Unsteady Supersonic Combustion

Mingbo Sun · Hongbo Wang · Zun Cai ·
Jiajian Zhu

Unsteady Supersonic Combustion

Springer

Mingbo Sun
Science and Technology on Scramjet Lab
National University of Defense Technology
Changsha, Hunan, China

Hongbo Wang
Science and Technology on Scramjet Lab
National University of Defense Technology
Changsha, China

Zun Cai
Science and Technology on Scramjet Lab
National University of Defense Technology
Changsha, China

Jiajian Zhu
Science and Technology on Scramjet Lab
National University of Defense Technology
Changsha, China

ISBN 978-981-15-3594-9 ISBN 978-981-15-3595-6 (eBook)
https://doi.org/10.1007/978-981-15-3595-6

This Springer imprint is published by the registered company Springer Nature Singapore Pte Ltd.
The registered company address is: 152 Beach Road, #21-01/04 Gateway East, Singapore 189721, Singapore

Dedication

The contributions from our colleagues and students are listed as follows:

Chapter 1 Introduction

Written jointly by Yanan Wang.

Chapter 2 Acoustic oscillation in supersonic combustor

Written jointly by Yongchao Sun, Dapeng Xiong and Fan Li.

Chapter 3 Flow dominating instability in supersonic flows

Written jointly by Tianyun Gao, Yuan Liu and Guiyun Gao.

Chapter 4 Cavity ignition in supersonic flows

Written jointly by Minggang Wan and Yanan Wang.

Chapter 5 Flame flashback in supersonic flows

Written jointly by Guoyan Zhao, Changhai Liang and Fei Li.

Chapter 6 Flame behaviors near blowoff in supersonic flows

Written jointly by Xiliang Song, Yanxiang Zhang and Liang Li

We would like to thank the staff at Springer for their help and support.

Preface

Hypersonic aircrafts that can fly at a speed above Mach 5 have revolutionary applications in national security and space exploration, and thus the development of hypersonic aircrafts has been regarded as 'highest technical priority' in many countries. Supersonic combustion ramjet (Scramjet) engines are the optimal propulsion system for hypersonic aircrafts within the atmosphere.

In the combustor of a Scramjet engine, the injected fuel experiences atomization, evaporation and mixing processes in an extremely short period of time, which significantly affects the subsequent combustion process. In addition, chemical reactions and heat release occurring in a supersonic and high-enthalpy stream during the combustion process also result in intense combustion instabilities. As a result, the combustion process in the Scramjet engine is a highly unsteady phenomenon. There are many issues affecting the unsteady supersonic combustion, and the characteristics of the unsteady combustion still need to be investigated, such as flow instability, acoustic oscillation, ignition and flame limit dynamic issues. Owing to great significance of both scientific research and engineering-related applications, unsteady supersonic combustion has attracted more and more attention among researchers and engineers.

Science and Technology on Scramjet Laboratory at National University of Defense Technology has carried out a huge amount of studies on fuel injection, mixing and combustion in Scramjet combustors, which significantly promotes the development of Scramjet engines in China. This book summarizes the research on unsteady supersonic combustion that has been carried out by our group in the past 15 years, and presents many state-of-the-art results and analyses in this subject.

The book is aimed at graduate students majoring in aeronautical and aerospace engineering, as well as researchers and engineers working in design of Scramjet engines. The prerequisite knowledge includes fluid mechanics, combustion principles, and computational fluid dynamics.

This book describes the unsteady phenomena for understanding the supersonic combustion. It is organized into six chapters. Chapter 1 introduces the basic thoughts and classical studies in the unsteady supersonic combustion fields. In Chaps. 2–6, the recent studies of the unsteady supersonic combustion are described, including interactions between acoustic wave and flame, flow dominating instability, ignition instability, flame flashback and combustion near blowout limits.

Changsha, China

Mingbo Sun
Hongbo Wang
Zun Cai
Jiajian Zhu

Acknowledgements

We would like to thank all the colleagues and students in our research group who have contributed to the understanding of the fundamentals of unsteady supersonic combustion. First and foremost, we express our deep gratitude to Prof. Zhenguo Wang, Prof. Weidong Liu, Prof. Jianhan Liang and Prof. Yuxin Zhao for the significant contributions they have made to the work presented here.

Particularly, we are grateful to Guoyan Zhao, Yanan Wang, Minggang Wan, Yongchao Sun, Dapeng Xiong, Yanxiang Zhang, Xiliang Song, Tianyun Gao, Changhai Liang, Yuan Liu, Liang Li, Fan Li, Fei Li and Guiyun Gao for their support in the writing up of the book.

Contents

About the Authors

Prof. Mingbo Sun is the Director of Science and Technology on Scramjet Laboratory at National University of Defense Technology (NUDT) in China. He was awarded a Doctorate in Aerospace Science and Technology from NUDT (2008) and a bachelor degree in Aerodynamic Engineering from NUDT (2002). His Ph.D. thesis entitled "Studies on Flow Patterns and Flameholding Mechanisms of Cavity Flameholders in Supersonic Flows" was rated as outstanding doctoral dissertation. He started his research career as a Lecturer at NUDT from 2008, and was promoted to a Professor in Science and Technology on Scramjet Laboratory in 2014. He has been working on experimental and numerical studies of the supersonic flow/combustion in scramjet engines in the past 15 years. He was awarded the National Science Fund for Distinguished Young Scholars for his outstanding research in supersonic combustion. He authored over 100 SCI-indexed journal papers and 16 patents.

Dr. Hongbo Wang is an Associate professor at National University of Defense Technology (NUDT) in China. He was awarded a Doctorate in Aerospace Science and Technology from NUDT (2012), Master of Science degree in Aerospace Science and Technology from NUDT (2007) and a bachelor degree in Aerodynamic Engineering from NUDT (2005). He used to be a visiting Ph.D. student in Aerospace Engineering at the University of Sheffield (UK) from 2009 to 2010. His Ph.D. thesis entitled "Combustion Modes and Oscillation Mechanisms of Cavity-Stabilized Jet Combustion in Supersonic Flows" was rated as outstanding doctoral dissertation. He started his Hypersonic Propulsion Technology research career working as a Lecturer at NUDT from 2012. He conducted research in the areas of scramjet combustor design, supersonic combustion and computational fluid/combustion dynamics. He authored over 50 publications in journals and several patents.

Dr. Zun Cai is an Associate professor at National University of Defense Technology in China. He was awarded a Doctorate in Aerospace Science and Technology from NUDT (2018), Master of Science degree in Aerospace Science and Technology from NUDT (2014) and a bachelor degree in Aerodynamic Engineering from NUDT (2012). He used to be a visiting Ph.D. student in Computational Fluid Dynamics at the University of Lund (Sweden) from 2015 to 2016, during which he developed a supersonic combustion solver based on the OpenFOAM. His Ph.D. thesis entitled "Investigation on the Flame Ignition and Stabilization Processes in a Cavity-based Scramjet Combustor with a Rear-wall-expansion Geometry" was rated as outstanding doctoral dissertation. He started his Hypersonic Propulsion Technology research career working as a Lecturer at NUDT from 2018 and promoted to an Associate professor in 2019. He conducted research in the areas of scramjet combustor design, supersonic combustion and RBCC propulsion issues. He authored over 20 publications in journals and several patents.

Dr. Jiajian Zhu is an Associate professor at National University of Defense Technology (NUDT) in China. He was awarded a Doctorate in Combustion Physics from Lund University in Sweden (2015), and a bachelor degree in Opto-electronic Engineering from NUDT in China (2009). He was employed as a Lecturer by the Science and Technology on Scramjet Laboratory, NUDT in 2015 and promoted to an Associate professor in 2018. His present research focus is supersonic combustion and laser-based combustion diagnostics. He was supported by the Huxiang Youth Talent Program and co-authored over 40 journal publications.

Nomenclature

a	Speed of sound
d	Jet exit diameter
e	Specific energy of the mixed gas
f	Oscillation frequency
h	Flame lift distance
k	Turbulent burning velocity
q	Heat-release
s	The distance from the leading edge of the cavity to the injector
ϕ	Equivalence ratio
α	Phase delay
β	Acoustic reflection coefficient
θ	Angle between fuel jet and axial direction
γ	Specific heat ratio
ψ	Dimensionless total heat release value
A	Cross-sectional area of the flow passage
D	Droplet diameter
E	Activation energy in Arrhenius equation
F	Non-viscous flux term
H	Source item
L	Cavity length
M	Mach number
Q	Chemical energy release in the ethylene-air mixture
U	Conservation variable
W	Width of the engine runner
Y	Fuel mass fraction
Z	Mixture fraction
Ω	Dimensionless frequency for admittance function
ρ_a	Density of air
ρ_e	Density of main stream
ρ_f	Density of fuel

ρ_j	Fuel density at the injector
d_q	The diameter equal to the quenching distance
f_j	The preferred mode frequency
f_n	Resonant frequencies
f_s	Stoichiometric contour
f_{if}	The characteristic frequency for the acoustic-convective feedback loop between the fuel injector and flame zone
f_{0_j}	Initial vortex shedding frequency
f_{sfi}	Oscillation frequencies
f_{up}	Mixing fraction of the upper edge of the shear layer at the flame base at the lean combustion limit
f_{RZ}	Stoichiometric in the cavity recirculation zone
h_c	Heat of combustion per unit mass of fuel
$h_{t,f}$	Total enthalpy of fuel
h_{fk}^0	Standard enthalpy of production for component k
k_O	Rate constant in Arrhenius equation
m_{air}	Air mass flow rate
m_f	Fuel mass flow rate
m_A^*	Effective air mass flow rate
m_F^*	Effective fuel mass flow rate
r_s	Stoichiometric fuel-air mass ratio
u_f	Effective flame propagation speed
u_j	Fuel jet velocity
x_a	Auto-ignition position
x_f	Locations of the flame
x_i	Location of the fuel injector
x_r	The relative distance between auto-ignition position and flame front position
x_s	Locations of the shock
p'	Pressure fluctuations
q'	Heat release fluctuations
u'	Velocity fluctuations
u_∞	Free-stream velocity
γ_∞	Free-stream specific heat ratio
M_∞	Free-stream Mach number
α_0	Thermal diffusivity of the fuel
ϕ_0	Effective equivalence ratio
S_0	Laminar combustion rate
c_p	Specific heat at constant pressure
$c_{p,a}$	Specific heat at constant pressure for air
d_l	Distance between the flame base and the cavity leading edge
ε_1	Dimensionless heat release rate
τ_1	Premixed gas residence time in the control volume
τ_2	Time for complete chemical reaction of premixed gas

τ_a	Time for the acoustic wave propagating from the flame zone to the injector
τ_c	Time for the fluctuating mixture convected from the fuel injector to the flame zone
τ_d	Dimensionless action time
τ_f	Flame time scale
τ_{if}	Time for the acoustic-convective feedback loop between the fuel injector and flame zone
τ_{sf1}	Characteristic times
τ_{sfi}	Oscillation period
δ_L	Thickness of the shear layer
δ^*	Displacement thickness
η_c	Combustion efficiency
η_m	Mixing efficiency
θ_j	The initial shear layer momentum thickness
A_d	Acoustic admittance
B_{st}	Stoichiometry mass transfer number
C_s	Correction factor
$E_{\min}$	A classical formula of minimum ignition energy
L_f	Length of the combustible zone
L_i	Energy loss of acoustic wave
L_{if}	Distance between the fuel injector and the flame
L_{sf}	Distance between the shock and the flame
L_{sep}	Time-averaged separated flow length
L_{RZ}	The length of the recirculation zone
K_v	Empirical constants corresponding to the average convection speed of disturbances in the shear layer
M_j	Atomic masses for elements i
S_t	Dimensionless frequency
T_{cav}	Mass-averaged temperature within the cavity
T_{ma}	Mass-averaged temperature
T_{AD}	Adiabatic combustion temperature
T_{0F}	Total temperature of the fuel
T_{0A}	Total temperature of the inflow
$T_{0,up}$	Total temperature of the upper edge of the shear layer at the flame base at the lean combustion limit
U_A	Upstream inflow velocity
U_e	Velocity of main stream
Y_j	Elemental mass fractions
Y_k	Mass fraction of the component
Y_{react}	Fuel mass fraction mixed in a proportion that can react
Y_{stoic}	Fuel stoichiometric mass fraction
$Y_{P,RZ}$	Mass fraction of combustion products in the cavity recirculation zone

ΔH	Reaction heat
ΔT	Temperature rise
ΔT_{st}	Stoichiometry temperature rise
S_L	Laminar burning velocity
S_T	Turbulent burning velocity
Da	Damkohler number
St	Strouhal number
u'	The velocity for acoustic admittance
p'	The pressure fluctuations for acoustic admittance
ρ'	Density fluctuations
$\dot{m}'$	Mass flow rate fluctuations
$\dot{m}'_f$	Fuel consumption rate fluctuations
$\dot{m}_f'''$	Rate of change in mass volume caused by fuel injection due to chemical reaction
u'	Velocity fluctuations in x direction
x'	Flame position pulsation
$\dot{Q}$	Overall heat-release rate
$\dot{m}$	Mass flow rate of the fuel/air mixture at a given axial location
$\dot{m}_f$	Fuel consumption rate
$\dot{\omega}_f$	Mass generation rate w of chemical reaction
$\dot{\omega}_k$	Net mass production rate of component k in the unit volume
$\bar{a}$	Mean speed of sound
$\bar{\dot{m}}_f$	Mean fuel consumption rate fluctuations
$\bar{\dot{m}}$	Mean mass flow rate
$\bar{p}$	Mean pressure
$\bar{\gamma}$	Mean specific heat ratio
$\underline{a}$	Speed of sound longitudinally averaged between the shock and flame
$\underline{u}$	Velocity in x direction longitudinally averaged between the shock and flame
$\bar{M}$	Mean Mach number
$\underline{M}$	Mach number longitudinally averaged between the shock and flame
Flux_{HRR}	Non-dimensional heat release rate
$(u_{CJ})_{deflagration}$	Chapman-Jouguet deflagration speed

List of Figures

List of Tables

Chapter 1
Introduction

Supersonic combustion in a scramjet engine has been investigated widely for decades [1]. With digging deeper into the supersonic combustion issues, the hot spot has been shifted gradually from quasi-steady state such as flame stabilization to unsteady state such as combustion fluctuations. Nowadays, unsteady supersonic combustion and its control strategy pose a big challenge for real scramjet engine applications.

In the supersonic combustor, chemical reaction and heat release occur in a high-speed and high-enthalpy stream, resulting in the intense combustion unsteadiness. There are many issues affecting the unsteady supersonic combustion that need to be investigated. The mechanisms of the unsteady supersonic combustion can be roughly divided into five categories based on the dominating factors, i.e., the interactions between acoustic wave and flame, flow dominating instability, ignition unsteadiness, flame flashback, and near-blowout combustion.

The combustion generally emerges in the subsonic regions generated by the flame-holder in the combustor. In the subsonic region, acoustic waves can propagate freely upstream and affect the mixing and reaction process. Once the acoustic waves couple with the heat release processes, the thermo-acoustic instabilities could be induced in the combustion.

Due to the pressure rise in the combustion region, boundary layer separation generally occurs upstream, and the shock train is also induced. Oscillations of the shock train along with the separated regions dominate the flow in the combustor and change the mixing and reaction conditions. The intrinsic unsteadiness in the flow field could become a certain cause of unsteady combustion.

Ignition is a transition process from the unreacted state to the consistently reacted state. As the beginning phase of combustion, it is vitally important to the combustion instability. The forced ignition methods are widely used in the scramjet combustor, and the influencing factors still need to be investigated. Under certain conditions, auto-ignition is significant and also affects both the ignition process and the consistently reacted state. Thus, the effects of the ignition on combustion instability deserve particular attention.

M. Sun et al., *Unsteady Supersonic Combustion*,
https://doi.org/10.1007/978-981-15-3595-6_1

During the combustion, transient flame flashback was observed under some occasions. The flow conditions rapidly change in the combustor when flame flashback emerges. It is an important sub-process of the combustion oscillation. The flame flashback is a complex unsteady combustion phenomenon coupled with deflagration–detonation transition, boundary layer separation, and thermal choking.

When the flow condition is near the blowout limits, the combustion characteristics are concerned specially and important for understanding the instability mechanism. The combustion instability is observed remarkably increasing when the flow condition approaching the blowout limits. These unsteady phenomena are frequently encountered and closely related to the combustion dynamics in supersonic combustors. They raise a great challenge in organizing the supersonic combustion.

This book describes the unsteady phenomena for understanding supersonic combustion. It is organized into five chapters. This chapter introduces the basic thoughts and important researches in the unsteady supersonic combustion. In Chaps. 2–5, the recent studies of the unsteady phenomena are described; such as the interactions between acoustic wave and flame, flow dominating instability, ignition instability, flame flashback, and near-blowout combustion.

1.1 Interactions Between Acoustic Wave and Flame

In many low-speed combustion systems, acoustic waves can be easily excited and sustained in the confined volume, and the frequency ranges of high-amplitude pressure oscillations are close to those of the natural acoustic modes in those systems. The high-amplitude pressure oscillations arise from the feedback loop between acoustic waves and unsteady heat release. It is a common assumption that acoustic waves cannot travel upstream in a supersonic flow, and any flow oscillations arising in the flame zone will simply travel downstream and exit from the engine without forming the feedback loop required to sustain combustion and flow instabilities. However, there are various subsonic flow regions in scramjets. In these subsonic flow regions, the acoustic wave can propagate towards upstream and lead to unsteady combustion ultimately. So, researches on oscillatory phenomena and mechanisms about thermoacoustic instabilities with experiment or numerical simulation play a decisive role in the development of supersonic combustion systems.

Before introducing the acoustic vibration modes of the combustion chamber, it is necessary to give a brief review and description of the related concept of the acoustic wave and acoustic oscillations.

1.1.1 Fundamentals of the Coupling Between Acoustic Wave and Combustion Process

The coupling between the acoustic wave and combustion process was discovered by Higgins in 1777. Many researches have been carried out to investigate the mechanism of the coupling between the acoustic wave and the combustion process.

Rayleigh criterion [2], as an universal explanation for the coupling between acoustic wave and combustion process, is used in many theoretical researches on thermoacoustic instabilities in combustion systems. The criterion replaces the burning process by a hypothetical heating process, and the combustion process is simplified to an interaction of heat release and acoustic field.

Rayleigh criterion gives the condition for thermoacoustic instability and is described as the following equation:

$$\iint_V p'(x,t)q'(x,t)dtdV \geq \iint_V \sum_{i=1} L_i(x,t)dtdV \tag{1.1}$$

where $p'(x,t)$ is the pressure fluctuations, $q'(x,t)$ is the heat release fluctuations, $L_i(x,t)$ is the energy loss of acoustic wave. The unsteady heat release delivered energy to acoustic filed is not necessarily to bring sustaining instability. Only when the rhythms of working fluids movements and fluctuations are in accordance with thermal processes, will the thermoacoustic oscillations be held.

The Rayleigh criterion shows the coupling between acoustic wave and heat release fluctuations. However, a variety of complex physical processes may be involved in the combustion process. Some of the interactions during the coupling between the acoustic wave and combustion process have been given in Fig. 1.1. It is important to understand the elementary processes of interaction between combustion and waves

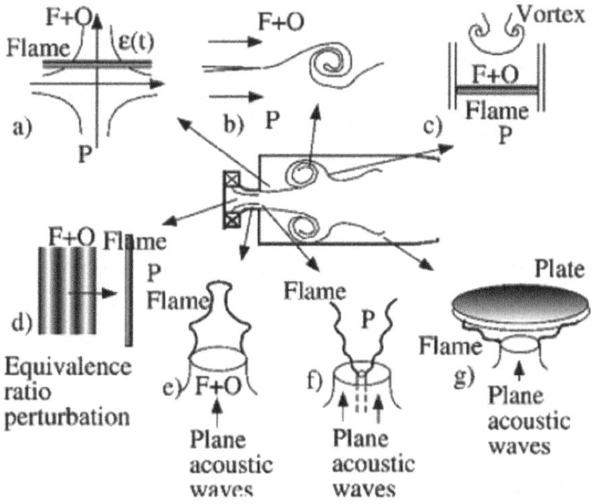

Fig. 1.1 Elementary processes: **a** unsteady strained diffusion flame, **b** unsteady strained premixed flame, **c** premixed flame/vortex interaction, **d** equivalence ratio perturbation interacting with a premixed flame, **e** acoustically modulated conical flame, **f** acoustically modulated V flame, and **g** perturbed flame interacting with a plate [3]

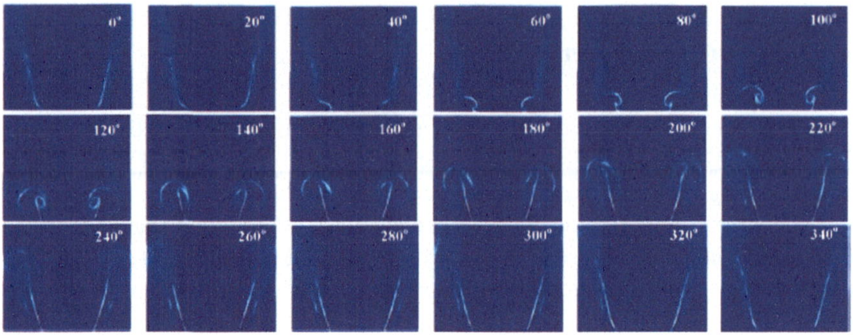

Fig. 1.2 Phase-averaged flame surface density image sequence under strong acoustic forcing [4]

or flow perturbations (acoustics, convective modes, injection inhomogeneities, etc.), which may become driving or coupling processes under unstable conditions. In this section, the influence of vortex (flow instabilities) and injection inhomogeneities will be introduced.

Experiments and theoretical analysis indicate that certain types of instabilities in lean premixed combustors may be driven by perturbations in the fuel and air ratio. This is illustrated here by assuming that pressure oscillations in the combustor interact with the fuel supply line and change the fuel flow rate. A positive pressure excursion produces a decrease in the fuel supply at a later instant. This causes a negative perturbation in the equivalence ratio, which is then convected by the flow to the flame zone. The interaction may also take place with the air supply, and this will also affect the equivalence ratio. The fluctuation of the equivalence ratio leads to a fluctuation of the heat release, and if the fluctuation of heat release is in phase with the pressure oscillation, energy may be fed into the resonant acoustic mode involved in the combustion process. Vortex structures drive various types of combustion instabilities, and the evolution of the vortex in one period can be observed in Fig. 1.2. In many premixed systems, the ignition and delayed combustion of these structures constitute the mechanism that feeds energy into the oscillation. The process involves at least two distinct mechanisms. In the first, the flame area is rapidly changing in the presence of a vortex, and the change of the flame area leads to a fluctuation of heat release. In the second, the vortex interacts with a wall, or another structure, inducing a sudden ignition of fresh material. Rollup by a vortex often controls the mixing of fresh gases into the burning regions, and this determines the unsteady rate of conversion of reactants in the flow and the amplitude of the pressure pulse resulting from the vortex burnout. The fluctuation of heat release will feed back to the combustion process, and the unsteady combustion will be intensified if these processes are in phase with the pressure fluctuation.

1.1.2 *Classification of Combustion Instability Related to Acoustic Wave*

Historically, the combustion instabilities are classified according to their frequency range, but between the so-called low frequency, intermediate frequency and high frequency, there is no clear borderline.

1.1.2.1 Low-Frequency Instability

The frequency of low-frequency combustion instability is usually below 200 Hz, mainly caused by the coupling of the combustion process in the combustion chamber and the flow process of the propellant feed system, and it is usually related to the ignition quality and injection speed of the propellant entering the combustion chamber. Ignition quality includes ignition delay time, flame propagation speed and flame stability characteristics. Combustion chamber, the scale of the propellant pipeline and the flow rate and mixing ratio of the propellant have a key role in the low-frequency oscillations. Coupling of the combustion process and the injector structure can also cause low-frequency instability: injector may work like a diaphragm, and produce an "Oiler"-type oscillation, causing inhomogeneous propellant injection and atomization, resulting in low-frequency instability. Some other situations can also result in the coupling between the combustion (or chamber pressure) and structure system and cause low-frequency instability. For example, the perturbation of chamber pressure makes the cooling jacket bend, causing pressure oscillations of the propellant contained in the cooling jacket. This coupling can lead to low-frequency instability.

When low-frequency combustion instability occurs, the wavelength of the gas oscillation is usually much larger than the characteristic length of the chamber or the supply system. Therefore, it can be considered that, the pressure oscillation of combustion chamber is uniformly distributed in any instantaneous, and it can be seen as the oscillations of the whole gas field in the combustion chamber; Meanwhile, the pipeline of propellant supply system or liquid collection chamber also exhibits oscillations. This instability is often a sine wave with low amplitude at the beginning, and then developed linearly into a higher amplitude.

In different types of combustion instabilities, low-frequency instability is probably the easiest one to deal with from a viewpoint of theoretical and experimental analysis or development. From the standpoint of theoretical analysis, the combustion chamber can be simulated by using a concentrated volume element, and the combustion is represented by a simple constant time delay, the resistance of propellant supply system is neglected, although the inertia and capacity of the supply system may be important in the analysis. Combustion time delay is defined as: the time required for the liquid propellant to be completely vaporized and consumed. An experiential average value can often be obtained for each propellant. The time delay usually referred to is the flight time of the component with the worst volatility from the injector surface to the impinging point. Because it is a major part of the total time delay. Methods to eliminate low-frequency instabilities include increasing the injector pressure drop, increasing fluid inertia, as well as reducing the volume of the combustion chamber, and so on. For approaches used to change the time delay, some are successful, but some are problematic since they may degrade system performance or cause high-frequency instability though they can successfully eliminate low-frequency instability.

1.1.2.2 High-Frequency Instability

High-frequency instability is a result of combustion processes coupled with the combustor acoustic oscillations, also known as resonant combustion or acoustic instability. The oscillation frequency is usually above 1000 Hz. When high-frequency combustion instability occurs, for the measured dynamic pressure in the combustion chamber at different locations, the relationship between the oscillation frequency and the phase of each point is often consistent with the natural modes of acoustic modes of the combustion chamber. Thus, according to the acoustic characteristics of the combustion chamber, high-frequency instability can be divided into the axial (longitudinal) or horizontal (radial and tangential) mode. The above various modes of high-frequency combustion instability can be divided according to their order of resonance into the first-order vibration mode, the second-order vibration mode, etc., such as first-order radial vibration mode, second-order longitudinal vibration mode and third-order tangential vibration mode. The heat release rate varies when the flame is under different vibration mode, which can be observed from Fig. 1.3.

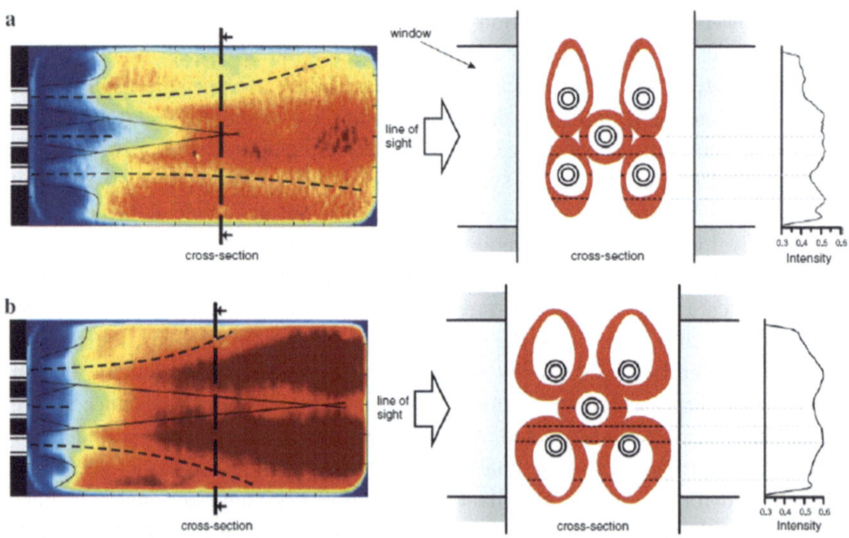

Fig. 1.3 Flame cross-sectional intensity profiles extracted from time-averaged OH* images during; **a** off-resonance, and **b** 1T mode excitation [5]

For the mechanism of high-frequency instability, current points of view include: ignition time lag, sensitive chemical time lag, physical time lag, detonation process, the changes of chemical reaction rate caused by the fluctuations of pressure or temperature, the "explosion" when the droplets are heated to beyond its critical temperature and critical pressure, and the jet flow, liquid fan or the crushing and mixing of liquid droplets.

To maintain high-frequency instability, firstly, there must be an oscillating energy to maintain the high-frequency instability of liquid rocket engine. The energy comes from the combustion of the propellant; secondly, the oscillation energy must be added at appropriate time phase related to oscillating pressure. Therefore, the methods to eliminate high-frequency instability usually have two categories: (1) change propellant spray combustion field or pressure wave characteristics, so that the energy released by the combustion fluctuations is less than the oscillation energy required to maintain oscillation, such as baffle devices; (2) change the dynamic energy loss or damping, making it greater than the energy obtained from combustion response, such as various different types of damping devices.

1.1.2.3 Intermediate-Frequency Instability

Intermediate-frequency combustion instability is the oscillation caused by the coupling between the combustion process in the combustion chamber and a portion of flow processes of the propellant supply system. The frequency range is usually 200–1000 Hz, lying between high and low-frequency oscillations.

When intermediate-frequency combustion instability occurs, it is often accompanied by a gradually increased combustion noise with a specific frequency, and its amplitude increases slowly. Besides the gas oscillations, fluctuations usually appear in the propellant supply system, the frequency and phase of gas oscillation are not consistent with the inherent acoustic modes of the combustion chamber, which is different from the high-frequency combustion instability. On the other hand, it is also different from the low-frequency combustion instability. Because its frequency is slightly higher, the wavelength of gas oscillation is close to or slightly larger than the characteristic length of the combustion chamber, so fluctuations in the combustion chamber and the supply system pipeline cannot be ignored; The pressure oscillation in combustion chamber will change spatially, and cannot be seen as a whole gas field like that in low-frequency combustion instability. Intermediate-frequency combustion instability may lead to oscillations of the propellant mixture ratio and decrease of engine performance.

1.1.3 Acoustic Induced Combustion Instabilities in Supersonic Flows

The experience with airbreathing propulsion systems and rocket engines suggests that combustion instability coupled to thermoacoustic may present an important obstacle in the development of scramjet engines. It is a common assumption that acoustic waves cannot travel upstream in a supersonic flow, any flow oscillations arising in the flame zone will simply travel downstream and exit from the engine without forming the feedback loop required to sustaining combustion and flow instabilities. In reality, with an experimental investigation, Stamp et al. [6] have found that acoustic waves can propagate upstream in various subsonic-flow regions and a scramjet combustor may be susceptible to acoustic-feedback type self-sustained combustion instabilities. Besides, the interactions between injector flows, shock waves, and boundary layers have strong unsteady characteristics and may cause instabilities in a supersonic combustor. The Acoustic-convective feedback loops in supersonic flow have been given in Fig. 1.4. Cavity flameholders increase the resonance of a sound and may introduce additional oscillation mechanisms to the combustor. The cavity-induced oscillations, on the one hand, can enhance the fuel-air mixing. On the other hand, they couple the internal and external regions, making the cavity flow fields highly complicated and even causing combustion instabilities [7].

Choi et al. [9, 10] carried out a comprehensive numerical analysis for both non-reacting and reacting flows in a scramjet engine combustor with and without a cavity. The results showed a wide variety of phenomena resulting from the interactions between the injector flows, shock waves, boundary layers, and cavity flows. Flow oscillations caused by the cavity overrode those induced by the interactions between shock waves and boundary layers, the captured high-frequency oscillations were associated with the cavity and flow unsteadiness. However, further investigations are required to achieve a better understanding of detailed fluid and flame dynamics and acoustic characteristic in a scramjet combustor.

Ma et al. [11] observed the low-frequency oscillations at 100–160 Hz for liquid JP-7 fuel and 300–350 Hz for gaseous ethylene in a dual-mode scramjet by using frequency pressure sensors. The low-frequency oscillations for liquid JP-7 can be observed in Fig. 1.5. A quasi-one-dimensional model to simulate the main features of the oscillatory flow fields in both the isolator and combustor was established, and the flow oscillations were reproduced by the numerical results. The mechanism responsible for driving flow oscillations was identified as the acoustic-convective

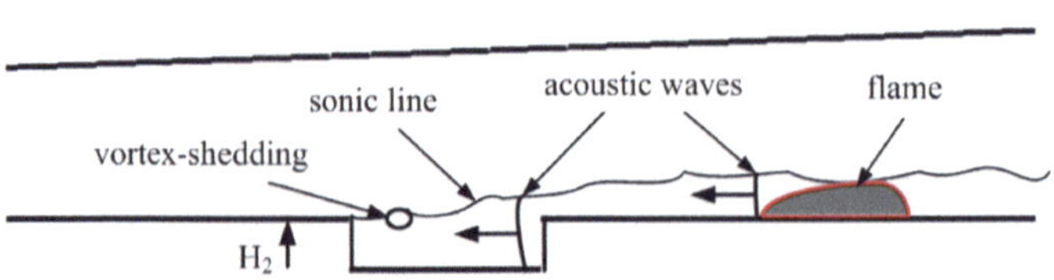

Fig. 1.4 Acoustic-convective feedback loops [8]

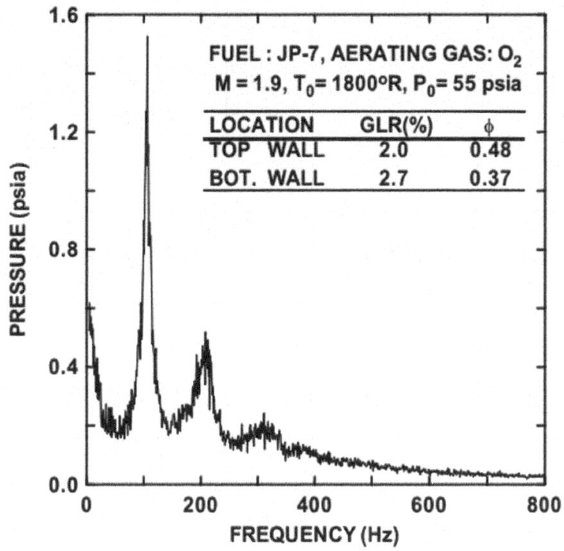

Fig. 1.5 Typical pressure power spectrum inside a scramjet combustor with liquid JP-7 fuel [11]

interactions between the fuel injector and the flame zone. Li et al. [12] carried out three-dimensional simulation of the ethylene-fueled scramjet combustor investigated by Ma et al. [11]. The results displayed the oscillations of the flame and fuel distribution.

Lin et al. [13] investigated acoustic oscillation instabilities inside an ethylene-fueled supersonic combustor with a recessed cavity flameholder. The schematic in Fig. 1.6 shows the flow-path with key combustor features identified. Under various flow conditions and flameholder geometries, the acoustic signals were recorded by high-speed pressure transducers positioned at the base and downstream of the cavity flameholder. The effects of fuel/air equivalence ratio, fueling scheme, cavity length, and simulated flight conditions on the stability characteristics of the combustor were examined systematically, and the results of the pressure oscillations for various fueling schemes and equivalence ratios can be observed in Fig. 1.7. In order

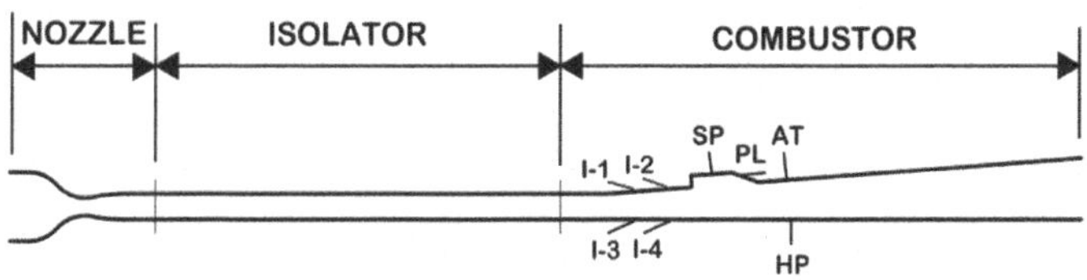

Fig. 1.6 Schematic of the combustor flowpath and key interior features (AT, Air throttle; HP, High-speed pressure transducer; I-1, Body side first row 15° gaseous injectors; I-2, Body side second row 15° gaseous injectors; I-3, Cowl side first row 15° gaseous injectors; I-4, Cowl side second row 15° gaseous injectors; PL, Pilot fuel injectors; SP, Spark plugs) [13]

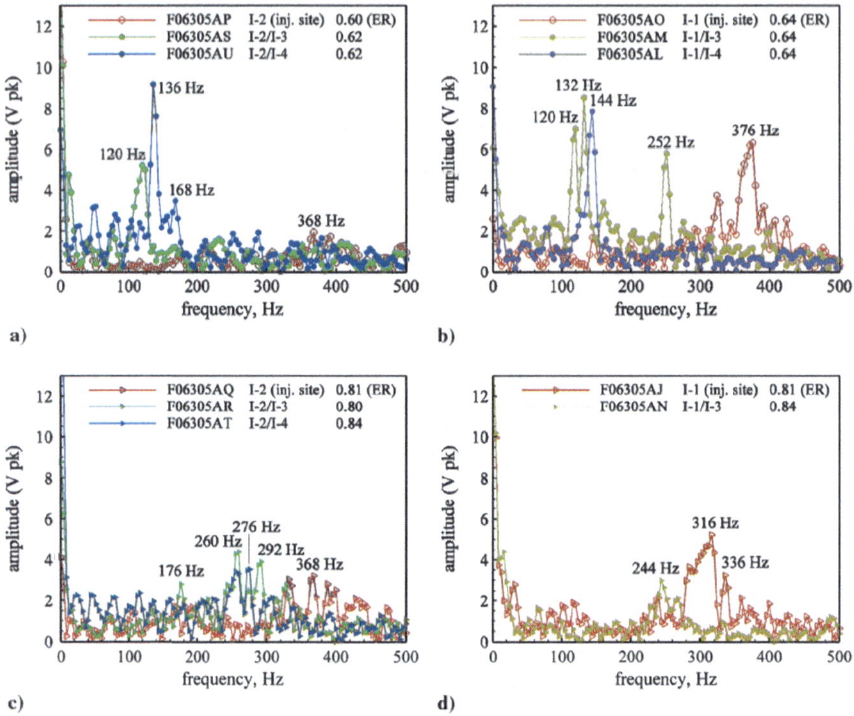

Fig. 1.7 Frequency spectra of pressure oscillations for various fueling schemes and equivalence ratios [13]

to explain these observed frequencies characteristic, three prospective mechanisms were identified in Lin's research [13].

The first and second mechanisms were concerned with the coupling between the terminal shock and flame zone. The shock-flame acoustic feedback loop was established by the upstream propagation of acoustic waves produced in the combustor and the interaction of these acoustic disturbances with the shock wave in the isolator. Then the perturbations traveled downstream as acoustic or entropy waves, enhancing the unsteady combustion in the flame zone.

The third mechanism described the interaction of acoustic waves and fluctuation happened in the region between the fuel injection and flame zone, where disturbances from the flame zone propagated upstream and caused an air mass flow-rate oscillation in the fuel injection region. For comparison, some acoustic admittance equations were used to estimate characteristic times and corresponding oscillation frequencies, and the measured oscillation frequencies agreed well with the characteristic frequencies related to each acoustic feedback loop between the shock and flame or the feedback loop between the fuel injector and flame.

As indicated by the feedback mechanisms, it is reasonable to believe that these instabilities basically occur in the ramjet mode rather than in the scramjet mode since the large subsonic regions behind the pre-combustion shock seem necessary for the acoustic waves to readily propagate upstream.

1.1.4 Summary

As this review of the research efforts on studying the interaction between acoustic wave and combustion process, many experimental, numerical simulation and theoretical works have been done to observe and measure the characteristic of combustion oscillation. It is found that acoustic wave can propagate to upstream in the channel with a supersonic main flow. Some feedback loop of acoustic wave between combustor and isolator are proposed to predict the characteristic time and corresponding oscillation frequencies for comparison with experimental data. Thus, the differences of configuration or coupling mechanism cause disparities on spectra of frequency, which demonstrate their association with acoustic wave.

1.2 Flow Dominating Instability

The shock dominated flow is typical in supersonic combustion, and the intrinsic unsteadiness of flowfield plays a key role under some occasions. Boundary layer separation often occurs in combustor due to combustion-induced pressure rise [14]. Oscillation of shock train along with the separated regions is a certain cause of unsteady combustion.

1.2.1 Low-Frequency Unsteadiness of Shock Wave/Turbulent Boundary Layer Interaction

Shock wave/boundary-layer interactions (SWBLI) represent complex flow phenomena that are associated with a wide range of flows, including transonic airfoils, supersonic inlets, over-expanded nozzles, etc. Often the shock induces significant boundary layer separation, which leads to a highly unsteady flow field [15].

The unsteadiness of SWBLI consists of a high-frequency component and a low-frequency component [16]. Figure 1.8 depicts pressure power spectra in a Mach 5 compression ramp from five different locations taken by Erengil and Dolling [17]. The high-frequency peak of pressure fluctuations from locations 1, 4 and 5 are around the outer-scale frequency, U_∞/δ_0. The high-frequency oscillation of location 1 is determined by fluctuations from the upstream boundary layer, and the unsteadiness of the separation bubble is affected by radiation from the turbulent shear layer above. However, the pressure power spectra of stations 2 and 3 represent a dominant peak

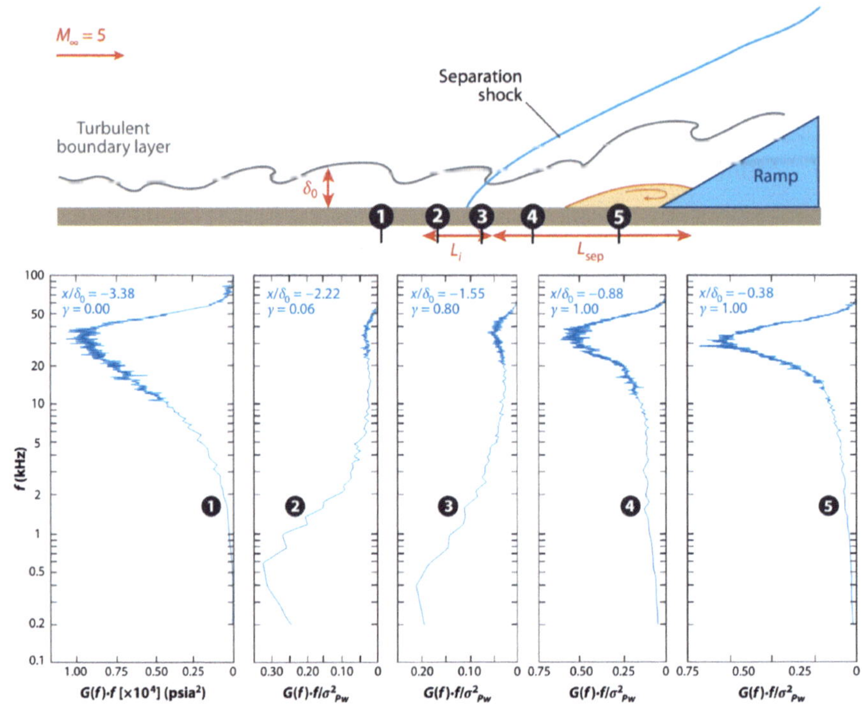

Fig. 1.8 Pressure power spectra underneath the interaction generated by a 28° compression ramp in a Mach 5 flow [23]

at a much lower frequency, which is of order 0.01 U_∞/δ_0. Such a low-frequency unsteadiness depicts the oscillation frequency of the separation shock foot. These basic trends in the power spectra of pressure fluctuations remain nearly the same for other traditional interactions, including those generated by ramps with sweep [18, 19], blunt [20] and sharp fins [21], and reflected shocks [22].

The shock foot unsteadiness in SWBLI is a typical low-frequency oscillation in supersonic flow, whose characteristics can be described as follows. The shock foot undergoes larger-scale motion and lower oscillation frequencies as the scale of separation increases. The dimensionless frequency Strouhal number however, is nearly the same [23] (Ma = 3: St = 0.09 and 0.11 [24], Ma = 2.3: St = 0.03, 0.035 and 0.04 [22]). The dimensionless frequency can be calculated as follows:

$$S_t = f L_{sep}/U_\infty \tag{1.2}$$

where f is the peak frequency of static pressure oscillation, L_{sep} is the time-averaged separated flow length, and U_∞ is the free stream velocity. Compared to the high-frequency unsteadiness, the mechanism of low-frequency oscillations remains not fully explained.

Previous researches mainly describe the unsteady flow by pressure data, which is usually not enough for mechanism study. In recent years, more and more studies try to uncover the mechanism of low-frequency unsteadiness with advanced experimental and computational methods. Humble et al. [25] carried out an experimental study to investigate the three-dimensional instantaneous structures of an incident SWTBLI at Mach 2.1. The large-scale coherent motions within the incoming boundary layer were observed using tomographic particle image velocimetry. As shown in Fig. 1.9, the instantaneous reflected shock wave pattern was found to be consistent with the streamwise-elongated low- and high-speed regions as they enter the interaction. Priebe et al. [26] characterized the low-frequency unsteadiness of a SWTBLI generated by a compression ramp at Mach 2.9. With the direct numerical simulation (DNS), the low-frequency streamwise oscillation of the shock wave was captured. The statistical relation between the low-frequency shock motion and the upstream/downstream flow was analysed. The changes in the velocity and vorticity profiles in the initial part of the interaction were found to be affected by an inherent instability in the downstream separated flow. On the other hand, the statistical relation of the shock motion and the upstream boundary layer was rather weak.

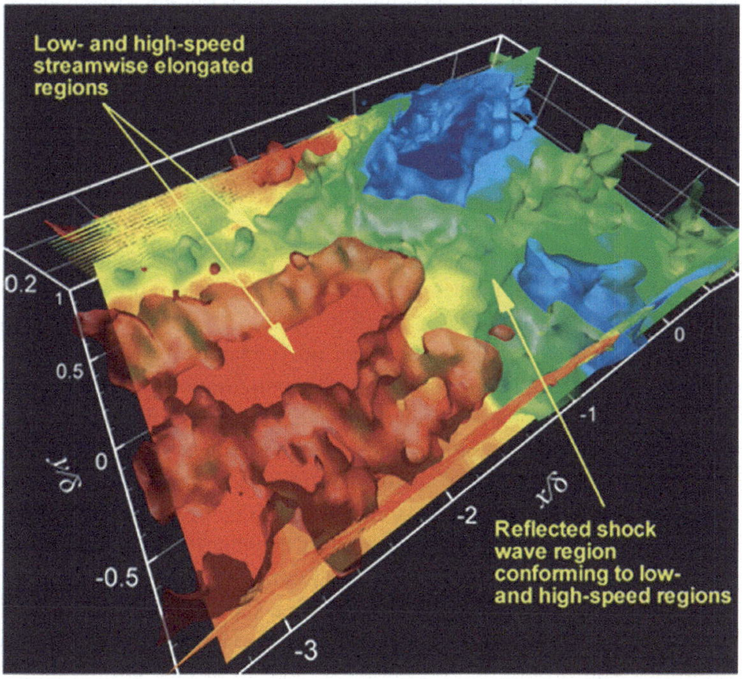

Fig. 1.9 Volumetric representations of the instantaneous flow organization of the interaction: lower region ($z/\delta = 0.1$–0.6). Iso-surfaces of streamwise velocity are shown: relatively high-speed in red ($0.9\ U_\infty$), intermediate velocity in green ($0.75\ U_\infty$), and relatively low-speed in blue ($0.55\ U_\infty$). Velocity vectors are shown flooded with instantaneous streamwise velocity [25]

Pasquariello et al. [27] analysed the low-frequency dynamics of a high Reynolds number impinging SWTBLI at Mach 3. The large-eddy simulation (LES) was performed for a very long integration time to obtain a trustworthy result for the strong separated flow. Consistent with experimental data, the simulated power spectral densities (PSD) of wall-pressure exhibited an energetic, broadband and low frequency component associated with the separation-shock unsteadiness. Sparsity-promoting dynamic mode decompositions (SPDMD) yielded a classical low-frequency breathing mode of the separation bubble, as well as a medium-frequency shedding mode responsible for reflected and reattachment shock corrugation.

Based on numerous researches implemented all over the world, some preliminary work on modeling the low-frequency unsteadiness in SWTBLI was established. Piponniau et al. [28] developed a model to describe the properties of fluid entrainment in the mixing layer generated downstream of the separation shock. The model well estimated the low-frequency shock unsteadiness observed in various shock-induced separation cases ranging from Mach 0 to 5. It was concluded that the main source of low-frequency unsteadiness was the dynamics of the separated bubble. Due to the complexity in low-frequency unsteadiness in SWTBLI, its mechanism is still not fully explained so far. In general, it is acknowledged [23] that the downstream mechanism dominates for strongly separated flows, and a combined mechanism (both upstream and downstream) dominates for weakly separated flows.

1.2.2 Unsteadiness of Shock-Induced Separation in Non-reacting Flow

Unsteadiness of flow separation has been widely investigated in non-reacting flows, such as the inlet of the unstart process, isolator under strong backpressure and nozzle in over-expanded condition.

Studies of the unsteady shock motions in inlets mainly focused on the unstart process. Koo et al. [29] studied an inlet-isolator configuration with large eddy simulation. The unstart dynamics were fully simulated under three different inlet ramp angles (0, 6, and 8°). It was found that the separated boundary layers on both walls played key roles in the initiation of the unstart process. Simulated results indicated that the unstart shock propagation was accelerating during the unstart process, as in Fig. 1.10. However, the propagation speed of the shock train from LES was 3–4 times larger than the experiment due to the simplified models for the boundary layers. Do et al. [30] injected a transverse jet into a supersonic inlet flow to induce unstart. They concluded that thick turbulent boundary layers in asymmetric boundary conditions would prompt the formation of unstart shocks. On the other hand, the symmetric boundary conditions led to the propagation of pseudo-shocks. Zhang et al. [31] carried out an experimental study in a hypersonic inlet with side compression at a freestream Mach number of 6.0. A flow plug was placed at the duct exit to simulate the combustion induced high pressure. During the retreating process of

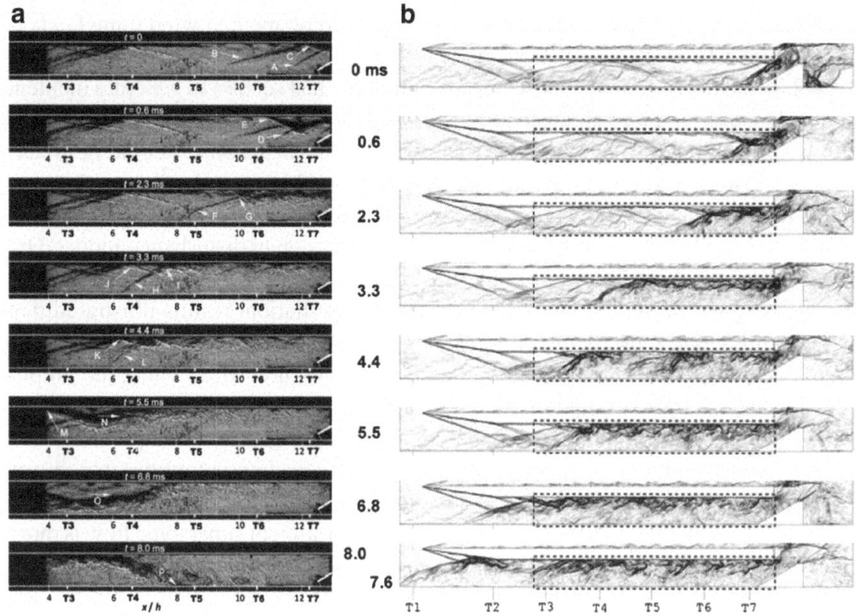

Fig. 1.10 **a** Experimental schlieren images and **b** density-gradient magnitude images from LES computations at corresponding scaled time for the 8 deg inlet unstart case. Experimental measurements are available corresponding to the boxed area in **b** [29]

the external unstart shock, they observed two kinds of secondary oscillations with high dominant frequencies of about 360 and 900–1300 Hz. It was found that these two secondary oscillations were both acoustic resonance modes formed in different parts of the duct.

Geerts et al. [32, 33] used background oriented schlieren (BOS) to study the shock train movement under slowly varying backpressure conditions in a rectangular isolator with Mach 2.5 upstream flow. It was observed that throughout the unstart process, the shock system behaved apparently oscillatory nature. Bruce et al. [34] studied a transonic duct with parallel walls at Ma 1.4, low-frequency downstream pressure perturbations (16–90 Hz) was enforced to the flowfield. It was concluded that the unsteady behaviors of relatively low frequency (40 Hz) could be captured well by the unsteady Reynolds-averaged Navier-Stokes scheme. However, the size of the interaction region was exaggerated by simulation. Bruce et al. [34] also found out that asymmetry existed in transonic channel flows, which was induced by the interaction of corner flows. Numerical simulations indicated that flowfield asymmetry occurred when the size of a corner interaction exceeds 35–40% of the channel width or height. Researches carried out by Su et al. [35, 36] focused on self-sustained and imposed oscillations of pseudoshock induced by back pressure. Their simulated results indicated that, an oscillation of 3107 Hz occurred when the ratio of backpressure to the freestream static pressure reached 70. Li et al. [37] implemented wind tunnel experiments to investigate the oscillation characteristics of the shock train in an isolator at Mach 2.7, and a wedge was mounted upstream of the test section to generate incident shocks. It was found that an unsteadiness appeared when the leading edge of shock train was travelling through the SWTBLI region. This unsteadiness could be weakened by a faster backpressure rising rate. Xiong et al. [38] carried out several experiments to investigate the flow unsteadiness in a constant-area rectangular isolator. In self-excited oscillation, it was found that the low-frequency disturbance induced by the upstream shock foot motions could travel downstream and the frequency would be magnified by the separation bubble. In forced oscillation, results illustrated that the separation shock oscillation frequencies increased and the intermittent region lengths decreased with the increasing steady backpressure. Meanwhile, the amplitude of the shock train oscillation increased with the decreasing excitation frequency of the fluctuating backpressure. An analytical model [39] was developed based on the 'relative Mach number' mechanism and a quasi-steady assumption, which was able to predict the unsteady motion of shock train quite well.

Compared with inlets and isolators, unsteady separation of supersonic flow is more widely studied in nozzles. Since the supersonic nozzles resemble combustors in shape (expanded flow path), the systematic studies of unsteady separation in nozzles are especially heuristic.

Several researchers have been studying the asymmetric and unsteady separation phenomena in a supersonic nozzle for a long time. Typical flow behaviors and the mechanism of unsteadiness are revealed step by step. Reijasse et al. [40] made a preliminary study on shock-induced separation in a planar two-dimensional nozzle. With the rapid shadowgraph and three-dimensional laser doppler velocimetry method, they concluded that the flow field transferred from symmetric to asymmetric as throat contraction ratio increased, and returned to symmetric due to the further increase of throat contraction ratio. Yu et al. [41] discussed the switch of separation modes in an over-expanded single expansion ramp nozzle. Research showed that separation patterns changed between restricted shock separation and free shock separation during the startup process. Meanwhile, a shock wave instability [42] occurred during the separation transition phenomenon.

Papamoschou et al. [43, 44] found out that asymmetric separation occurred in a convergent-divergent nozzle which worked in over-expanded condition. Wall pressure measurements indicated that a low-frequency, piston-like unsteady shock motion exists without any resonant tones. Xiao et al. [45] used RANS to study the same nozzle as Papamoschou investigated, numerical results captured asymmetric separation under moderate NPR (nozzle pressure ratio) 1.6–2.3. Johnson et al. [46] made further study based on their previous work. They concluded that enhanced shear layer instability was strongly coupled to shock motion unsteadiness, while the wave pattern itself was not a cause of enhanced mixing. Olson et al. [47, 48] carried out LES simulation on the same case as Papamoschou and Johnson worked on. Based on directional artificial fluid properties method developed for wall-bounded flow, their simulation fully described the process of shock unsteady motions, as shown in Fig. 1.11. A reduced-order model was proposed based on the quasi 1d flow equations.

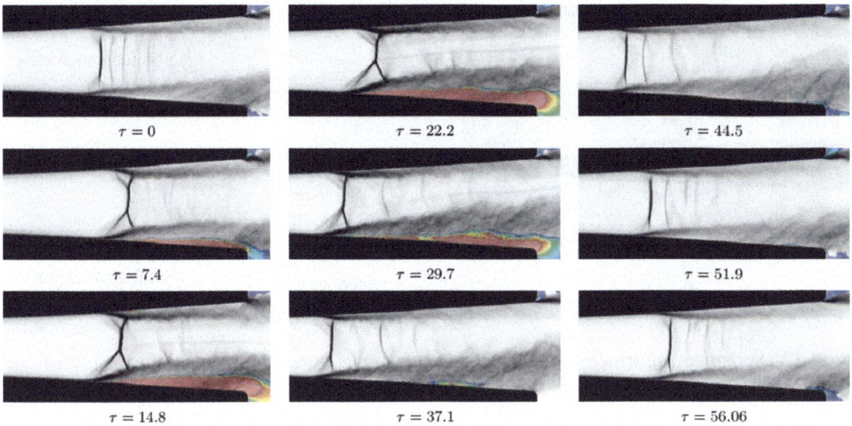

Fig. 1.11 Shock wave motion and the corresponding separated shear layer over one low-frequency process (Contours of $\|\nabla\rho\|$ are shown in grayscale and colored regions depict negative U velocity. Red represents a Mach number of approximately 0.1 and blue is 0) [48]

With systematic studies, their conclusions can be summed up as follows:

1 The asymmetry and unsteadiness of shock train become obvious when NPR is high, and the amplitude of oscillation is about half of the nozzle's height [43].
2 The shock foot and the shear layer of large separation side behave strong unsteadiness [49], which forms large vortices downstream. The shear layer at the other side develops at a normal rate. Such an unsteadiness is helpful in mixing enhancement.
3 The main component of oscillation is low frequency, which do not have an obvious peak. The back and forth motion of shock train results from shear layer instability, which has nothing to do with acoustic effects [50]. The oscillation is broadband, the high frequency component is controlled by the turbulent boundary layer and separated shear layer, while the low-frequency component is affected by shock intensity. Stronger shock leads to unsteady behavior more obvious [49].
4 Correlation analysis indicates that the total pressure of the shear layer at the large separation side has a positive correlation [46] with shock motion, which is opposite to normal one-dimensional shock theory.
5 Alternating wave pattern downstream the separation shock is not the main cause of unsteady process [46]. A wavy wall is implemented in the separate experiment to study the isolated effect of the alternating compression and expansion waves. Results suggest no increases in RMS pressure from both walls, which means the imposed wave pattern does not increase instability.
6 The source of the unsteady process probably results from the interaction of unsteady waves generated past the main separation shock with the shear layer of the large separation region [44], as shown in Fig. 1.12.

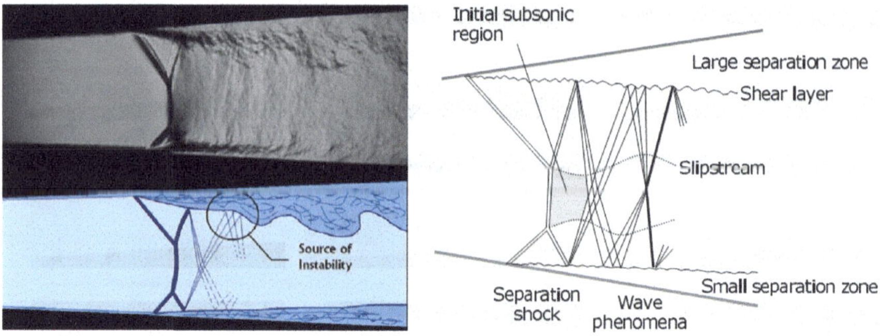

Fig. 1.12 Schematic of principal phenomena in supersonic nozzle flow separation [44]

1.2.3 Unsteady Combustion Dominated by Flow Instability

Unlike other parts, the flow dominating instabilities in combustor have not drawn enough attention. The separation induced unsteadiness in inlet and nozzle usually appears in off-design conditions (such as unstart and over-expanded). However, the large-scale separated region often exists in combustors which work under normal operating conditions (especially for high fuel equivalence ratio). Therefore, studying the separation-related phenomenon in supersonic combustor maybe even more important than the other parts of a scramjet. Moreover, due to the different configuration (converging inlet, constant-area isolator, and diverging combustor) and flow structures (normal shock train in the nozzle and oblique shock train in the combustor), the conclusions acquired from other parts of scramjet are probably problematic in combustors.

Up to now, only a few of researches have been focused on unsteady supersonic combustion processes which are dominated by flow instabilities. Laurence et al. [51] performed a series of experiments in the High Enthalpy Shock Tunnel Göttingen, in order to investigate the response of the HyShot II scramjet combustor to equivalence ratios close to the critical value at which the onset of thermal choking occurred. For the case with an equivalence ratio of 0.41, flow separated on the injector-side wall, which leads to the presence of large oscillations on the cowl-side wall. Based on the analysis from schlieren images, the high-frequency oscillation of shock train was observed. Fotia et al. [52] made contributions in searching out the mechanism of flame/shock-train interactions during the ram-scram transition. Under certain conditions of ramjet mode, they observed that some periodic low-frequency flame oscillations occurred. During the unsteady process, the oscillation of flame correlated well with pressure fluctuations. They concluded that the mechanism of flame oscillation was induced by a self-sustained shear-layer instability, associated with the flameholding cavity, as in Figs. 1.13 and 1.14. Yuan et al. [53] studied the flame stabilization characteristics in a dual-mode scramjet combustor with inflow Mach number of 2.5. It was found that the flame oscillated between the shear layer and the jet wake mode if the thermal

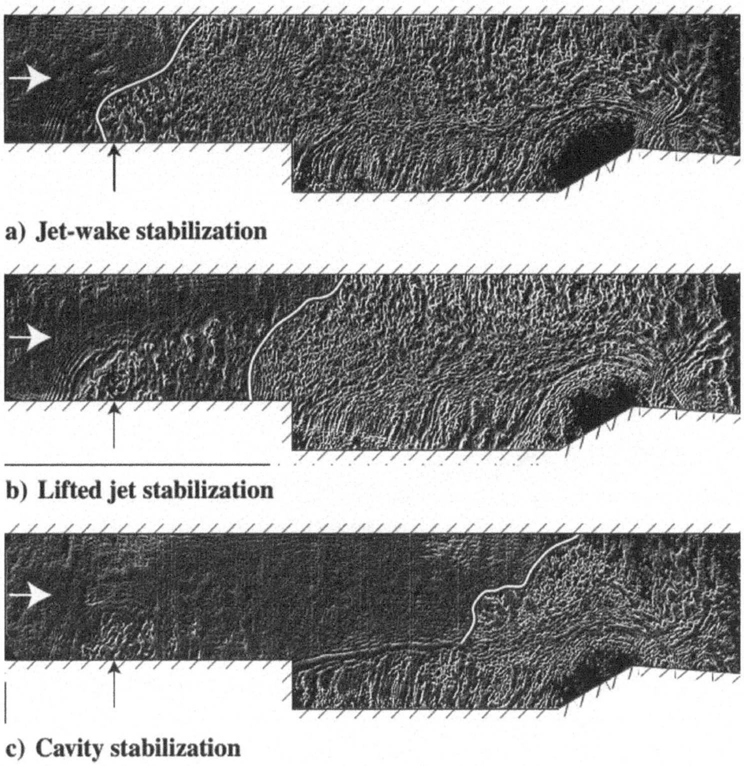

Fig. 1.13 Shearing interferograms of the **a** jet wake, **b** lifted jet, and **c** cavity flame stabilization modes in unsteady ramjet operation [52]

choke occurred around the injection location. They concluded that the short-lived aerodynamic throat was probably the cause of the flame oscillation.

Combustion is mainly affected by the mixing process when the combustor operates near the lean extinction limit. Instabilities accompanied by the flow features such as vortex, shear layer, shock wave, and boundary layer may strengthen the combustion unsteadiness. Flameholders and reasonable fuel injection patterns can improve combustion steadiness and make it more resistant to flow disturbances.

The flow in the combustor interacts with the combustion mainly through ignition and mixing processes. Therefore, stable combustion is often achieved by using recirculation zones to provide continuous sources of ignition, by well mixing the combustion products with fresh fuel and oxidant reactants [54]. Conventionally, swirl vanes [55], bluff-bodies [56, 57] and rearward-facing [58] steps are used as effective approaches to establish a recirculation zone for flame stabilization [55, 56, 58, 59]. These flameholders also introduce flow instabilities. In subsonic combustor, a bluff body could separate the incoming flow and develop shear layer instabilities. The alternating array of vortices shed from the trailing edge of the bluff body. These instabilities in most circumstances are responsible for initiating a blowout [60–62].

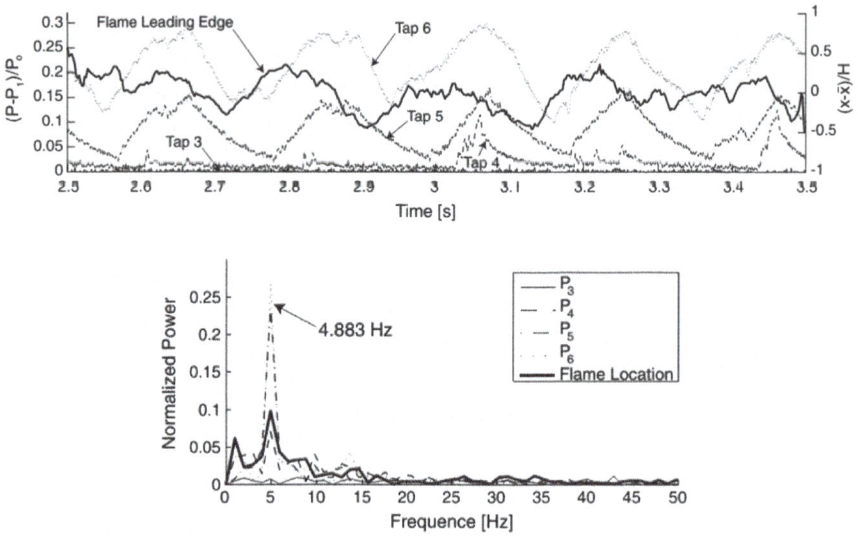

Fig. 1.14 Isolator shock-train static pressure traces (above) and spectra showing the dominant frequencies of the flame position and the pressure transducers (below) [52]

Gas turbines and aero-engines are used to apply swirl stabilizer to provide circulation zones [55, 63]. But the spinning speed of the swirler is limited to the velocity of coming flow. Considering the stringent NO_x emissions, industrial combustors must be operated near the lean extinction limit [64]. The combustion stability is more sensitive to the flow patterns.

Instead of aerodynamic features in swirl stabilizer, the trapped vortex combustor (TVC) uses geometric features to ignite the incoming fuel-air mixture and is less sensitive to unstable combustion [65, 66]. This concept was proposed by AFRL (Air Force Research Laboratory) in the 1990s or late 1980s [67] which is similar to the cavity used in scramjet. The conventional swirl-stabilized combustor and TVC are schematically shown in Fig. 1.15. A bluff or forebody is located upstream of a

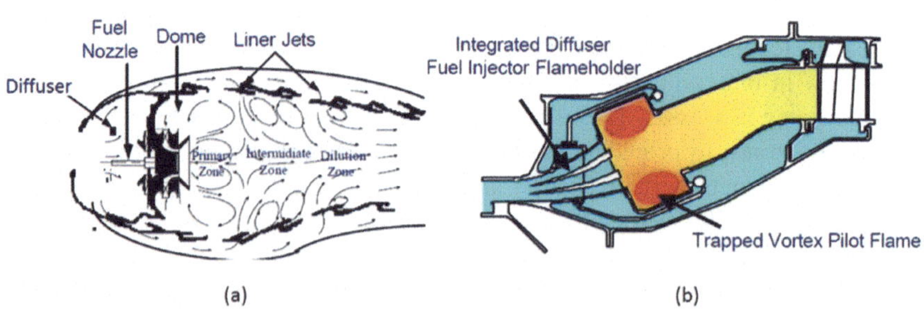

Fig. 1.15 Comparison between a conventional swirl-stabilized combustor and trapped vortex one. Adapted from Ref. [69]

smaller bluff body in TVC, and the vortices are trapped or locked between the two bodies [68].

In a TVC, air and fuel injection should be strategically-placed in the forward and rear walls of the cavity to drive the vortex contained. As the fuel is injected into the cavities, it is quickly mixed and burned in the stable trapped-vortex flow structure. The residence time of the supersonic and subsonic flow inside a cavity depends on the mass exchange rate in and out of the cavity. In the open cavities, mass and momentum transfer mechanisms are determined by the vortex structure inside the cavity and the longitudinal oscillations. Numerical results demonstrated [70] that there was one large vortex stationed near the trailing edge of the cavity and a secondary vortex near the upstream wall. The large trailing vortex interacts with the unstable shear layer and determines the mass exchange of the cavity. As the trailing edge vortex occupies a larger volume inside the cavity, the mass exchange is increased and the flow residence time inside the cavity is decreased. To minimize the combustion instability, the vortex must be "safely locked" in the cavity [71, 72].

There are also interests in whether spinning motion can improve the fuel-air mixing and combustion performance in TVCs. 3D streamlines in combusting flows with spinning motions are shown in Fig. 1.16a. Due to the swirling flows, strong tangential motion is introduced into the cavity vortex, and vortex breakdown is established in the sudden expansion region of the TVC. Turbulence kinetic energy and turbulence intensity are significantly increased. This indicates the fuel-air mixing can be dramatically improved. The spinning motion is found to lead to an increased combustion efficiency close to the spinning disc illustrated in Fig. 1.16b. And a study with a sudden change in swirl number is made. The transient results show that the cavity vortex is quite resistant to the flow disturbances. The vortex is trapped well in the cavity during the changing process.

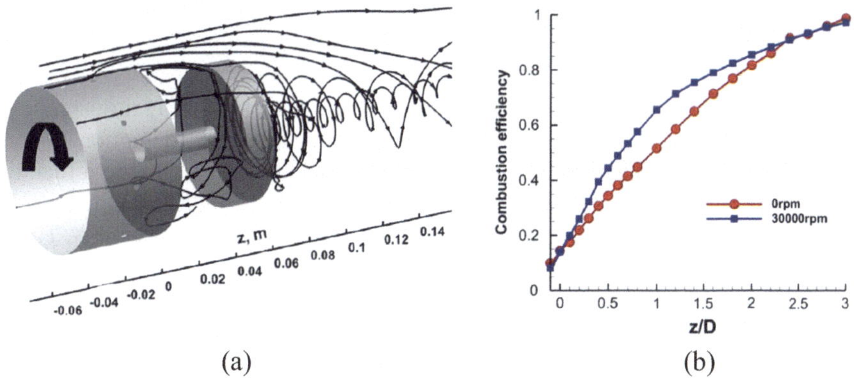

Fig. 1.16 3D streamlines of the combusting flows in the combustor: **a** 30,000 rpm, **b** Combustion efficiency for the non-spinning and spinning combustor [73]

1.2.4 Summary

Shock induced separation is a traditional problem in scramjet. Characteristic structures of reacting flow field are usually dominated by separation patterns when the heat release is strong enough. As the development of the experimental facility and computational capability, complex unsteady phenomena in non-reacting supersonic flow have been investigated by more and more researchers. The flow unsteadiness is believed to occur when the separation is severe, and is usually accompanied by asymmetric behaviors. The unsteadiness of separated flow is broadband, and main components of oscillation concentrate in the low-frequency band. Some studies suggest that interactions between separated shocks and shear layer instabilities account for the unsteadiness, while the majority of investigations only focus on the description of unsteady phenomena. It is believed that the unsteadiness in supersonic non-reacting flow has a strong relationship with the low-frequency unsteadiness in shock wave/boundary-layer interactions (SWBLI). Since the driving force for low-frequency unsteadiness in SWBLI is still controversial (upstream, downstream or combine), the cause of low-frequency oscillation in non-reacting flow remains not fully explained. In general, it is acknowledged that the downstream mechanism dominates for strongly separated SWBLI flows, and a combined mechanism (both upstream and downstream) dominates for weakly separated SWBLI flows. Meanwhile, the unsteadiness in supersonic combustion dominated by flow separation has not gained enough concern. Due to intrinsic complexities in shock wave/boundary-layer interactions, the unsteady combustion driven by back pressure induced separation is a challenging problem. Further studies are required to shed light on the unsteadiness of separation in reactive flows.

1.3 Ignition

Ignition is the beginning phase of combustion. The mechanism of the ignition process is widely investigated, and the effects of the parameters of the forced ignition system are clarified. In the scramjet combustor, the supersonic flow increases the difficulty of the ignition. Auto-ignition cloud also be induced with the high enthalpy flow. It changes the traditional ignition method and affects the combustion instability. Thus, the ignition process deserves extra attention.

1.3.1 Basic Concepts for the Forced Ignition

In low speed inflow conditions, the researches were concentrated on both spontaneous ignition (auto-ignition) and forced ignition (spark ignition). Many classical combustion articles have been widely reported. The technological applications in low speed inflow conditions are closely related to our daily lives, such as spark ignition in gas turbines and auto-ignition in diesel engines. A classical formula of minimum ignition energy ($E_{\min}$) [74, 75] is proposed to calculate the minimum energy required for forced ignition, as seen in Eq. (1.3). In the equation, c_p is defined as specific heat at constant pressure, ρ is the gas density, ΔT is the temperature rise due to combustion and d_q is the diameter equal to the quenching distance.

$$E_{\min} = c_p \rho \Delta T \left(\frac{1}{6}\pi\right) d_q^3 \tag{1.3}$$

Under the conditions of low turbulence, the equation of minimum ignition energy is given as

$$E_{\min} = c_p \rho \Delta T \left(\frac{1}{6}\pi\right) \left[\frac{Ak}{c_p \rho (S_L - 0.16u')}\right]^3 \tag{1.4}$$

Under the conditions of high turbulence, the equation of minimum ignition energy is given as

$$E_{\min} = c_p \rho \Delta T \left(\frac{1}{6}\pi\right) \left[\frac{Bk}{c_p \rho (S_T - 0.63u')}\right]^3 \tag{1.5}$$

In Eq. (1.4), k is the thermal conductivity, S_L is the laminar burning velocity, S_T is the turbulent burning velocity, u' is the value of fluctuating velocity, A and B are constants. According to Eq. (1.5), it is concluded that the minimum ignition energy increases with the increase in the thermal diffusivity and turbulence intensity, and it decreases with the increase in density and burning velocity. When the turbulence level is high, they also proved that larger turbulence scale increases the quenching distance and causes a significant enhancement in the minimum ignition energy accordingly, which can be easily understood from Eq. (1.5) when u' increases much faster than S_T. Therefore, for the gaseous fuel ignition process, a much larger ignition energy is needed to achieve a successful ignition process under severe turbulent flow fields.

The equation of minimum ignition energy for liquid fuels [75] as

$$E_{\min} = \left[\frac{\left(\frac{1}{6}\pi\right) c_{p,a} \Delta T_{st} D^3}{\rho_a^{\frac{1}{2}}}\right] \left[\frac{\rho_f}{\phi \ln(1 + B_{st})}\right]^{\frac{3}{2}} \tag{1.6}$$

where $c_{p,a}$ and ρ_a are the above mentioned nomenclatures for air, ρ_f is the density of fuel, D is the droplet diameter, ΔT_{st} is the stoichiometry temperature rise, B_{st} is the stoichiometry mass transfer number, and ϕ is the equivalence ratio. From Eq. (1.6), it can be seen that the minimum ignition energy of liquid fuel is strongly influenced by the drop size, and to a less extent affected by the equivalence ratio and fuel density. As a result, fuel evaporation process is the key factor affecting the ignition of quiescent liquid fuels.

The above spark ignition model has been extended to include the effects of finite chemical reaction rates and the presence of fuel vapour in the mixture flowing into the ignition zone. In the new model, different mixtures were represented by using different quenching distance equations. The detailed modified minimum ignition energy equations could be referred to Ref. [76].

The most important concept of the ignition energy is the minimum ignition energy (MIE) which is widely studied in low speed flows. From these equations of minimum ignition energy discussed above, fundamental requirements of ignition energy can be known based on sample calculations and easily achieved in many industry related combustion operations. However, for the flame kernel needs much more energy to resist the severe turbulent dissipation, many ignition methods in supersonic flows could provide ignition energy much more than MIE during the ignition process. Therefore, MIE is less focused in supersonic flows whereas flame behaviors with different ignition energy are concentrated.

The biggest challenge for the ignition process is the flame propagation at the initial ignition phase, which is highly affected by the fuel/air mixing as well as the turbulent flow field. During the past decades, effects have been gained on investigating the ignition flame propagation process, and it is found that turbulent-chemistry interactions are of vital importance to the above process. Considering the complexity of turbulent-chemistry interactions, however, accurate flame propagation models are rarely reported. Mastorakos [77] did comprehensive reviews on both the auto-ignition and the spark ignition processes in turbulent non-premixed flames under low speed inflow conditions and emphasized fundamental turbulent-chemistry interactions. Detailed descriptions about the research progress on the turbulent-chemistry interactions during ignition process can be referred to their research.

1.3.2 Effects of the Forced Ignition Methods

In a cavity-based scramjet combustor under a supersonic inflow condition, the inflow velocity is typically over 1000 m/s and the recirculation flow velocity inside the cavity is varied from approximately 0–200 m/s [78]. The velocity gradient from inside the cavity to the core flow brings a significant challenge to the flame propagation, which is likely to extinguish the weak flame during the initial ignition phase. It can be indicated that a much severe turbulent dissipation will occur during the ignition process in the supersonic flow. On the other hand, the stagnation temperature of the supersonic flow (such as, over 1500 K in Mach 6 condition and over 800 K in Mach 4 condition) is much higher than that of the atmosphere. The static temperature of the recirculation flow inside the cavity under supersonic inflow conditions is also estimated over about 700 K [79], which in turn creates a more chemical friendly environment. Thus, the turbulent-chemistry interactions of the cavity ignition process will be much more complicated than that reviewed by Mastorakos [77]. Nevertheless, considering the complex flow phenomenon occurring in supersonic flows, such as the fuel/air mixing, boundary layer separation and air stream shearing, the cavity ignition becomes a much more complicated physical process. As a result, much harder predicted ignition probability and more complex influencing factors are presented in the cavity ignition process in supersonic flows, leading to a relatively slow research progress in the supersonic research fields during the past decades.

As mentioned in the open literature, many new ignition methods have been developed to achieve successful cavity ignition in a supersonic flow, however, the most widely used igniter is still the spark ignition system. Although the spark plug is widely applied in our daily life, the mechanism of the spark ignition process is still complex and rarely focused especially in a supersonic flow. McNeil [80] studied the ignition process by electrical sparks and classified the spark ignition process into five stages as breakdown, electron heating, relaxation of internal plasma energy, shock wave generation and propagation, and subsonic flows. Brieschenk et al. [81, 82] conducted an experimental investigation of a capacitive-discharge spark ignition system in a cavity based combustor by utilizing Schlieren and luminescence imaging techniques in a Mach 6.6 flight condition. In their study, three ignition coils were compared with different spark plug gaps in order to evaluate the ethylene ignition performance and it was demonstrated that the spark plug gap is a vital factor affecting the ignition process. It was found that the ignition system parameters can be set to cause sufficient heating of the electrodes to obtain a successful ignition. Denman et al. [83] also applied spark plug to test a cavity flameholder in a Mach 8 shock tunnel. In their experiments, successful ignition and flameholding of ethylene and hydrogen were observed at an equivalence ratio ranging from 0.58 to 0.71. Although the inflow Mach number is relatively high with a favorable chemical reaction temperature, however, methane did not ignite at any tests. The spark ignition is mainly used in igniting gaseous hydrocarbon fuels under supersonic conditions. Spark ignition in liquid kerosene fueling cavity is rarely reported without the aid of other methods. The reason causing the above limitation is the shortage of spark ignition energy.

Pulse detonation igniter is recently developed and proved favorable for the ignition process in supersonic flows. Compared to spark discharge providing a small pure electrical energy addition for ignition, the pulse detonation provides a chemical heat release technique with a high pressure, temperature and radical-rich plume. Besides, detonation is a superior combustion form owing to the coupling between the shock and flame front which can provide rapid heat release and elevated pressure. Ombrello et al. [84, 85] did comprehensive and frontier investigations on the pulse detonation ignition method, and their pulse detonation igniter system is shown in Fig. 1.17. They selected propane (C_3H_8) and nitrous oxide (N_2O) as the fuel and oxidizer, respectively, because of their reliability of igniting and transitioning to detonation under the conditions needed for the experiments. They studied the cavity ignition processes in a supersonic flow by using two different ignition method, spark discharge and pulse detonation, and also discussed the effects of inflow distortion and mixing enhancement on the above ignition processes [84, 86]. Compared to the spark discharge, the pulse detonation can create an environment with a higher pressure and temperature which not only broaden the ignition limits but also cause a significant disruption to the cavity flow field. In addition, they emphasized the importance of cavity flow field

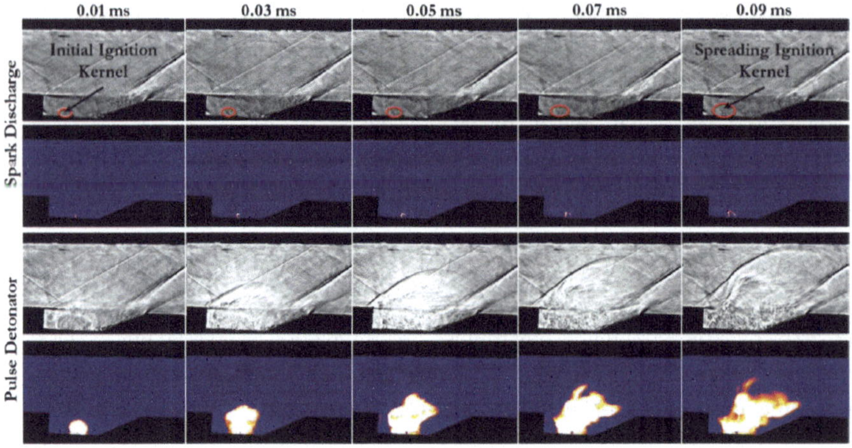

Fig. 1.17 Shadowgraph and CH* chemiluminescence images of cavity ignition processes excited by spark discharge and pulse detonation, respectively [84]

dynamics and fueling rate for a successful ignition. Recently, they applied simultaneous 100 kHz formaldehyde planar laser-induced fluorescence imaging in combination with CH* chemiluminescence imaging to investigate the transient ignition process in a cavity-based supersonic combustor. According to their research, it was indicated that there exists a strong correlation between the delay time from the onset of ignition to flame stabilization and the cavity fueling rate [87]. Furthermore, they studied the ignition mechanisms involved in transferring a detonation to a deflagrating scramjet cavity by clarifying the decoupling process of the detonation recently [88]. They proved that the shedding of high-temperature intermediate species is the primary mechanism governing successful ignition in the scramjet cavity. The pulse detonation is a promising ignition method with adequate ignition energy, and it will be more practical after decreasing the size of the igniter as well as the detonation fuel delivery system.

In the fields of cavity laser induced plasma (LIP) ignition, Yang et al. [89–91] performed a lot of experiments in hydrocarbon fueled cavity based scramjets in Mach 6 flight conditions. They applied LIP method to investigate single-pulse ignition [90], dual-pulse ignition [92], dual-point ignition [93] and also ignition mechanisms [94, 95]. According to their research, the energy of an individual laser pulse can be reduced by half via a dual-pulse LIP method as compared with a single-pulse LIP with the same total energy. Besides, a pulse interval shorter than 40 μs is suggested for dual-pulse LIP method. Even though they tried dual-point LIP ignition method, it is not suggested to apply in a real scramjet combustor owing to significant radical loss and heat loss as a result of spatial distribution of the plasma by dual-point LIP method. They revealed that the ignition process can be characterized into four stages: an initial plasma ignition stage, followed by the plasma-quenching stage, the re-ignition stage, finally, the stable flame stage. Despite detailed experimental observations mentioned above, numerical simulations were also performed to further study the mathematical model of dual-pulse laser ignition [96] and also the ignition process [97]. The significant advantages of LIP to other ignition methods are precisely controlled excitation energy, frequency as well as ignition location. In addition, it won't cause any disturbance to the flow field due to its non-intrusive characteristics. In the current LIP applications, laser system is rather big which is only suitable to use in a lab. In the future, LIP ignition method will become more and more practical after minimizing the laser system.

Electrical discharge ignition method is proposed to enhance the ignition ability of spark discharge and it belongs to the plasma-assisted ignition. In addition, the concept of plasma-assisted ignition not only includes ignition acceleration, but also the mixing enhancement and flame stabilization. Firsov et al. [98] conducted optimization of electrical discharge (10 kW) geometry in a Mach 2 supersonic inflow. It was reported the combined mixing/ignition geometry in which plasma penetrates into the fuel injector demonstrates a significant advantage in terms of ignition and flameholding. They also conducted experimental and numerical study on long spark plasma actuator for mixing enhancement in a supersonic flow [99]. It was demonstrated that there exists a strong correlation of jet instability with a local curvature of the discharge channel and the long spark discharge stretches the interface between fuel jets and supersonic inflow [100]. Savelkin et al. [101] investigated the ignition and flame stabilization processes in a Mach 2 supersonic inflow by an electrical discharge combined with an ethylene injector. In their study, a wall-fuel injector and a high-voltage electric discharge were installed into a single module, which demonstrated a significant advantage in terms of ignition and flameholding limits. Recently, Leonov et al. [102] conducted experiments to further study the ethylene ignition and flame stabilization by electrical discharge (15.7 kW) in a scramjet combustor and explored the sensitivity of the ignition dynamics to the plasma power. In the field of plasma-assisted ignition caused by electrical discharge, gliding arc plasma is also recognized to expand ignition and extinction limit with low energy consumption in the recent years. Wu et al. [103] experimentally demonstrated that gliding arc plasma (2 kW) could achieve combustion enhancement in a cavity based scramjet due to heating and chemical effects. In their study, it was concluded that the ignition limit is nearly equal to the blow-off limit of ethylene flame owing to the gliding arc. To achieve reliable and fast ignition in the scramjet combustor, Wu et al. [104] also proposed multichannel plasma igniter (MCPI) to induce a relatively larger ignition kernel at the beginning phase. They reported that the lean ignition limit of ethylene flame in a scramjet combustor through MCPI is expanded by 20–26% than that through spark ignition and the ignition time is reduced by about 50%.

To achieve a successful ignition in the scramjet engine, one of the ignition aids which is known as air throttling is to modulate the flow structures in the isolator and combustor in order to reduce the local flow velocity and increase the pressure, then the established shock train can facilitate ignition and flame stabilization. Li et al. [105] did comprehensive studies regarding air throttling ignition in an ethylene fueled scramjet combustor at flight Mach 5 condition. In their experiments, air throttling is activated after the cavity fueling is steady and the igniter is turned on. After

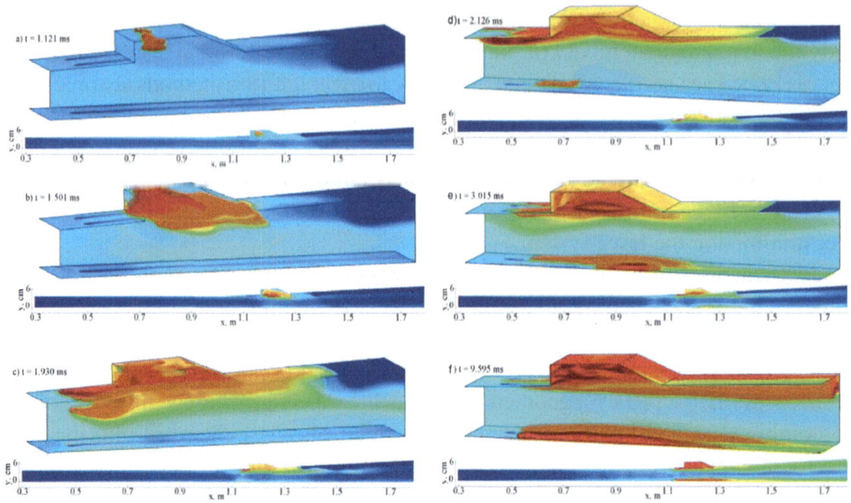

Fig. 1.18 Evolution of the temperature field in the cavity region during the ignition transient on the bodyside wall with air throttling ($\dot{m}_{throttle} = 20\% \ \dot{m}_{air}$) [106]

flame stabilization is achieved in the combustor, air throttling is then terminated to minimize the total pressure loss of the core supersonic flow. They numerically investigated the non-reacting and reacting flows during the air throttling working process [79, 89]. It was found that chemical reactions are intensified which produce sufficient heat release to maintain a flow environment conducive to flame stabilization. The temperature distribution is given in Fig. 1.18. As a result, a self-sustaining mechanism is established between the flow and flame development, and flame stabilization is achieved in the combustor even after the deactivation of air throttling [106]. In addition, they also studied control and optimization issues of the air throttling ignition process based on quasi-one-dimensional analysis, and a parametric investigation was conducted [107]. Noh et al. [108] numerically studied the auto-ignition process in an ethylene fueled cavity based combustor with air throttling. It was shown that air throttling increases the flow temperature and pressure as well as mixing enhancement in the pre-combustion region, which greatly improves the ignition efficiency and leads to a stable flame. Yang et al. [109] and Tian et al. [110] also conducted both numerical and experimental studies on the air throttling process in a kerosene fueled cavity based dual-mode scramjet combustor. They found that the combustion mode would change from supersonic combustion to subsonic combustion with throttling air injected into the combustor. The mode transition is mainly affected by the mass flow rate of throttling air and the throttling-off time [111]. They also classified the combustion into different parts to further study the mechanism controlling the air throttling process [112].

Obviously, self-ignition method is not suitable to work at low flight Mach number condition, and it still needs another igniter system to initiate the combustion of vaporized kerosene at high flight Mach number condition. It is also not favorable for the real scramjet applications.

Representative investigations of the above ignition methods discussed in this section are summarized as listed in Table 1.1, which presents recent research progress on the cavity ignition field. It is obvious that the ignition methods with consistent and newest concentration are pulse detonation, laser-induced plasma, electrical discharge and air throttling ignition methods. The advanced optical observation measurements as well as detailed simulations are conducted in these investigations. In addition, it is also indicated that the research routine is changed from experimental tests discussions in the early 2000s to detailed reacting flow field analysis nowadays with the aid of the development of experimental and numerical measurements. From the practical point of view, however, only laser-induced plasma and air throttling ignition methods of the above four are of real interest to the future scramjet applications owing to reported successful kerosene ignitions as listed in Table 1.1. Meanwhile, all the other ignition methods are still of great significance to understand the complex cavity ignition mechanism.

1.3.3 Effects of Auto-Ignition

The auto-ignition is expected to occur at high stagnation temperature conditions when the order of the ignition delay time is less than the order of the flow residence time in real scramjet operations.

Since flame propagation has significant influences on the supersonic combustion, it is difficult to clarify the effects of auto-ignition. Fureby et al. [122] investigated hydrogen-vitiated-air flames with two stage and alternating-wedge injection struts. The temperature of vitiated air was 830 K with Mach number 2.5 in the combustor. The combustion region is consisted of auto-ignition zones enfolded by self-igniting fronts embedded in the background of non-premixed flames. By introducing planar laser-induced fluorescence (PLIF) of OH and CH_2O, auto-ignition is observed in a jet flame with a vitiated co-flow of 1355 K by Gordon et al. [123]. Cabra [124] experimentally studied a lifted methane-air jet flame in a vitiated co-flow with total temperature of 1350 K, and Domingo et al. [125, 126] suggested that in the experiment, the turbulent flame base began with auto-ignition to provide the every first ignition point. When the temperature of co-flow air is 1550 K, Yoo et al. [127] suggested that auto-ignition provides every first ignition point in the turbulent flame base. Lu et al. [128] indicated that auto-ignition could be the controlling factor determining the lifted-off height under the temperature of 1150 K.

The unsteady phenomena are more complicated in the transverse jets. Turbulent structures carry fluid packets that are further broken down for efficient mixing. Shock wave discontinuities and their interactions with the boundary layer and recirculation zones create distinct regions of different mixing qualities. It increases the difficulty to investigate the effects of auto-ignition. Both auto-ignition and thickened flamelets were observed by Micka et al. [129, 130]. As shown in Fig. 1.19, a strong

Table 1.1 Information for the cavity ignition methods in partial literature

Authors	Ignition method	Technique	Inflow Mach number	Year	ER
Sun et al. [113]	Spark	Schlieren; high speed photography	1.92	2012	Hydrogen; 0.17, 0.34
Denman et al. [83]	Spark	Wall-pressure measurement; RANS	3.0–5.0	2016	Ethylene, hydrogen; 0.58–0.71
Ombrello et al. [87, 88]	Pulse detonation	Formaldehyde PLIF (100 kHz); Schlieren; CH* chemiluminescence	2.0	2017, 2018	Ethylene; 0.5–1.5
Yang et al. [90, 95]	Laser-induce plasma	CH* and OH* chemiluminescence	2.92	2018, 2017	Ethylene, 0.15; Kerosene, 0.23
Brieschenk et al. [81, 82]	Laser-induce plasma	PLIF; Schlieren	5.7	2014, 2013	Hydrogen
Leonov et al. [102]	Electrical discharge	Schlieren; High speed photography	2.0	2018	Ethylene; 0–0.2
Savelkin et al. [101]	Electrical discharge	Schlieren; High speed photography	2.0	2015	Ethylene; 0–0.16
Kim et al. [114]	Plasma torch	Schlieren; Wall-pressure measurement	2.0	2011	Hydrogen; 0–0.1
Matsubara et al. [115, 116]	Plasma torch	Wall-pressure measurement	1.0, 2.0, 2.5	2011, 2013	Hydrogen; 0.06
Li et al. [79, 89]	Air throttling	RANS	2.2	2015	Ethylene; 0.6
Tian et al. [110–112]	Air throttling	RANS; Schlieren; Wall-pressure measurement	2.0	2015–2017	Kerosene; 0.3, 0.6, 0.8
Xi et al. [117]	Piloted	Schlieren; High speed photography	2.52	2014	Kerosene (piloted ethylene)
Situ et al. [118, 119]	Hot jet	Photography; Wall-pressure measurement	2.15	2001, 2002	Kerosene; 1.11–1.51
Li et al. [120]	Hot jet	High speed photography; RANS	1.92	2012	Kerosene; 0.3, 0.5
Sung et al. [121]	Self	Wall-pressure measurement	2.5	1999	Hydrogen; 0.6

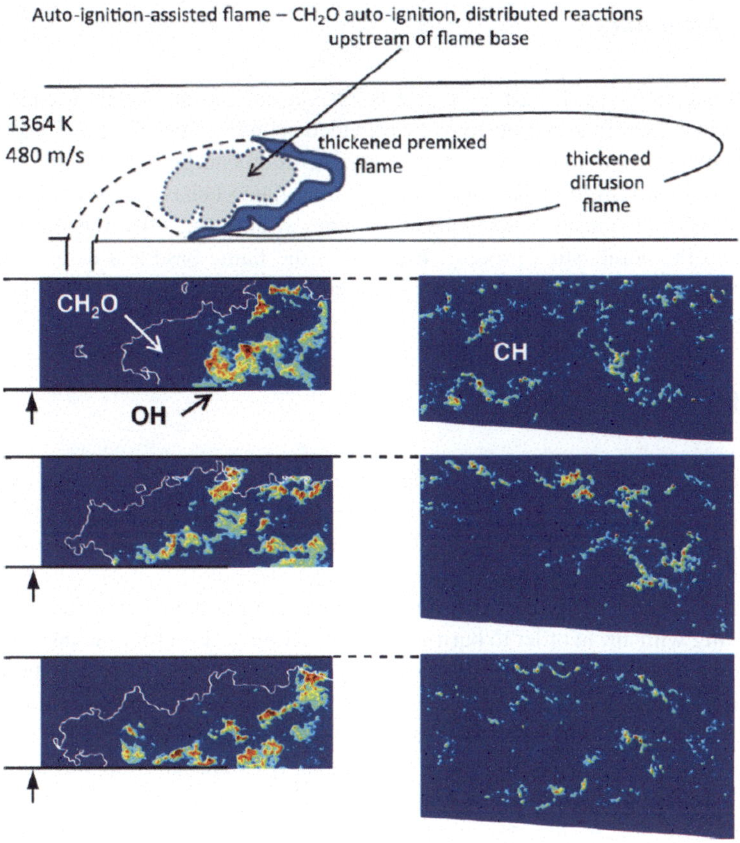

Fig. 1.19 Simultaneous formaldehyde-OH PLIF images [130]

formaldehyde signal was distributed in the lift-off region indicated that auto-ignition was presented. The auto-ignition existed upstream of the major heat release zone identified by OH signal, indicating the auto-ignition effected the flame base. They suggested that the combustion could be classified to an "auto-ignition assisted flame" under stagnation temperature of 1450 K.

Wang et al. [131] further reported that the flame front could be partially affected by the auto-ignition process of the combustible structure formed around the jet mixing layer. Under a very high stagnation temperature of 3750 K, auto-ignition was observed close to the bow shock, and combustion was a mixing-limited process mainly affected by auto-ignition [65].

1.3.4 *Summary*

Basic concepts for the forced ignition was introduced, and the widely used ignition methods in the scramjet combustor were provided in this section. The ignition process will be enhanced obviously with an increased ignition energy. Besides, the ignition method and the ignition location could affect the flame behaviors. When the flight Mach number increases, auto-ignition becomes significant, effecting the ignition phase and the combustion process. It provides the flame base and determines the lifted-off height and the flame may be stabilized as "auto-ignition assisted flame".

1.4 Flame Flashback

Flashback is the condition of the flame propagating down the hoses of an oxy-fuel welding and cutting system. The flame burns backwards into the hose, causing a popping or squealing noise, which affects combustion both in internal combustion engines and ramjets.

A number of studies [131–135] have been carried out concerning fuel injection and mixing with air, in order to burn completely within a short time. Transient flame flashback which is an indispensable key sub-process of combustion oscillation have been neglected for a long time because of the general thought that acoustic waves cannot propagate upstream in supersonic flowfield, and any flow oscillation resulting from an unsteady combustion process will be simply exhausted from the engine exit and will not interact with the flame zone [11, 12]. The first study to systematically investigate flashback limits is that of Lewis and von Elbe [136], whose model has remained as the state of the art for order-of-magnitude flashback predictions.

The existence of unsteady combustion process has been unfolded by many experiments. So far, the following reasons, such as DDT (deflagration–detonation transition) [134, 137], auto-ignition [138, 139], boundary layer separation [14, 140–142], and thermal choking [52, 143–146], have been considered responsible for the combustion oscillation in scramjet by different researchers.

1.4.1 *Flashback Due to DDT (Deflagration–Detonation Transition)*

When the fuel injection upstream of the cavity flameholder produces a premixed region with sufficiently high global equivalence ratio, a rapid flame flashback occurs against the incoming supersonic flow. The flashback develops explosively from the cavity pilot flame at regular intervals. Analysis of the experimental data suggests that the flame flashback is related to flame acceleration similar to deflagration-to-detonation transition. With high equivalence ratio, the immediate flame re-ignition

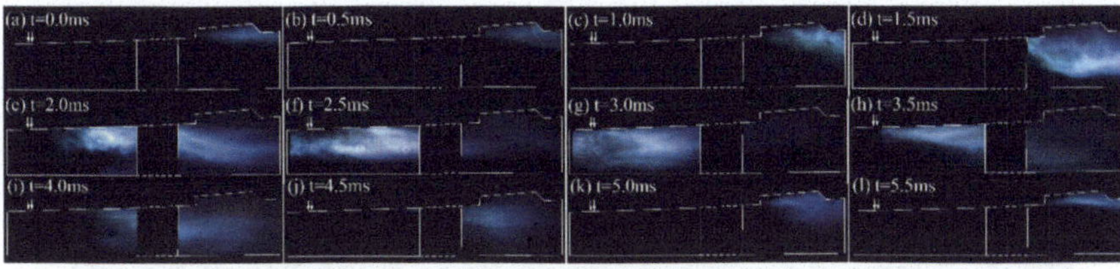

Fig. 1.20 Typical luminosity movie of flame flash-forward and blow-off event between cavity stabilized location and injection location [134]

and intense flame flashback decreases the flame blow off duration and finally leads to a more efficient heat release compared to the cases with lower equivalence ratio [132]. Many experiments shows that the flashback is caused by DDT (deflagration–detonation transition) [134, 147, 148]. O'Byrne et al. [140] also observed the flame flashback phenomenon, which is attributed to transition from diffusive to pre-mixed combustion. Aiming to compare the combustion stabilities of various mixing conditions, Wang et al. [134] designed several different injection schemes to investigate combustion instabilities inside an ethylene-fueled scramjet combustor. Figure 1.20 shows the high-speed flame luminosity images of the combustion oscillations. The images demonstrate the details of a flame flash-forward (from C0 cavity to I31 injection location) and flash-back. From Fig. 1.20d–g it is seen that the propagation of the flame from the cavity to the fuel jet occurs in the main flow, not just in the boundary layer. The flame base moves forward very quickly (Fig. 1.20c–e) until it reaches the fuel jet location (Fig. 1.20f).

Zhu et al. [149] investigated flame stabilization and propagation inside a kerosene-fueled two-stage strut dual-mode scramjet combustor experimentally, as shown in Fig. 1.21. The flame flashback equivalence ratio is much higher than the flame blowout limit, and the upstream strut can get a stabilized flame once it is reignited. A higher stagnation temperature and a lower inflow Mach number are advantageous to trigger flame flashback.

1.4.2 Flashback Due to Boundary Layer Separation

As previous references show [14, 140–142], the boundary layer separation can lead to the flame flashback. Usually, a disturbance is observed to upstream. This disturbance is interpreted as being due to separation of the boundary layers caused by the adverse pressure gradient to which the boundary layers are subjected. The separation reduces the effective flow area, thereby restricting the flow and pushing a shock (or shock trains) upstream to eventually unstart the duct [14]. Detailed shadowgraph images showed that the injection behind the pylon removed all the liquid from the wall surface, which is important for the elimination of potential flashback. These studies indicated that addition of fuel either in the inlet or in the scramjet's isolator

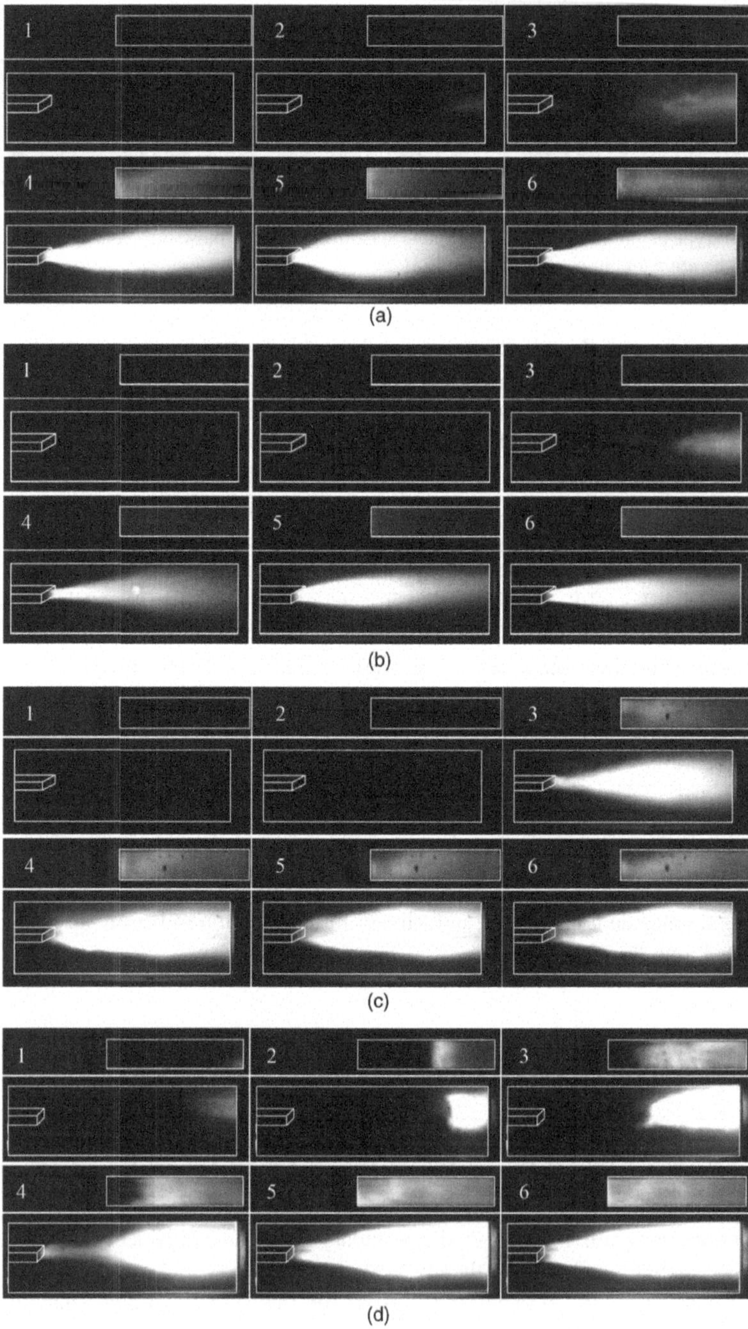

Fig. 1.21 Flame flashback captured near the upstream strut [149]

led to enhanced efficiency of the combustor operation. Flashback or inlet unstart can be avoided if careful consideration is given to fuel penetration, residence time, operational conditions, and fuel physicochemical properties. The main results indicated that thin pylons with sharp leading edges do not introduce significant pressure losses or flow distortion in the isolator airflow, but they create a region of low pressure, which has a great impact on fuel penetration. Even at moderate dynamic pressure ratios, the presence of these pylons promotes substantially higher penetration in comparison with simple wall injection. Because of this increased penetration, the entire liquid jet is lifted from the wall, eliminating the danger of flashback through seeding of the boundary layers with a combustible mixture [142].

Ducruix et al. [3] experimentally explored the combustion oscillation mechanism and control methods in an air-breathing engine. They found processes involving acoustic/flame coupling, unsteady strain rates, flame response to inhomogeneities, interactions of flames with boundaries, and flame/vortex interactions. Tian et al. [112] and Yang et al. [109, 150] found an intensive combustion in the whole cavity and wall boundary layer could be achieved with a careful air throttling method, or else leading to the combustion oscillation. The two processes might interact with each other, which would render the oscillations more complicated. As Fig. 1.22 shows, the shock waves kept moving upstream from the cavity region into the isolator, and then the flame near the cavity ramp propagated upstream along the cavity wall.

The investigations of Gruber et al. [151], which involved high-resolution experimental measurements and direct numerical simulations, focused on the characterisation of flame flashback for premixed and preheated hydrogenated flames in turbulent boundary layers. Those researchers showed that the near-wall speed fluctuation pattern found in turbulent boundary layers causes wrinkling of the initially flat flame sheet as it starts propagating against the direction of main flow, and that the structure of the characteristic streaks of the turbulent boundary layer has an important impact on the resulting flame shape and its propagation mechanism. They also indicated that flame flashback should be attributed to thermal choking, which is caused by downstream combustion and an adverse pressure gradient.

Figure 1.23 illustrates several stages of the unsteady flame propagation along the channel walls. The surfaces visualized represent:

(i) the streamwise velocity normalized by the friction velocity $u^+ = u/u_\tau$ on the $y^+ = 5$ plane (greyscale flooded contours on the plane parallel to the wall);
(ii) the wall-normal vorticity ω_y on the $y^+ = 5$ plane denoting the streamwise vorticity structures of the boundary layer (white lines on the plane parallel to the wall, solid and dashed lines denote opposite sign);
(iii) the flame surface as described by the progress variable isosurface, $C = 0.7$ (red isosurface);
(iv) back-flow regions characterized by negative streamwise velocity and located upstream of the flame surface portions that are convex towards the reactants (blue isosurfaces).

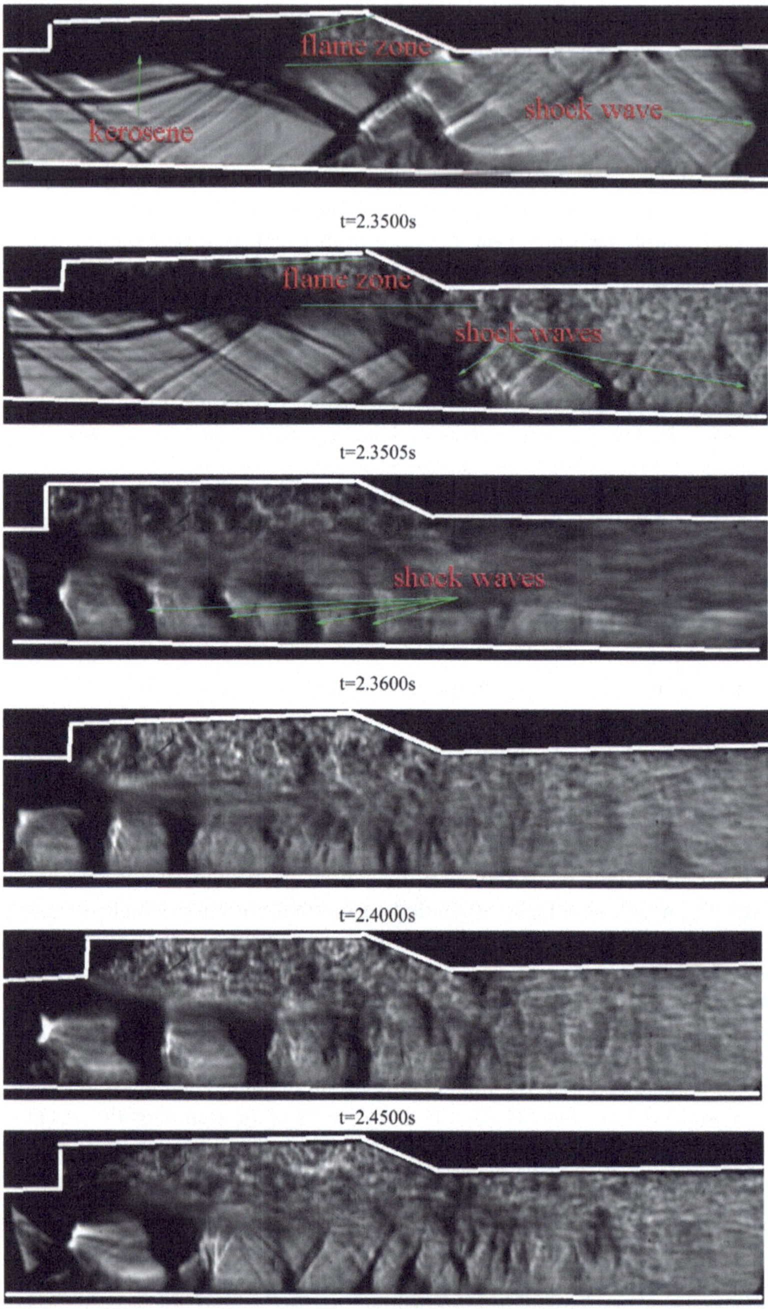

Fig. 1.22 The high-speed schlieren images of part-c and part-d combustion process [112]

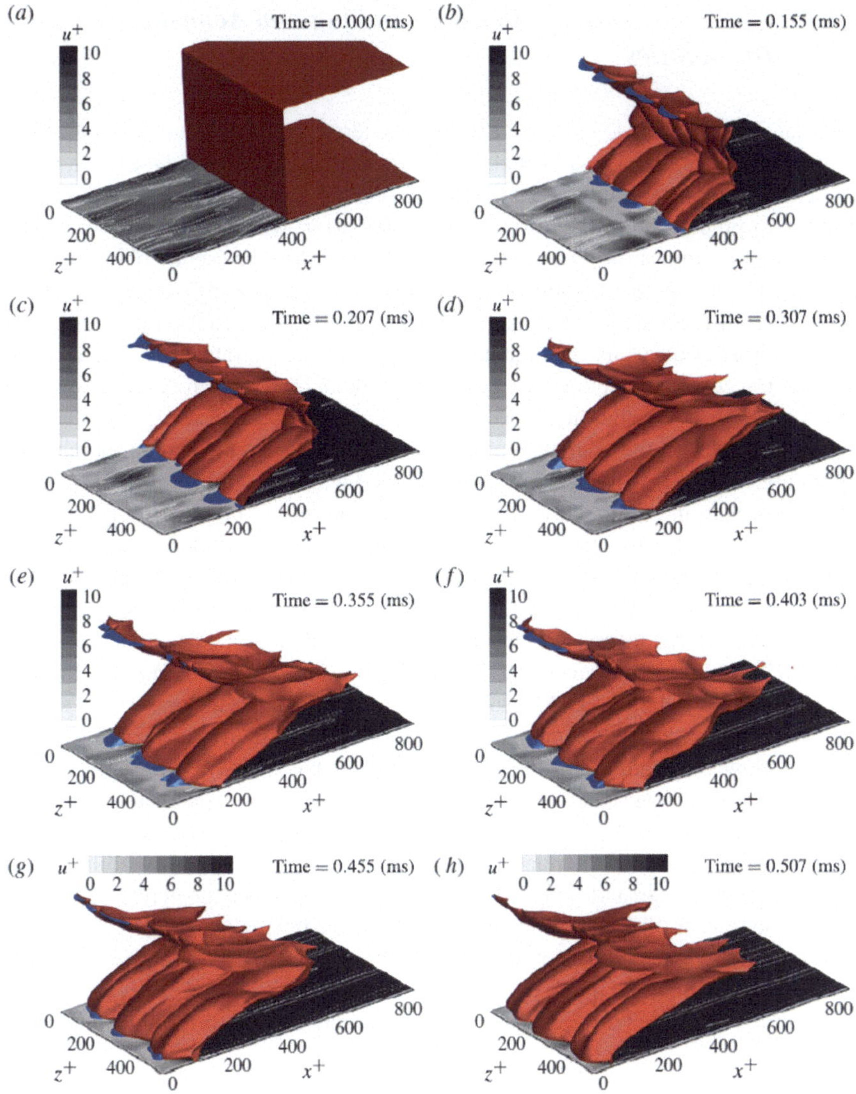

Fig. 1.23 Temporal evolution of the premixed flame (C = 0.7), red isosurface, and of the backflow regions, blue isosurfaces, from the beginning to the end of the reactive simulation. The non-dimensional streamwise velocity (greyscale flooded contours) is shown on the $y^+ = 0.5$ plane together with the trace of Y vorticity (white lines, solid and dashed patterns represent opposite sign) [151]

1.4.3 Flashback Due to Thermal Choking and Acoustic Instabilities

Thermal choking [52, 143–146] have been considered responsible for the combustion oscillation in scramjet. Larsson et al. [144] numerically investigated the mechanism of thermal choking using large eddy simulations (LES). The first objective of their study is to predict the pressure-rise in the combustor, which is within the experimental bounds, and shows reasonable grid-convergence. The second objective is to study the flow for increased fuel/air equivalence ratios. They also estimated the effect on the overall combustor performance. Karl et al. [145] focused on the numerical investigations of unsteady phenomena at large equivalence ratios. The main result was that the combustion efficiency decreases with an increasing fuel mass flow rate which coincides with significant flow separation. In the past decades, the effects of fueling schemes on the combustion stability characteristics in scramjet combustor equipping cavity flame holders have been studied widely. For high fuel-equivalence ratios, however, the flame base or combustion zone might be pushed upstream of the cavity intermittently due to the large separation of the upstream boundary layer and enlarged cavity recirculation, resulting from the intense heat release around the cavity and the interaction between the jet and the cavity shear layer. Laurence et al. [143] concluded that the primary mechanism responsible for the development of the transient shock system is thermal choking by OH* visualizations (as Fig. 1.24 shows) and numerical simulations. OH* visualizations did not indicate the presence of strong separation features propagating upstream with the shock train near its point of formation, suggesting that the driving mechanism for the transient development was thermal choking. Nevertheless, boundary-layer separation was observed to develop on the injector-side wall when the shock train had moved further upstream.

Lin et al. [13] studied the relationship between thermo-acoustic instabilities and the flame flashback. In addition, Ma et al. [11] obtain the mechanism responsible for driving the flow oscillation was identified as the acoustic-convective. The details are introduced in chapter 1.1.3. Rossiter et al. [152] developed semi-empirical formula to predict the resonant frequency of the compressible flow-induced cavity oscillation based on the coupling between the acoustic radiation and the vortex shedding.

1.4.4 Summary

There are several factors which causes flame flashback under different conditions, including DDT (deflagration–detonation transition), boundary layer separation, and thermal choking and acoustic instabilities.

Despite all the investigations performed, the physical mechanisms of the low-frequency oscillations in scramjet engines remain unclear. Additional studies are required to clarify the origin of this important phenomenon which controls scramjet performance and efficiency. The uncertainty and disagreement of these opinions

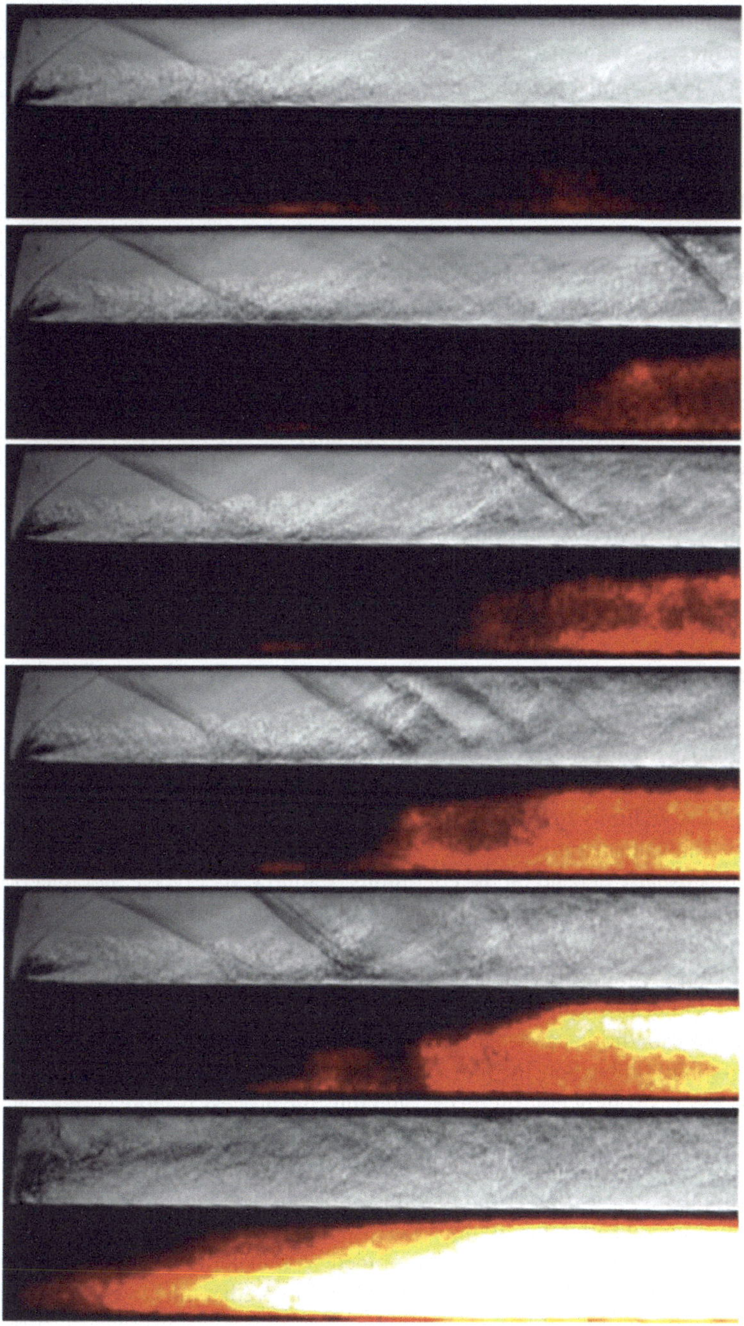

Fig. 1.24 Sequences of quasi-synchronous schlieren and OH* chemiluminescence images of the flow near the injector (x = 56–136 mm) for an equivalence ratio of $\varphi \approx 0.66$ [143]

require continuous and further research on this unsteady process. According to the public literatures, the influence of injection parameters on the combustion oscillation and flashback in a scramjet combustor has barely been researched so far.

1.5 Combustion Near Blowout Limits

The unsteady nature of turbulent combustion has been observed over a wide range of conditions including subsonic flow [56, 153], supersonic flow [131, 135], premixed combustions [154], and non-premixed combustions [155]. While in this section, the unsteady characteristics of combustions near blowout limits in supersonic flows are concerned specially, since it has been observed that the flame unsteadiness in a supersonic flow is remarkably increased when approaching blowout limits [156]. Since the combustion behaviors near lean and rich blowout limits are considerably different from those in moderate combustion, it is necessary to study the blowout limits and the corresponding flame behaviors of supersonic combustion. The existing relevant literatures are selected and summarized below.

1.5.1 Blowout Limits

Blowout limits, including rich blowout limit (RBO) and lean blowout limit (LBO), are the boundaries between stable combustion and flame blowout. Experimentally, the rich or lean blowout limits are respectively defined as the maximum or minimum fuel flow rates or equivalence ratios that are able to sustain continuous combustion. The blowout limits usually depend on the conditions of inflow, fuel types, fuel injection patterns, and the configuration of the cavity [157]. Most of the previous researches concentrated on premixed flames, Ozawa curve [158] gave a parabolic correlation of the blowout limits for a premixed system, which indicated that the flame can only be stabilized within a certain range of conditions thus either excess or insufficiency of any dominant condition can lead to unsteadiness.

For supersonic combustion, the fuel is injected into supersonic air inflow, the combustion process is strongly non-premixed, and thus the process is more complicated than premixed combustion [159]. A number of previous experiments [160–162] have demonstrated that non-premixed lifted jet flame is stabilized on the stoichiometric contour. Although fuel and air are non-premixed at the beginning, the condition at flame base is premixed after the mixing process within the lift distance. Based on the view of non-premixed combustion and combined with experimental data, Driscoll and Rasmussen [163] developed a correlation model to predict the blowout limits of a cavity in supersonic flow. The model avoided the assumption of perfectly stirred reactor (PSR) in premixed flames, and could be applied to cavities, steps, and struts in supersonic flows.

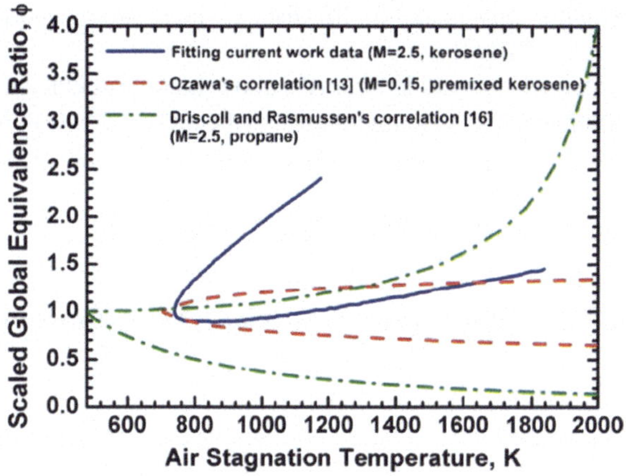

Fig. 1.25 Comparison of three existing correlations [157]

In order to study the effects of different conditions on the blowout limits, Zhang et al. [157] performed several sets of experiments in supersonic combustors. The stagnation temperature of the inflow and the injection pattern were observed to be the dominant parameters, while the impact of air stagnation pressure and the combustor divergence angle were negligible. Figure 1.25 shows the comparison of the three different correlations mentioned above. Experiments performed by Donohue et al. [164] also gave the same pattern of blowout limit correlations.

Retaureau et al. [165] also observed the significant impact of the free stream temperature on the stability domain. In addition, it was reported that the combustion process was barely stable when the static pressure in the combustor was low, thus a certain level of combustor pressure was also crucial in the blowout mechanisms. Different fuel types were also compared, hydrogen was preferred for a wider range of stability domain while ethylene was less sensitive when the static pressure was low.

The mechanism of flame blowout was associated with combustion modes by several researchers. Le et al. [166] divided the combustion stabilization mode into two different types: shear layer stabilization mode and recirculation zone stabilization mode. In the former mode, the flame base was anchored downstream of the cavity leading edge. While in the latter mode, the flame base was attached to the top of the cavity leading edge, and the flame was deep in the cavity, which lifted the shear layer up, expanded the mixing area of fuel and air, and thus more fuel could enter into the cavity. The flame was stabilized in the shear layer when near LBO limit, as the equivalence ratio decreased, less fuel was in the shear layer, and the flame base moved downstream. When the flame base shifted out of the shear layer, blowout occurred. As for RBO cases, the flame was stabilized in the shear layer at the beginning, the spread velocity of the flame at flame base was decided by the temperature of upstream unburnt gas. When fuel flow rate rose up to RBO limit, more cold fuels entered into the recirculation zone, lowering the heat release, the spread velocity then dropped, which finally made the lift distance increase beyond the length of shear layer and reaches RBO limit.

It can be concluded from the above literatures that the blowout limits are essentially dominated by the temperature and the fuel distribution. When combustion takes place near blowout limits, the change of the local conditions lead to different behaviors of the margin states.

1.5.2 Combustion Behaviors Near Blowout Limits

The combustion behaviors near blowout limits are remarkably different from those in stable states, and have considerable characteristics of unsteadiness. As concluded above, when the flow and reaction conditions approach critical points, some of the parameters intermittently overstep the boundary of stable combustion, which can couple with other sources of oscillations and enhance the degree of unsteadiness. The combustion behaviors near blowout limits, therefore, have drawn some attentions of the researchers.

The distributions of the reaction zones were initially observed to be different between stable states and margin states. Rasmussen et al. [167–169] reported the flame location within the cavity at different fueling rates, and analyzed the qualitative effect of heat release in the cavity on flow oscillations. Figure 1.26 shows the location of reaction zone near LBO and RBO limits. Near LBO, reaction zone moved into cavity volume for both wall injection and floor injection; while wall fueled flame moves to rear of cavity and floor fueled flame extended to length of cavity when RBO is approached.

Lin et al. [170, 171] investigated the flame structures and operating limits of an ethylene-fueled recessed cavity flameholder both experimentally and numerically. Figure 1.27 shows the instantaneous images from a high-speed video camera for cavity flames with various independent cavity fuel flow rates, and near LBO the

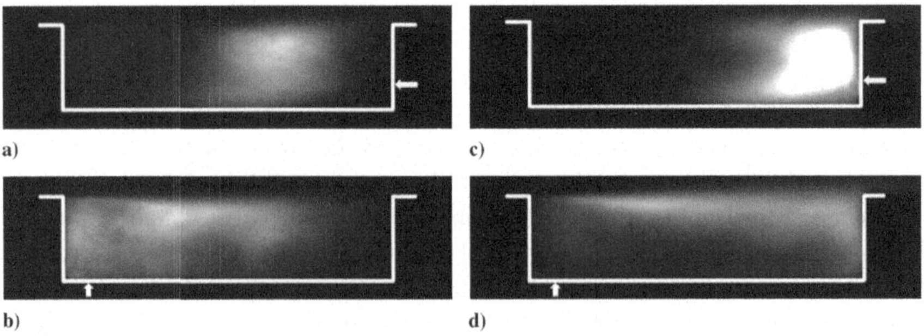

Fig. 1.26 Near lean blowout, reaction zone moves into cavity volume for both **a** wall injection and **b** floor injection; at high fueling rates approaching RBO, wall fueled flame **c** moves to rear of cavity and floor fueled flame **d** extends to length of cavity [168]

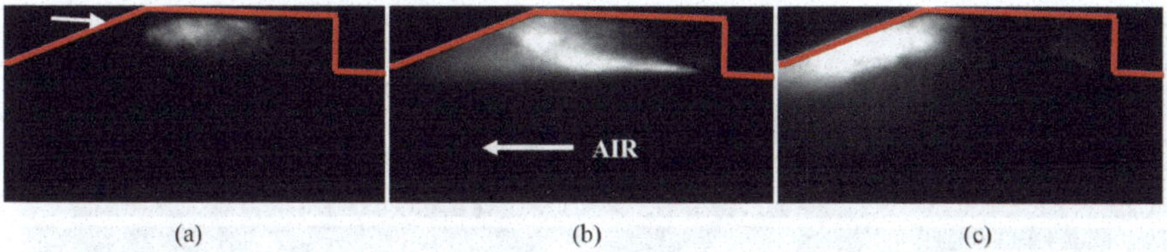

Fig. 1.27 Instantaneous images from a high-speed video camera for cavity flames with various independent cavity fuel flow rates for Mflight = 5.0 and q = 1000 psf. **a** Near lean blowout limit, 0.0011 lb/s, **b** Near rich ignition limit, 0.0047 lb/s, **c** Near rich blowout limit, 0.0089 lb/s [169]

flame was anchored near cavity floor and near RBO the flame moves downstream to cavity trailing edge. The numerical study also discovered the same characteristics near LBO.

The difference in the flame locations is partially due to fuel transport and convection. Choi et al. [172] employed Linear Eddy Mixing (LEM) Model as a subgrid closure for turbulent combustion in Large Eddy Simulation (LES), and numerically investigated stable combustion and blowout limits. The results showed that the fuel (a blended mixture of methane and hydrogen) was mostly found in the injection region. Some methane was convected along the shear layer and mixed with oxygen in the aft region of the cavity, but not enough hydrogen was present. It was observed that in the blowout case the flow structure inside the cavity was substantially different from the stable case.

Further researches indicated that the flow field would also change significantly when approaching blowout limits. According to Ghodke et al. [173], in stable combustion, a large vortical flow was formed in the aft region of the cavity that provided an effective mechanism to transport the hot products and to enhance mixing of the fuel-air mixture. While small localized vortical flow structures were seen in the near-blowout case that did not seem to be as efficient in mixing the hot products and/or in transporting the premixed mixture. Tuttle et al. [174] also observed increased streamwise oscillations in the flow field near rich blowout limit through particle image velocimetry (PIV).

Researches employing high frequency detection techniques further discovered the enhanced unsteady behaviors of combustions near blowout limits. Allen et al. [175] used standard deviation images to characterize the flame unsteadiness, and reported fluctuations of the flame distribution. Gruber et al. [156] reported increased emission fluctuations near lean blowout limit. The frequency spectra showed that the oscillations near lean blowout limit were governed by low frequencies below 250 Hz, and it was mainly caused by the reignition processes.

Hammak et al. [155] employed high-repetition-rate OH PLIF to observe the unsteady phenomena corresponding to the supersonic combustion in a cavity flameholder near lean blowout limit. As shown in Fig. 1.28, the averaged result showed that the flame was mainly stabled close to the aft wall of the cavity, while the sequential images showed that the flame was periodically convected and the combustion was extremely weak.

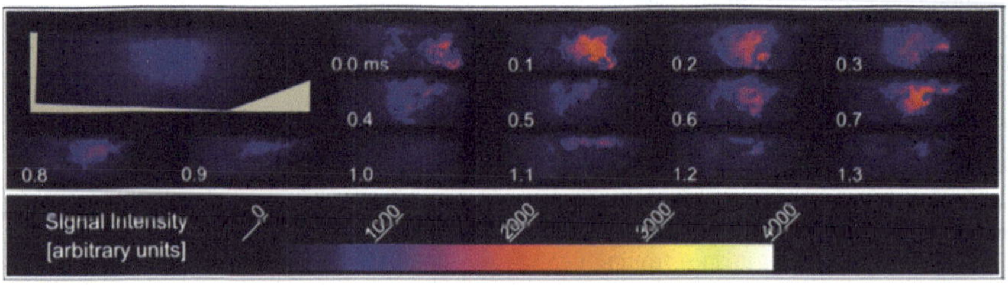

Fig. 1.28 Average and sequential images for lean blowout condition [155]

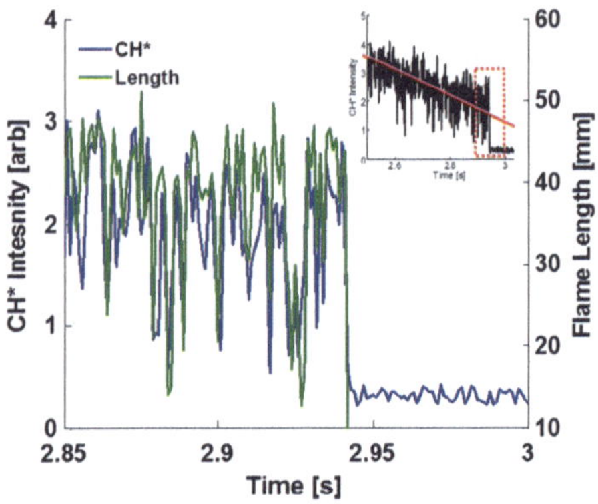

Fig. 1.29 Temporal comparison of integrated CH* signal and flame length in the moments before LBO showing that shortening of the flame is correlated with decreased heat release. The inset image depicts a general downward trend in overall CH* signal as the equivalence ratio is reduced [154]

Allison et al. [154] detected the dynamic characteristics of premixed ethylene flame near LBO by high-speed CH* chemiluminescence. In Fig. 1.29 the integrated CH* signal intensity is compared with the flame length, which reveals that the strong oscillations of the flame near LBO are correlated with the heat release.

1.5.3 Summary

Blowout limits are the boundaries between stable combustion and flame blowout, including rich blowout limit and lean blowout limit. The combustion behaviors near blowout limits are distinctly different from stable combustion, the flame unsteadiness in a supersonic flow is remarkably increased when approaching blowout limits. Supersonic combustion is strongly non-premixed, and the process is more complicated than premixed combustion. The change of the local conditions lead to different behaviors of the margin states. The existing literatures observed the unsteady characteristics near blowout limits, and quantitative analyses indicated that the unsteady nature could be associated with the flow field oscillations, the reignition process, and the heat release fluctuations in the margin state.

The behavior of the flame that is actually experiencing the blowout process is still observed in recent studies, which provide more information in characterizing the unsteady behaviors of combustion near blowout limits. In order to describe the unsteady behavior near blowout limits, it is necessary to further study the flame mechanism near blowout limits.

1.6 Discussion

A review of unsteady combustion physics in supersonic flows has been presented, trying to construct a framework for unsteady supersonic combustion and provide a guideline for future study. The topics cover several fundamental aspects, including interactions between acoustic wave and flame, Flow Dominating Instability, ignition, flame flashback and near-blowout combustion. The conclusions are drawn as below.

It is a common assumption that acoustic waves cannot travel upstream in a supersonic flow, and any flow oscillations arising in the flame zone will simply travel downstream and exit from the engine without forming the feedback loop required to sustaining combustion and flow instabilities. However, the researches indicates that there are various subsonic flow regions in scramjets. Acoustic wave induced by perturbations can propagate upstream in subsonic regions embedded in supersonic flows. Interactions among flame, shock wave and fuel injection may give rise to acoustic feedback loops, and pay a contribution to the sustainability of unsteadiness in supersonic reacting flows. Interactions of combustion and acoustic wave in scramjet combustors still need to be further explored so that the thermoacoustic effects may be controlled efficiently. In this work, the researches about the combustion oscillation characteristics in the scramjet combustor will be introduced, which are shown in Chap. 2.

The flow dominating instability is believed to occur when separation is severe, and is usually accompanied by asymmetric flow field structures. The frequency feature of unsteady separated flow is broadband, and major frequency components of oscillation concentrate in low frequency. Interactions between separated shocks and shear layer instabilities may be account for a certain type of unsteadiness. Due to the intrinsic complexities in shock wave-boundary layer interactions, the mechanism of unsteady combustion induced by backpressure forced separation remains unclear. Further studies are required to study the unsteadiness of separation in reacting flows. Chapter 3 focuses on the study of combustion unsteadiness dominated by shock-induced separation. For a rectangular supersonic combustor with dual parallel cavities and near-cavity fuel injections, flowfield structures generally transform from symmetry into asymmetry as the increment of equivalence ratio. Under an intermediate equivalence ratio, intermittent dynamic combustion occurs with a high-amplitude pseudo-shock oscillation in the streamwise direction. A decoupling analysis is carried out to discover the key impact factor of this unsteady combustion, it is found that the flame, fuel jet as well as cavity flameholder do not play a key role in this issue. Thus, cold flow analysis is applied, and the typical flame structures are fully reproduced by the backpressure induced separated flowfield. Specifically speaking, a symmetric separation under low backpressure generates symmetric combustion, while an asymmetric separation under high backpressure results in asymmetric combustion. Based on the detached-eddy simulation, the whole process of symmetric/asymmetric separation transition (occurs under threshold backpressure) is captured. Boundary layer separation tendency analysis shows that an interlaced shape factor distribution of the boundary layers from both walls accounts for the switch of separation modes.

Although significant research progress regarding cavity ignition in a supersonic flow has been achieved in the recent years, the understanding of the cavity ignition mechanism is still far from being satisfactory. The effects of auto-ignition on combustion stabilization have also been analyzed preliminary, and qualitative results have been obtained. The recent research of our work on the characteristics of the ignition process in the cavity-based supersonic combustion will be introduced in Chap. 4 typical ignition processes under diverse ignition methods will be presented, involving the spark ignition, the piloted ignition, the gliding-arc-discharge ignition and LIP ignition. This section aims to first introduce general ignition methods with their features, and second provide an intuitive observation on the dynamic ignition process. Tthe flame behaviors during ignition is emphasized from two aspects, namely flame kernel formation and initial flame propagation. The studies of flame kernel formation and flame propagation are conducted in single-cavity combustors, showing the effects of ignition energy, ignition position and rear wall height. The analysis of the ignition mechanism is introduced and the effects of the auto-ignition in the combustor with transverse injection is primarily explained.

Due to relatively long premixing length, a lean premixed gas mixture often induces the combustion oscillation in scramjet engines, which can directly degrade the engine performance and reduce its life cycle. Hence, it is important to explore the mechanism of the flame flashback, and devise possible control methods. Chapter 5 would show the experimental results and discusses the factors inducing combustion oscillation, such as the global equivalence-ratio, the pre-mixing distance, the injection angle, and the jets number. A simplified combustion opening system model will be constructed for mechanism analysis.

Studying the mechanism of blowout limits is of great value for understanding the unsteady characteristics of turbulent combustion and the flame behaviour near blowout. The existing literatures observed the unsteady characteristics near blowout limits, and quantitative analyses indicated that the unsteady nature could be associated with the flow field oscillations, the re-ignition process, and the heat release fluctuations in the margin state. In Chap. 6, blowout limits of cavity flame in supersonic flows is firstly analysed and modelled. Then, mixing and combustion characteristics with different injection schemes in low equivalence ratio conditions are investigated experimentally and numerically. At last, the flame blowout and re-ignition of premixed flame and non-premixed flame are studied systematically.

References

1. Curran, E. T. (2001). Scramjet engines: The first forty years. *Journal of Propulsion and Power, 17,* 1138–1148.
2. Rayleigh, J. (1878). The explanation of certain acoustical phenomena. *Nature, 18*(455), 319–321.
3. Ducruix, S., Schuller, T., Durox, D., & Candel, S. (2003). Combustion dynamics and instabilities: Elementary coupling and driving mechanisms. *Journal of Propulsion and Power*, *19*, 722–734.
4. Balachandran, R., Ayoola, B. O., Kaminski, C. F., Dowling, A. P., & Mastorakos, E. (2005). Experimental investigation of the nonlinear response of turbulent premixed flames to imposed inlet velocity oscillations. *Combustion and Flame, 143,* 37–55.
5. Hardi, J., Oschwald, M., & Dally, B. (2011). Flame response to acoustic excitation in a rectangular rocket combustor with LO_x/H_2 propellants. *Ceas Space Journal, 2,* 41–49.
6. Stamp, G., Ghosh, A., Zang, A., & Yu, K. (2005). Experimental characterization of acoustic wave propagation in a supersonic duct. In *41st AIAA/ASME/SAE/ASEE Joint Propulsion Conference & Exhibit* (p. 4144). AIAA.
7. Wang, H., Sun, M., Qin, N., Wu, H., & Wang, Z. (2013). Characteristics of oscillations in supersonic open cavity flows. *Flow, Turbulence and Combustion, 90,* 121–142.
8. Lin, K.-C., & Jackson, K. (2007). Acoustic characterization of an ethylene-fueled scramjet combustor with a recessed cavity flameholder. In *43rd AIAA/ASME/SAE/ASEE Joint Propulsion Conference & Exhibit* (p. 5382). AIAA.
9. Choi, J.-Y., Ma, F., & Yang, V. (2005). Dynamic combustion characteristics in scramjet combustors with transverse fuel injection. In *41st AIAA/ASME/SAE/ASEE Joint Propulsion Conference & Exhibit* (p. 4428). AIAA.
10. Choi, J.-Y., Ma, F., Yang, V., Won, S.-H., & Jeung, I.-S. (2007). Detached eddy simulation of combustion dynamics in scramjet combustors. In *43rd AIAA/ASME/SAE/ASEE Joint Propulsion Conference & Exhibit* (p. 5027). AIAA.
11. Ma, F., Li, J., Yang, V., Lin, K., & Jackson, T. (2005). Thermoacoustic flow instability in a scramjet combustor. In *41st AIAA/ASME/SAE/ASEE Joint Propulsion Conference & Exhibit* (p. 3824). AIAA.
12. Li, J., Ma, F., Yang, V., Lin, K.-C., & Jackson, T. (2007). A comprehensive study of combustion oscillations in a hydrocarbon-fueled scramjet engine. In *45th AIAA Aerospace Sciences Meeting and Exhibit* (p. 836). AIAA.
13. Lin, K. C., Jackson, K., Behdadnia, R., Jackson, T. A.,Ma, F. H., & Yang, V. (2010). Acoustic characterization of an ethylene-fueled scramjet combustor with a cavity flameholder. *Journal of Propulsion and Power, 26*, 1161–1170.
14. Frost, M. A., Gangurde, D. Y., Paull, A., & Mee, D. J. (2009). Boundary-layer separation due to combustion-induced pressure rise in a supersonic flow. *AIAA Journal, 47,* 1050–1053.
15. Knight, D., Yan, H., Panaras, A. G., & Zheltovodov, A. (2003). Advances in CFD prediction of shock wave turbulent boundary layer interactions. *Progress in Aerospace Sciences, 39,* 121–184.
16. Dolling, D. S. (2001). Fifty years of shock-wave/boundary-layer interaction research: What next? *AIAA Journal, 39,* 1517–1531.
17. Dolling, D. S., & Erengil, M. E. (1991). Unsteady wave structure near separation in a Mach 5 compression rampinteraction. *AIAA Journal, 29,* 728–735.
18. Schmisseur, J. D., & Dolling, D. S. (1994). Fluctuating wall pressures near separation in highly swept turbulent interactions. *AIAA Journal, 32,* 1151–1157.
19. Erengil, M. E., & Dolling, D. S. (1993). Effects of sweepback on unsteady separation in Mach 5 compression ramp interactions. *AIAA Journal, 31,* 302–311.
20. Brusniak, L., & Dolling, D. S. (1994). Physics of unsteady blunt-fin-induced shock wave/turbulent boundary layer interactions. *Journal of Fluid Mechanics, 273,* 375–409.
21. Dolling, D. S., & Brusniak, L. (1989). Separation shock motion in fin, cylinder, and compression ramp—Induced turbulent interactions. *AIAA Journal, 27,* 734–742.

22. Dupont, P., Haddad, C., & Debiève, J. F. (2006). Space and time organization in a shock-induced separated boundary layer. *Journal of Fluid Mechanics, 559,* 255–277.
23. Clemens, N. T., & Narayanaswamy, V. (2014). Low-frequency unsteadiness of shock wave/turbulent boundary layer interactions. *Annual Review of Fluid Mechanics, 46,* 469–492.
24. Dolling, D. S., & Murphy, M. T. (1983). Unsteadiness of the separation shock wave structure in a supersonic compression ramp flowfield. *AIAA Journal, 21,* 1628–1634.
25. Humble, R. A., Elsinga, G. E., Scarano, F., & Oudheusden, B. W. V. (2009). Three-dimensional instantaneous structure of a shock wave/turbulent boundary layer interaction. *Journal of Fluid Mechanics, 622,* 33–62.
26. Priebe, S., & Martín, M. P. (2012). Low-frequency unsteadiness in shock wave-turbulent boundary layer interaction. *Journal of Fluid Mechanics, 699,* 1–49.
27. Pasquariello, V., Hickel, S., & Adams, N. A. (2017). Unsteady effects of strong shock-wave/boundary-layer interaction at high Reynolds number. *Journal of Fluid Mechanics, 823,* 617–657.
28. Piponniau, S., Dussauge, J. P., Debiève, J. F., & Dupont, P. (2009). A simple model for low-frequency unsteadiness in shock-induced separation. *Journal of Fluid Mechanics, 629,* 87–108.
29. Koo, H., & Raman, V. (2012). Large-eddy simulation of a supersonic inlet-isolator. *AIAA Journal, 50,* 1596–1613.
30. Do, H., Im, S. K., Mungal, M. G., & Cappelli, M. A. (2011). The influence of boundary layers on supersonic inlet flow unstart induced by mass injection. *Experiments in Fluids, 51,* 679–691.
31. Zhang, Q. F., Tan, H. J., Sun, S., Bu, H. X., & Rao, C. Y. (2016). Unstart of a hypersonic inlet with side compression caused by downstream choking. *AIAA Journal, 54*(1), 28–38.
32. Jonathan, G., & Kenneth, Y. (2012). Experimental characterization of isolator shock train propagation. In *18th AIAA/3AF International Space Planes and Hypersonic Systems and Technologies Conference* (p. 5891). AIAA.
33. Jonathan, S. G., & Kenneth, H. Y. (2013). Visualization of shock train-boundary layer interaction in Mach 2.5 isolator flow. In *43rd Fluid Dynamics Conference* (p. 3102). AIAA.
34. Bruce, P. J. K., Babinsky, H., Tartinville, B., & Hirsch, C. (2011). Corner effect and asymmetry in transonic channel flows. *AIAA Journal, 49,* 2382–2392.
35. Su, W.-Y., & Zhang, K.-Y. (2012). Back-pressure effects on the hypersonic inlet-isolator pseudoshock motions. *Journal of Propulsion and Power, 29,* 1391–1399.
36. Su, W.-Y., Ji, Y.-X., & Chen, Y. (2016). Effects of dynamic backpressure on pseudoshock oscillations in scramjet inlet-isolator. *Journal of Propulsion and Power, 32,* 516–528.
37. Li, N., Chang, J.-T., Xu, K.-J., Yu, D.-R., Bao, W., & Song, Y.-P. (2018). Oscillation of the shock train in an isolator with incident shocks. *Physics of Fluids, 30,* 116102.
38. Xiong, B., Wang, Z. G., Fan, X. Q., & Wang, Y. (2017). Response of shock train to high-frequency fluctuating backpressure in an isolator. *Journal of Propulsion and Power, 33,* 1–9.
39. Xiong, B., Fan, X.-Q., Wang, Z.-G., & Tao, Y. (2018). Analysis and modelling of unsteady shock train motions. *Journal of Fluid Mechanics, 846,* 240–262.
40. Reijasse, P., Corbel, B., & Soulevant, D. (1999). Unsteadiness and asymmetry of shock-induced separation in a planar two-dimensional nozzle—A flow description. In *30th Fluid Dynamics Conference* (p. 3694). AIAA.
41. Yu, Y., Xu, J., Mo, J., & Wang, M. (2014). Principal parameters in flow separation patterns of over-expanded single expansion RAMP nozzle. *Engineering Applications of Computational Fluid Mechanics, 8,* 274–288.
42. Yu, Y., Xu, J., Yu, K., & Mo, J. (2015). Unsteady transitions of separation patterns in single expansion ramp nozzle. *Shock Waves, 25,* 623–633.
43. Dimitri, P., & Andreas, Z. (2004). Fundamental investigation of supersonic nozzle flow separation. In *42nd AIAA Aerospace Sciences Meeting and Exhibit* (p. 1111). AIAA.
44. Dimitri, P., & Andrew, J. (2006). Unsteady phenomena in supersonic nozzle flow separation. In *36th AIAA Fluid Dynamics Conference and Exhibit* (p. 3360). AIAA.

45. Xiao, Q., Tsai, H.-M., & Papamoschou, D. (2007). Numerical investigation of supersonic nozzle flow separation. *AIAA Journal, 45,* 532–541.
46. Johnson, A. D., & Papamoschou, D. (2010). Instability of shock-induced nozzle flow separation. *Physics of Fluids*, *22*, 016102.
47. Olson, B., & Lele, S. (2012). Low frequency unsteadiness in nozzle flow separation. In *42nd AIAA Fluid Dynamics Conference and Exhibit* (p. 2974). AIAA.
48. Olson, B., & Lele, S. (2011). Large-eddy simulation of an over-expanded planar nozzle. In *41st AIAA Fluid Dynamics Conference and Exhibit* (p. 3908). AIAA.
49. Andrew, J., & Dimitri, P. (2008). Shock motion and flow instabilities in supersonic nozzle flow separation. In *38th Fluid Dynamics Conference and Exhibit* (p. 1111). AIAA.
50. Papamoschou, D., Zill, A., & Johnson, A. (2009). Supersonic flow separation in planar nozzles. *Shock Waves, 19,* 171–183.
51. Laurence, S. J., Lieber, D., Martinez Schramm, J., Hannemann, K., & Larsson, J. (2015). Incipient thermal choking and stable shock-train formation in the heat-release region of a scramjet combustor. Part I: Shock-tunnel experiments. *Combustion and Flame*, *162*, 921–931.
52. Fotia, M. L., & Driscoll, J. F. (2013). Ram-scram transition and flame/shock-train interactions in a model scramjet experiment. *Journal of Propulsion and Power, 29,* 261–273.
53. Yuan, Y., Zhang, T., Yao, W., & Fan, X. (2015). Study on flame stabilization in a dual-mode combustor using optical measurements. *Journal of Propulsion and Power, 31,* 1524–1531.
54. Zhao, D., Gutmark, E., & de Goey, P. (2018). A review of cavity-based trapped vortex, ultra-compact, high-g, inter-turbine combustors. *Progress in Energy and Combustion Science, 66,* 42–82.
55. Huang, Y., & Yang, V. (2009). Dynamics and stability of lean-premixed swirl-stabilized combustion. *Progress in Energy and Combustion Science, 35,* 293–364.
56. Santosh, S., Sajjad, H., & Tim, L. (2009). Lean blowoff of bluff body stabilized flames—Scaling and dynamics. *Progress in Energy and Combustion Science, 35,* 98–120.
57. Penner, S. S., & Williams, F. (1957). Recent studies on flame stabilization of premixed turbulent gases. *Applied Mechanics Review, 10.*
58. Schadow, K. C., & Gutmark, E. (1992). Combustion instability related to vortex shedding in dump combustors and their passive control. *Progress in Energy and Combustion Science, 18,* 117–132.
59. Lefebvre, A. H.,& Whitelaw, J. H. (1984). Gas turbine combustion. *International Journal of Heat and Fluid Flow*, *5*(4), 228.
60. Lee, J. G., Armstrong, J. P., & Santavicca, D. A. (2011). Experimental on lean blowout and Nox emissions of a premixed trapped vortex combustor with high G-loading. *Proceedings of ASME Turbo Expo 2011, GT2011-46396.* British Columbia, Canada: Vancouver.
61. Xing, F., Wang, P., Zhang, S., Zou, J., Zheng, Y., & Zhang, R., et al. (2012). Experiment and simulation study on lean blow-out of trapped vortex combustor with various aspect ratios. *Aerospace Science and Technology, 18,* 48–55.
62. Xing, F., Fan, W., & Yang, M. (2009). Experiments and simulation study on flow field and LBO of trapped vortex combustor with H/L changed. In *45th AIAA/ASME/SAE/ASEE Joint Propulsion Conference and Exhibit.*
63. Biagioli, F., Güthe, F., & Schuermans, B. (2008). Combustion dynamics linked to flame behaviour in a partially premixed swirled industrial burner. *Experimental Thermal and Fluid Science, 32,* 1344–1353.
64. McManus, K. R., Poinsot, T., & Candel, S. M. (1993). A review of active control of combustion instabilities. *Progress in Energy and Combustion Science, 19,* 1–29.
65. Ben-Yakar, A., & Hanson, R. K. (2001). Cavity flame-holders for ignition and flame stabilization in scramjets: An overview. *Journal of Propulsion and Power, 17*(4), 869–878.
66. Sturgess, G. J., Zelina, J., Shouse, D. T., & Roquemore, W. M. (2005). Emissions reduction technologies for military gas turbine engines. *Journal of Propulsion and Power, 21,* 193–217.
67. Hsu, K., Gross, L., Trump, D., Roquemore, W. (1995). Performance of a trapped-vortex combustor. In *33rd Aerospace Sciences Meeting and Exhibit* (p. 810). AIAA.

68. Mongia, H., Gore, J., Grinstein, F., Gutmark, E., Jeng, S. M., McDonell, V., Menon, S., Samuelsen, G., Santavicca, D., & Santoro, R. (2001). Combustion research needs for helping development of next-generation advanced combustors. In *37th Joint Propulsion Conference and Exhibit* (p. 3853). AIAA.
69. Shouse, D. T. (2000). *Trapped vortex combustion technology MITE workshop*. Dayton Ohio, USA: US Air Force Research Laboratory, Wright-Patterson Air Force Base.
70. Gruber, M. R., Baurle, R. A., Mathur, T., & Hsu, K. Y. (2001). Fundamental studies of cavity-based flameholder concepts for supersonic combustors. *Journal of Propulsion and Power, 17,* 146–153.
71. Yu, K. H., Wilson, K. J., & Schadow, K. C. (2001). Effect of flame-holding cavities on supersonic-combustion performance. *Journal of Propulsion and Power, 17,* 1287–1295.
72. Ben-Yakar, A., Hanson, R. K. (1998). *Cavity flame-holders for ignition and flame stabilization in scramjets: Review and experimental study*. In *34th AIAA/ASME/SAE/ASEE Joint Propulsion Conference and Exhibit* (p. 3122). AIAA.
73. Chen, S. (2016). *Numerical study of trapped vortex combustors characteristics in small Ramjets*. Singapore: Nanyang Technological University.
74. Ballal, D. R., & Lefebvre, A. H. (1977). Ignition and flame quenching in flowing gaseous mixtures. *Proceedings of the Royal Society A, 357,* 163–181.
75. Ballal, D. R., & Lefebvre, A. H. (1978). Ignition and flame quenching of quiescent fuel mists. *Proceedings of the Royal Society A, 364,* 277–294.
76. Ballal, D. R., & Lefebvre, A. H. (1981). A general model of spark ignition for gaseous and liquid fuel-air mixtures. *Symposium (International) on Combustion, 18*(1), 1737–1746.
77. Mastorakos, E. (2009). Ignition of turbulent non-premixed flame. *Progress in Energy and Combustion Science, 35,* 57–97.
78. Zhao, Y., Liang, J., Zhao, Y., & Duan, J. (2017). Research on interactions between cavity and upstream transverse jet in supersonic combustor. In *21st AIAA International Space Planes and Hypersonics Technologies Conference, AIAA 2017–2368.*
79. Li, J., Zhang, L., Choi, J. Y., Yang, V., & Lin, K.-C. (2014). Ignition transients in a scramjet engine with air throttling part 1: Nonreacting flow. *Journal of Propulsion and Power, 30,* 438–448.
80. McNeill, D.H. (2005). Minimum ignition energy for laser spark ignition. *Proceedings of the Combustion Institute, 30*(2), 2913–2920.
81. Brieschenk, S., O'Byrne, S., & Kleine, H. (2013). Laser-induced plasma ignition studies in a model scramjet engine. *Combustion and Flame, 160,* 145–148.
82. Brieschenk, S., O'Byrne, S., & Kleine, H. (2014). Ignition characteristics of laser-ionized fuel injected into a hypersonic crossflow. *Combustion and Flame, 161,* 1015–1025.
83. Denman, Z. J., Chan, W. Y. K., Brieschenk, S., Veeraragavan, A., Wheatley, V., & Smart, M. K. (2016). Ignition experiments of hydrocarbons in a Mach 8 shape-transitioning scramjet engine. *Journal of Propulsion and Power, 32*(6), 1462–1471.
84. Ombrello, T., Peltier, S., & Carter, C. (2015). Effects of inlet distortion on cavity ignition in supersonic flow. In *53rd AIAA Aerospace Sciences Meeting* (p. 0882). AIAA.
85. Ombrello, T., Carter, C., Tam, C.-J., & Hsu, K.-Y. (2014). Cavity ignition in supersonic flow by pulse detonation. In *AIAA SciTech Forum 52nd Aerospace Sciences Meeting* (p. 0286). AIAA.
86. Ombrello, T., Carter, C., McCall, J., & Schauer, F. (2015). Enhanced mixing in supersonic flow using a pulse detonator. *Journal of Propulsion and Power, 31,* 654–663.
87. Miller, J. D., Peltier, S. J., Slipchenko, M. N., Mance, J. G., Ombrello, T. M., & Gord, J. R., et al. (2017). Investigation of transient ignition processes in a model scramjet pilot cavity using simultaneous 100 kHz formaldehyde planar laser-induced fluorescence and CH* chemiluminescence imaging. *Proceedings of the Combustion Institute, 36,* 2865–2872.
88. Cuppoletti, D., Ombrello, T., & Rein, K. (2019). Energy coupling mechanism for pulse detonation ignition of a scramjet cavity. *Proceedings of the Combustion Institute, 37,* 3453–3460.

89. Li, J., Zhang, L., Choi, J. Y., Yang, V., & Lin, K.-C. (2015). Ignition transients in a scramjet engine with air throttling part II: Reacting flow. *Journal of Propulsion and Power, 31,* 79–88.
90. An, B., Wang, Z., Yang, L., Li, X., & Zhu, J. (2017). Experimental investigation on the impacts of ignition energy and position on ignition processes in supersonic flows by laser induced plasma. *Acta Astronautica, 137,* 444–449.
91. Yang, L., Li, X., Liang, J., Yu, Y., Yu, X. (2015). Laser-induced plasma ignition of hydrocarbon fuel in supersonic flows. In *20th AIAA International Space Planes and Hypersonic Systems and Technologies Conference* (p. 3544). AIAA.
92. Yang, L., An, B., Liang, J., Li, X., & Wang, Z. (2018). Dual-pulse laser ignition of ethylene-air mixtures in a supersonic combustor. *Optics Express, 26,* 7911–7919.
93. An, B., Wang, Z., Yang, L., Li, X., & Liu, C. (2019). The ignition characteristics of the close dual-point laser ignition in a cavity based scramjet combustor. *Experimental Thermal and Fluid Science, 101,* 136–140.
94. Li, X., Liu, W., Pan, Y., Yang, L., & An, B. (2017). Experimental investigation on laser-induced plasma ignition of hydrocarbon fuel in scramjet engine at takeover flight conditions. *Acta Astronautica, 138,* 79–84.
95. Li, X., Liu, W., Pan, Y., Yang, L., An, B., & Zhu, J. (2018). Characterization of ignition transient processes in kerosene-fueled model scramjet engine by dual-pulse laser-induced plasma. *Acta Astronautica, 144,* 23–29.
96. Tropina, A. A., Miles, R. B., & Shneider, M. N. (2018). Mathematical model of dual-pulse laser ignition. *Journal of Propulsion and Power, 34,* 408–414.
97. Gibbons, N., Gehre, R., Brieschenk, S., & Wheatley, V. (2018). Simulation of laser-induced plasma ignition in a hypersonic crossflow. *AIAA Journal, 56*(8), 3047–3059.
98. Firsov, A. A., Shurupov, M. A., Yarantsev, D. A.,& Leonov, S. B. (2014). Plasma-assisted combustion in supersonic airflow: Optimization of electrical discharge geometry. In *52nd Aerospace Sciences Meeting*. AIAA.
99. Leonov, S. B., Firsov, A. A., Yarantsev, D. A., Falempin, F.,& Miller, A. (2009). Flow control in model supersonic inlet by electrical discharge. In *16th AIAA/DLR/DGLR International Space Planes and Hypersonic Systems and Technologies Conference* (p. 7367). AIAA.
100. Firsov, A. A., Dolgov, E. V., Rakhimov, R. G., Shurupov, M. A., & Leonov, S. B. (2018). Mixing enhancement by electrical discharge in supersonic airflow. In *2018 AIAA Aerospace Sciences Meeting*. AIAA 2018-1195.
101. Savelkin, K. V., Yarantsev, D. A., Adamovichb, I. V., & Leonov, S. B. (2015). Ignition and flameholding in a supersonic combustor by an electrical discharge combined with a fuel injector. *Combustion and Flame, 162,* 825–835.
102. Leonov, S., Houpt, A., Elliott, S., & Hedlund, B. (2018). Ethylene ignition and flameholding by electrical discharge in supersonic combustor. *Journal of Propulsion and Power, 34,* 499–509.
103. Feng, R., Li, J., Wu, Y., Zhu, J., Song, X., & Li, X. (2018). Experimental investigation on gliding arc discharge plasma ignition and flame stabilization in scramjet combustor. *Aerospace Science and Technology, 79,* 145–153.
104. Huang, S., Wu, Y., Song, H., Zhu, J., Zhang, Z., Song, X., et al. (2018). Experimental investigation of multichannel plasma igniter in a supersonic model combustor. *Experimental Thermal and Fluid Science, 99,* 315–323.
105. Li, J., Ma, F., Yang, V., Lin, K.-C., & Jackson, T. A. (2007). A comprehensive study of ignition transient in an ethylene-fueled scramjet combustor. In *43rd AIAA/ASME/SAE/ASEE Joint Propulsion Conference & Exhibit* (p. 5025). AIAA.
106. Yang, V., Li, J., Choi, J. Y., & Lin, K.-C. (2010). Ignition Transient in an Ethylene Fueled Scramjet Engine with Air Throttling Part II: Ignition and Flame Development. In *48th AIAA Aerospace Sciences Meeting Including the New Horizons Forum and Aerospace Exposition* (p. 410). AIAA.
107. Li, J., Ma, F., Yang, V., Lin, K.-C., & Jackson, T. A. (2006). Control and optimization of ignition transient in scramjet engine using air throttling. In *44th AIAA Aerospace Sciences Meeting and Exhibit* (p. 1028). AIAA.

108. Noh, J., Choi, J.-Y., Byun, J.-R., Lim, J.-S., & Yang, V. (2010). Numerical simulation of auto-ignition of ethylene in a scramjet combustor with air throttling. In *46th AIAA/ASME/SAE/ASEE Joint Propulsion Conference & Exhibit* (p. 7036). AIAA.
109. Yang, S., Tian, Y., & Le, J. (2017). Flow oscillation in a scramjet combustor with air-throttling. In *21st AIAA International Space Planes and Hypersonics Technologies Conference* (p. 2301). AIAA.
110. Tian, Y., Xiao, B., Zhang, S., & Xing, J. (2015). Experimental and computational study on combustion performance of a kerosene fueled dual-mode scramjet engine. *Aerospace Science and Technology, 46,* 451–458.
111. Tian, Y., Yang, S., & Le, J. (2016). Numerical study on effect of air throttling on combustion mode formation and transition in a dual-mode scramjet combustor. *Aerospace Science and Technology, 52,* 173–180.
112. Tian, Y., Yang, S., Le, J., Zhong, F., & Tian, X. (2017). Investigation of combustion process of a kerosene fueled combustor with air throttling. *Combustion and Flame, 179,* 74–85.
113. Sun, M. B., Gong, C., Zhang, S. P., Liang, J. H., Liu, W. D., & Wang, Z. G. (2012). Spark ignition process in a scramjet combustor fueled by hydrogen and equipped with multi-cavities at Mach 4 flight condition. *Experimental Thermal and Fluid Science, 43,* 90–96.
114. Kim, C.-H., Jeung, I.-S., Choi, B., Kouchi, T., Takita, K., & Masuya, G. (2011). Effect of fuel injection location on a plasma jet assisted combustion with a backward-facing step. *Proceedings of the Combustion Institute, 33,* 2375–2382.
115. Matsubara, Y., Takita, K., & Masuya, G. (2013). Combustion enhancement in a supersonic flow by simultaneous operation of DBD and plasma jet. *Proceedings of the Combustion Institute, 34,* 3287–3294.
116. Matsubara, Y., & Takita, K. (2011). Effect of mixing ratio of N_2/O_2 feedstock on ignition by plasma jet torch. *Proceedings of the Combustion Institute, 33,* 3203–3209.
117. Xi, W., Wang, Z., Sun, M., Liu, W., & Li, Q. (2014). Experimental investigation of ignition transient phase in model supersonic combustor. *Journal of Aerospace Engineering, 27,* 04014009.
118. Situ, M., Wang, C., Lu, H. P., Yu, G., & Zhang, X. Y. (2001). Hot gas piloted energy for supersonic combustion of kerosene with dual-cavity. In *39th AIAA Aerospace Sciences Meeting & Exhibit* (p. 0523). AIAA.
119. Situ, M., Wang, C., & Zhuang, K. (2002). Investigation of supersonic combustion of kerosene jets with hot gas piloted energy and dual-cavity. In *40th AIAA Aerospace Sciences Meeting & Exhibit* (p. 13938). AIAA.
120. Qing, L., Wenxiong, X., Lin, L., Yu, P., Meng, D., & Jianhan, L. (2012). Research on hot gas jet ignition process of scramjet combustor fueled with kerosene. In *18th AIAA/3AF International Space Planes and Hypersonic Systems and Technologies Conference* (p. 5947). AIAA.
121. Sung, C. J., Li, J. G., Yu, G., & Law, C. K. (1999). Chemical kinetics and self-ignition in a model supersonic hydrogen-air combustor. *AIAA Journal, 37,* 208–214.
122. Fureby, C., Nordin-Bates, K., Petterson, K., Bresson, A., & Sabelnikov, V. (2015). A computational study of supersonic combustion in strut injector and hypermixer flow fields. *Proceedings of Combustion Institute, 35,* 2127–2135.
123. Gordon, R. L., Masri, A. R., & Mastorakos, E. (2008). Simultaneous rayleigh temperature, OH- and CH_2O-LIF imaging of methane jets in a vitiated coflow. *Combustion and Flame, 155,* 181–195.
124. Cabra, R., Chen, J. Y., Dibble, R. W., Karpetis, A. N., & Barlow, R. S. (2005). Lifted methane-air jet flames in a vitiated coflow. *Combustion and Flame, 143,* 491–506.
125. Domingo, P., Vervisch, L., & Veynante, D. (2008). Large-eddy simulation of a lifted methane jet flame in a vitiated coflow. *Combustion and Flame, 152,* 415–432.
126. Domingo, P., Vervisch, L., & Veynante, D. (2006). Auto-ignition and flame propagation effects in LES of burned gases diluted turbulent combustion. *Proceedings of the Summer* Program 2006, 337-348.
127. Yoo, C. S., Richardson, E. S., Sankaran, R., & Chen, J. H. (2011). A DNS study on the stabilization mechanism of a turbulent lifted ethylene jet flame in highly-heated coflow. *Proceedings of the Combustion Institute, 33*(1), 1619–1627.

128. Lu, S., Fan, J., & Luo, K. (2012). High-fidelity resolution of the characteristic structures of a supersonic hydrogen jet flame with heated co-flow air. *International Journal of Hydrogen Energy, 37,* 3528–3539.
129. Micka, D. J., & Driscoll, J. F. (2012). Stratified jet flames in a heated (1390 K) air cross-flow with autoignition. *Combustion and Flame, 159,* 1205–1214.
130. Micka, D. J., & Driscoll, J. F. (2011). Stratified jet flames in a heated (1364 K) cross-flow with auto-ignition. In *49th AIAA Aerospace Sciences Meeting including the New Horizons Forum and Aerospace Exposition* (P. 321). AIAA.
131. Wang, H., Wang, Z., Sun, M., & Qin, N. (2013). Combustion characteristics in a supersonic combustor with hydrogen injection upstream of cavity flameholder. *Proceedings of the Combustion Institute, 34,* 2073–2082.
132. Sun, M. B., Cui, X. D., Wang, H. B., & Bychkov, V. (2015). Flame flashback in a supersonic combustor fueled by ethylene with cavity flameholder. *Journal of Propulsion and Power, 31,* 976–981.
133. Huang, Z. W., He, G. Q., Qin, F., Xue, R., Wei, X. G., & Shi, L. (2016). Combustion oscillation study in a kerosene fueled rocket-based combinedcycle engine combustor. *Acta Astronautica, 129,* 260–270.
134. Wang, Z. G., Sun, M. B., Wang, H. B., Yu, J. F., Liang, J. H., & Zhuang, F. C. (2015). Mixing-related low frequency oscillation of combustion in an ethylene-fueled supersonic combustor. *Proceedings of the Combustion Institute, 35,* 2137–2144.
135. Wang, H., Wang, Z., Sun, M., & Qin, N. (2013). Large-Eddy/Reynolds-averaged Navier-stokes simulation of combustion oscillations in a cavity-based supersonic combustor. *International Journal of Hydrogen Energy, 38,* 5918–5927.
136. Lewis, B., & von Elbe, G. (1943). Stability and structure of burner flames. *The Journal of Chemical Physics, 11,* 75–97.
137. Kim, C.-H., Jeung, I.-S., Choi, B., Kouchi, T., & Masuya, G. (2012). Flowfield characteristics of a hypermixer interacting with transverse injection in supersonic flow. *AIAA Journal, 50,* 1742–1753.
138. Micka, D. J. (2010). *Combustion stabilization, structure, and spreading in a laboratory dual-mode scramjet combustor.* The University of Michigan.
139. Mitani, T., & Kouchi, T. (2005). Flame structures and combustion efficiency computed for a Mach 6 scramjet engine. *Combustion and Flame, 142,* 187–196.
140. Sean, O. B., Ingo, S., Andrew, N., Russell, B., Neil, M., & Frank, H. (2005). OH PLIF imaging of supersonic combustion using cavity injection. In *AIAA/CIRA 13th International Space Planes and Hypersonics Systems and Technologies Conference* (p. 3357). AIAA.
141. Tarun, M., Mark, G., Kevin, J., Jeff, D., & Wayne, D. (2001). Supersonic combustion experiments with a cavity-based fuel injector. *Journal of Propulsion and Power, 17,* 1305–1312.
142. Vinogradov, V. A., Shikhman, Y. M., & Segal, C. (2007). A review of fuel pre-injection in supersonic, chemically reacting flows. *Applied Mechanics Reviews, 60,* 139–148.
143. Laurence, S. J., Karl, S., Schramm, J. M., & Hannemann, K. (2013). Transient fluid-combustion phenomena in a model scramjet. *Journal of Fluid Mechanics, 722,* 85–120.
144. Larsson, J., Laurence, S., Bermejo-Moreno, I., Bodart, J., Karl, S., & Vicquelin, R. (2015). Incipient thermal choking and stable shock-train formation in the heat-release region of a scramjet combustor. Part II: Large eddy simulations. *Combustion and Flame, 162,* 907–920.
145. Karl, S., Laurence, S., Martinez Schramm, J., & Hannemann, K. (2012). CFD analysis of unsteady combustion phenomena in the HyShot-ii scramjet configuration. In *18th AIAA/3AF International Space Planes and Hypersonic Systems and Technologies Conference* (p. 5912). AIAA.
146. Laurence, S., Ozawa, H., Lieber, D., Martinez Schramm, J., & Hannemann, K. (2012). Investigation of unsteady/quasi-steady scramjet behavior using high-speed visualization techniques. In *18th AIAA/3AF International Space Planes and Hypersonic Systems and Technologies Conference* (p. 5913). AIAA.

147. Sunami, T., Itoh, K., Satoh, K., & Komuro, T. (2006). Mach 8 ground tests of the hypermixer scramjet for HyShort-Iv flight experiment. In *14th AIAA/AHI Space Planes and Hypersonic Systems and Technologies Conference* (p. 8062). AIAA.
148. Sunami, T., & Kodera, M. (2012). Numerical investigation of a detonation wave system in a scramjet combustor. In *18th AIAA/3AF International Space Planes and Hypersonic Systems and Technologies Conference* (p. 5861). AIAA.
149. Zhu, S.-H., & Xu, X. (2017). Experimental study on flame transition in a two-stage struts dual-mode scramjet. *Journal of Aerospace Engineering, 30,* 06017002.
150. Yang, V., Li, J., Choi, J. Y., & Lin, K.-C. (2010). Ignition transient in an ethylene fueled scramjet engine with air throttling part I: non-reacting flow development and mixing. In *48th AIAA Aerospace Sciences Meeting Including the New Horizons Forum and Aerospace Exposition* (p. 409). AIAA.
151. Gruber, A., Chen, J. H., Valiev, D., & Law, C. K. (2012). Direct numerical simulation of premixed flame boundary layer flashback in turbulent channel flow. *Journal of Fluid Mechanics, 709,* 516–542.
152. Rossiter, J. E. (1964). *Wind tunnel experiments on the flow over rectangular cavities at subsonic and transonic speeds*. RAE Farnborough: Ministry of Aviation; Royal Aircraft Establishment.
153. Tim, L., Santosh, S., Sachin, K., & Clifford, S. (2007). Dynamics of bluff body flames near blowoff. In *45th AIAA Aerospace Sciences Meeting and Exhibit* (p. 169). AIAA.
154. Allison, P., Frederickson, K., Lempert, W., Sutton, J., Kirik, J., Rockwell, R., & Goyne, C. (2016). Investigation of flame structure and combustion dynamics using CH_2O PLIF and high-speed CH chemiluminescence in a premixed dual-mode scramjet combustor. In *54th AIAA Aerospace Sciences Meeting* (p. 0441). AIAA.
155. Hammack, S. D., Lee, T., Hsu, K.-Y., & Carter, C. D. (2013). High-repetition-rate OH planar laser-induced fluorescence of a cavity flameholder. *Journal of Propulsion and Power, 29,* 1248–1251.
156. Gruber, M. R., Donbar, J. M., Carter, C. D., & Hsu, K. Y. (2004). Mixing and combustion studies using cavity-based flameholders in a supersonic flow. *Journal of Propulsion and Power, 20,* 769–778.
157. Zhang, T., Wang, J., Qi, L., Fan, X., & Zhang, P. (2014). Blowout limits of cavity-stabilized flame of supercritical kerosene in supersonic combustors. *Journal of Propulsion and Power, 30,* 1161–1166.
158. Ozawa, R. I. (1971). *Survey of basic data on flame stabilization and propagation for high speed combustion systems*. OH: U. S. Air Force.
159. Barnes, F. W., & Segal, C. (2015). Cavity-based flameholding for chemically-reacting supersonic flows. *Progress in Aerospace Sciences, 76,* 24–41.
160. Namazian, M., Kelly, J. T., & Schefer, R. W. (1988). Near field instantaneous flame and fuel concentration structures. *Proceedings of the Combustion Institute, 22,* 627–634.
161. Schefer, R. W., Namazian, M., & Kelly, J. T. (1990). CH, OH, and CH_4 concentration measurements in a lifted turbulent jet flame. *Proceedings of the Combustion Institute, 23,* 669–676.
162. Donbar, J. M., Driscoll, J. F., & Carter, C. D. (2000). Reaction zone structure in turbulent nonpremixed jet flames—From CH-OH PLIF images. *Combustion and Flame, 122,* 1–19.
163. Driscoll, J. F., & Rasmussen, C. C. (2005). Correlation and analysis of blowout limits of flames in high-speed airflows. *Journal of Propulsion and Power, 21,* 1035–1044.
164. Donohue, J. M. (2014). Dual-mode scramjet flameholding operability measurements. *Journal of Propulsion and Power, 30,* 592–603.
165. Ghislain, R., Suresh, M. (2010). Experimental studies on flame stability of a fueled cavity in a supersonic crossflow. In *46th AIAA/ASME/SAE/ASEE Joint Propulsion Conference & Exhibit* (p. 6718). AIAA.
166. Le, J., Yang, S. & Li, H. (2012). Analysis and correlation of flame stability limits in supersonic flow with cavity flameholder. In *18th AIAA/3AF International Space Planes and Hypersonic Systems and Technologies Conference* (p. 5948). AIAA.

167. Rasmussen, C. C., Dhanuka, S. K., & Driscoll, J. F. (2007). Visualization of flameholding mechanisms in a supersonic combustor using PLIF. *Proceedings of the Combustion Institute, 31*, 2505–2512.
168. Rasmussen, C. C., Driscoll, J. F., Carter, C. D., & Hsu, K.-Y. (2005). Characteristics of cavity-stabilized flames in a supersonic flow. *Journal of Propulsion and Power, 21*(4), 765–769.
169. Rasmussen, C., & Driscoll, J. (2008). Blowout limits of flames in high-speed airflows: Critical damkohler number. In *44th AIAA/ASME/SAE/ASEE Joint Propulsion Conference & Exhibit* (p. 4571). AIAA.
170. Lin, K.-C., Tam, C.-J., Boxx, I., Carter, C., Jackson, K., & Lindsey, M. (2007). Flame characteristics and fuel entrainment inside a cavity flame holder of a scramjet combustor. In *43rd AIAA/ASME/SAE/ASEE Joint Propulsion Conference & Exhibit* (p. 5381).
171. Lin, K.-C., Tam, C.-J., & Jackson, K. (2009). Study on the operability of cavity flameholders inside a scramjet combustor. In *45th AIAA/ASME/SAE/ASEE Joint Propulsion Conference & Exhibit* (p. 5028). AIAA.
172. Choi, J., Ghodke, C., & Menon, S. (2010). Large-eddy simulation of cavity flame-holding in a Mach 2.5 cross flow. In *48th AIAA Aerospace Sciences Meeting Including the New Horizons Forum and Aerospace Exposition* (p. 414). AIAA.
173. Ghodke, C., Retaureau, G., Choi, J., & Menon, S. (2011). Numerical and experimental studies of flame stability in a cavity stabilized hydrocarbon-fueled scramjet. In *17th AIAA International Space Planes and Hypersonic Systems and Technologies Conference* (p. 2365). AIAA.
174. Tuttle, S. G., Carter, C. D., & Hsu, K.-Y. (2014). Particle image velocimetry in a nonreacting and reacting high-speed cavity. *Journal of Propulsion and Power, 30*, 576–591.
175. Allen, W., King, P., Gruber, M., Carter, C., & Hsu, K.-Y. (2005). Fuel-air injection effects on combustion in cavity-based flameholders in a supersonic flow. In *41st AIAA/ASME/SAE/ASEE Joint Propulsion Conference & Exhibit* (p. 4105). AIAA.

Chapter 2
Acoustic Oscillation in Supersonic Combustor

In the supersonic combustors, there exist intense exothermic reactions between fuel and oxidizer. In many cases the combustor does not work as expected, where the pressure oscillations occur, especially for the combustor with cavity. The oscillation frequency varies from hundreds of Hz to 15,000 Hz or higher. This phenomenon is called combustion instability.

Combustion instability is the oscillatory combustion phenomenon caused by the coupling of combustion processes and system fluid-dynamic processes or acoustic oscillations, accompanied by periodic oscillation of the gas pressure, temperature and velocity. This chapter focuses on the instability related to acoustic oscillations.

2.1 High Frequency Acoustic Oscillations of Cavity

High frequency acoustic oscillations are found in the cavity-based supersonic flows and supersonic combustion. The dominant types of high frequency acoustic oscillation of cavity-based supersonic flow are Rossiter and wake modes, while the cavity-based supersonic combustion is influenced by the interaction of the jet-with-cavity shear layer and the auto-ignition of the combustible fluid packets formed around the fuel jet accompanied by the generation of the hairpin-like vortices.

2.1.1 Characteristics of Oscillations in Supersonic Cavity Flows

Cavity flow under the influence of an external cross-stream can exhibit several kinds of dynamics–bulk mode (Helmholtz mode), acoustic mode, Rossiter mode (shear-layer mode) and wake mode [1]. In general, the nature of the cavity flow mainly depends on the external flow Mach number and L/D ratio. As illustrated by Sahoo

M. Sun et al., *Unsteady Supersonic Combustion*,
https://doi.org/10.1007/978-981-15-3595-6_2

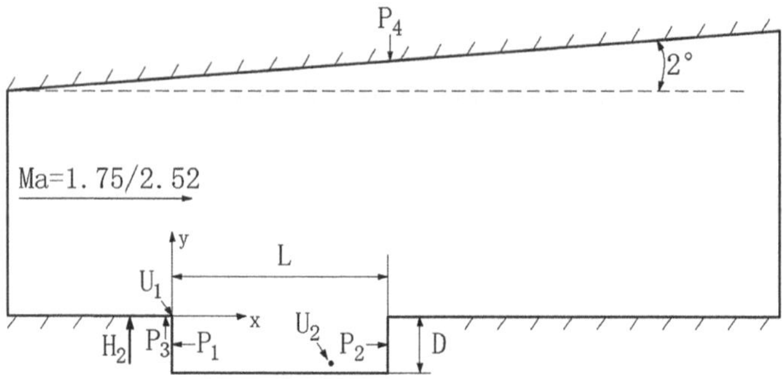

Fig. 2.1 Schematic of the cavity [3]

Table 2.1 Inflow conditions for cavity flows [3]

Parameter	Case 1	Case 2
M_∞	1.75	2.52
P_∞, Pa	100,000	86,200
T_∞, K	820.0	753.8
Flight Mach number	5.0	6.0

et al. [2], low speed cavities typically exhibit fluid-resonant (acoustic) mode oscillations if they are very shallow or deep; otherwise, they exhibit bulk mode oscillations. Higher speed cavities exhibit Rossiter and wake modes, which are also the most dominant types of dynamics in the typical regions of operation for aircraft applications.

There are currently several interpretations for the driving mechanism of the cavity oscillations under Rossiter mode. Researchers appear to agree that an oscillating shear layer exists, that the primary and secondary vortices residing within the cavity are driven by the shear layer, that a mass breathing effect occurs within the cavity, and that pressure oscillations exist.

In order to deepen the understanding of the characteristics of oscillations in supersonic cavity flows, Wang et al. [3] numerically investigated the oscillating regimes and driving mechanisms of the cavity oscillations under the flow conditions corresponding to scramjet applications based on the combustor shown in Fig. 2.1

Additionally, the influences of Mach number and upstream injection on the cavity oscillations are also studied. The conditions of the researches are shown in Table 2.1.

2.1.1.1 Oscillation Regimes

For the acoustics-vortex resonant mechanism of cavity oscillations proposed by Rossiter [4], Heller and Bliss [5] developed the modified Rossiter's semi-empirical formula to predict the resonant frequencies as

$$f_n = \frac{u_\infty}{L} \frac{n - \alpha}{M_\infty \Big/ \sqrt{1 + [(\gamma_\infty - 1)/2]M_\infty^2} + 1/K_v} \tag{2.1}$$

where u_∞ is free-stream velocity, n is the number of the acoustic mode, γ_∞ is the specific heat ratio of the gas, L is the cavity length, M_∞ is free-stream Mach number, K_v and α are empirical constants corresponding to the average convection speed of disturbances in the shear layer and a phase delay. Under the condition of $Ma_\infty = 1.75$ of Wang's research, the first three shear-layer modes of cavity oscillation, as predicted by Rossiter's semi-empirical formula, are 6200 Hz, 14,460 Hz and 22,720 Hz, respectively.

Figure 2.2 shows the pressure histories and FFT results of points P1 and P2, located at the middle of the front and aft wall, respectively. There seem to be five dominant oscillation frequencies: 1034, 7383, 14,766, 22,297 and 44,446 Hz. The three frequencies, 7383, 14,766 and 22,297 Hz, captured in the calculation correspond to the first three shear-layer modes. It is notable that the two higher modes agree well with the theoretical prediction while the first mode has an apparent departure from the predicted value. The possible reason is that Rossiter's semi-empirical formula uses a uniform convection-speed ratio for all modal disturbances developing in the shear layer, which may be questionable as studied by Kegerise et al. [6]. Also, one can see that the oscillation energy is mainly concentrated in the third shear-layer mode. The highest frequency of 44,446 Hz is the quadratic coupling of the first three shear-layer modes since it is exactly the sum of the three frequencies. However, the oscillation frequency 1034 Hz falls well below the first shear-layer mode prediction, which indicates there might be other dynamics responsible for the cavity oscillations.

As shown by the pressure histories, there seems to be two different oscillation modes that occur alternately. For the first mode, the amplitude of the pressure oscillation is relatively small, as denoted by stage 'S' in Fig. 2.2a. For the second mode, however, the amplitude becomes several times larger than that in the first mode, as

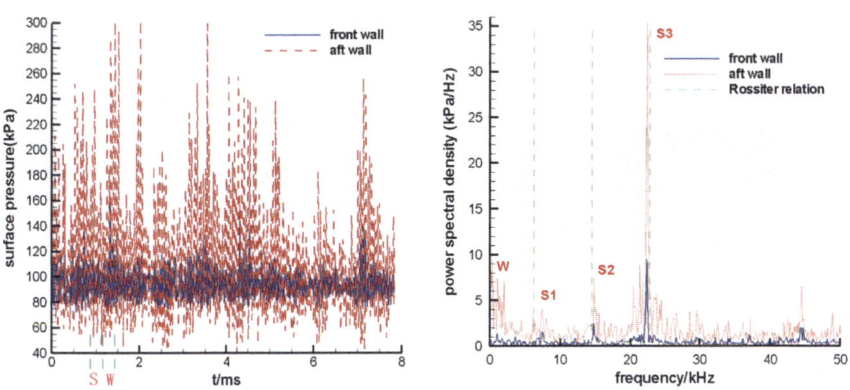

(a) Pressure histories (b) FFT results

Fig. 2.2 Pressure histories and FFT results of points P1 and P2 [3] denoted by stage 'W'. As pointed out by Colonius et al. [7] and Rowley et al. [8], the shed vortex has dimensions of nearly the cavity size in the wake mode, and as it is forming, irrotational free-stream fluid is directed into the cavity, impinging on the cavity base. The vortex is shed from the leading edge and ejected from the cavity in a violent event. The vortex is large enough to cause flow separation upstream of the cavity during its forming. Based on these discussions and the calculated pressure oscillations, it is inferred that the present supersonic cavity flow behaves a mixed shear-layer/wake oscillation mode. Furthermore, it is conjectured that stage 'S' is dominated by the shear-layer mode and stage 'W' by the wake mode, which will be carefully analyzed below.

Figure 2.3 shows the streamwise velocity histories of point U1 and U2, located around the leading edge ($x/D = -0.01,\ y/D = 0.002$) and in the cavity ($x/D = 2.8,\ y/D = -0.8$), respectively. It can be seen that the upstream boundary layer always attaches and the backflow in the cavity is rather weak during stage 'S'. When stage 'W' comes, however, the upstream boundary layer separates apparently and the backflow in the cavity becomes several times larger (more than 30% of the free-stream velocity). These results appear to support that stages 'S' and 'W' correspond to shear-layer and wake mode, respectively.

Figures 2.4 and 2.5 show the pressure histories and FFT results of points P3 and P4, located upstream of the leading edge ($x/D = -0.2,\ y/D = 0.0$) and at the upper wall ($x/D = 3.88$), respectively. From Fig. 2.4, it can be seen that the amplitude of pressure oscillation is rather small during stage 'S' while it becomes several times larger during stage 'W'. Additionally, it seems that the upstream boundary sees a great adverse pressure gradient during stage 'W', which may be responsible for the intense separation upstream of the cavity. The pressure oscillations occur at the upper wall may attribute to the intermittent generation of shock wave around the leading edge. During stage 'S', the shock wave is generated due to the oscillating shear layer and shedding vortex around the leading edge, and the shock wave is rather weak.

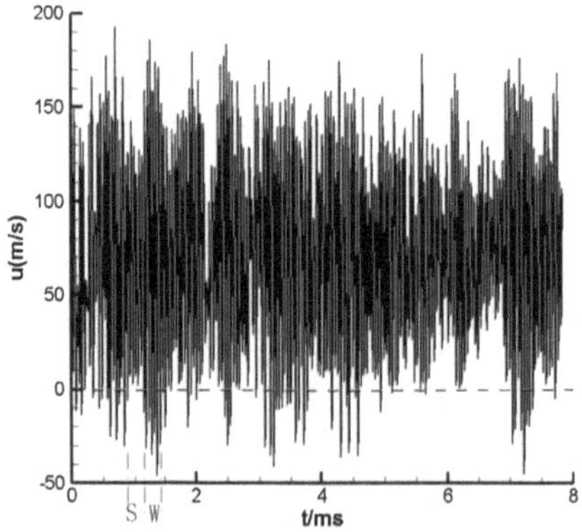

(a) History at point U1

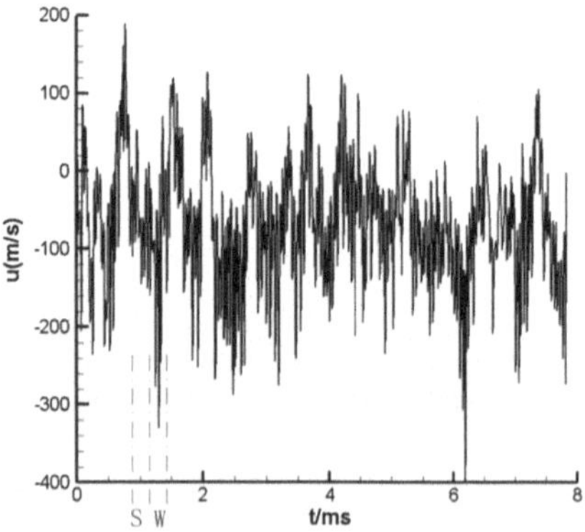

(b) History at point U2

Fig. 2.3 Velocity histories of points U1 and U2 [3]

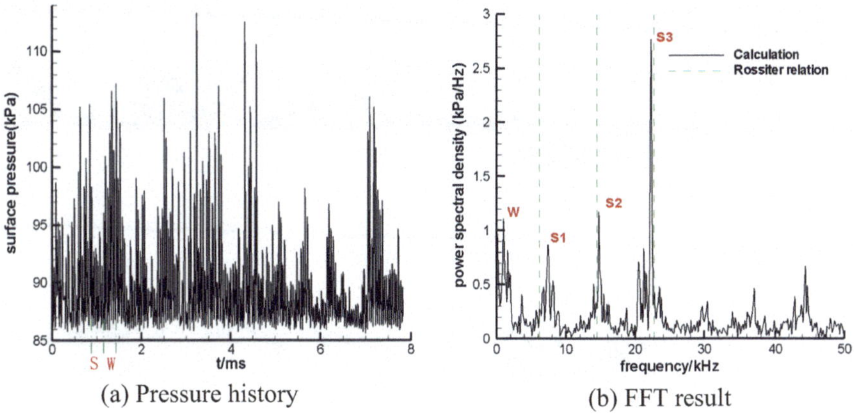

(a) Pressure history (b) FFT result

Fig. 2.4 Pressure history and FFT result of point P3 [3]

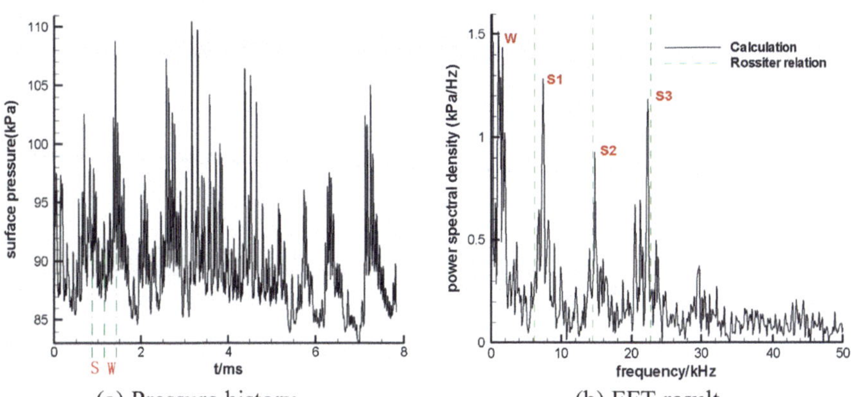

(a) Pressure history (b) FFT result

Fig. 2.5 Pressure history and FFT result of point P4 [3]

During stage 'W', however, the shock wave is generated due to the forming of large vortex and the separation of the upstream boundary layer, and the shock wave is strong enough to induce intense pressure oscillations around its incidence area at the upper wall.

Figures 2.6 and 2.7 show the instantaneous vorticity contours in stages 'S' and 'W', respectively. During stage 'S', the vortex forms mainly due to the Kelvin-Helmholtz instability [4] developing in the shear layer and the size of the vortex grows slowly along the shear layer. One can see that the vortex is far from saturation after developing over the cavity and it continues to grow downstream of the cavity. During stage 'W', however, the vortex forms mainly due to the absolute instability resulted from the strong backflow within the cavity. The shed vortex has dimensions of nearly the cavity size, and as it is forming, irrotational free-stream fluid is directed

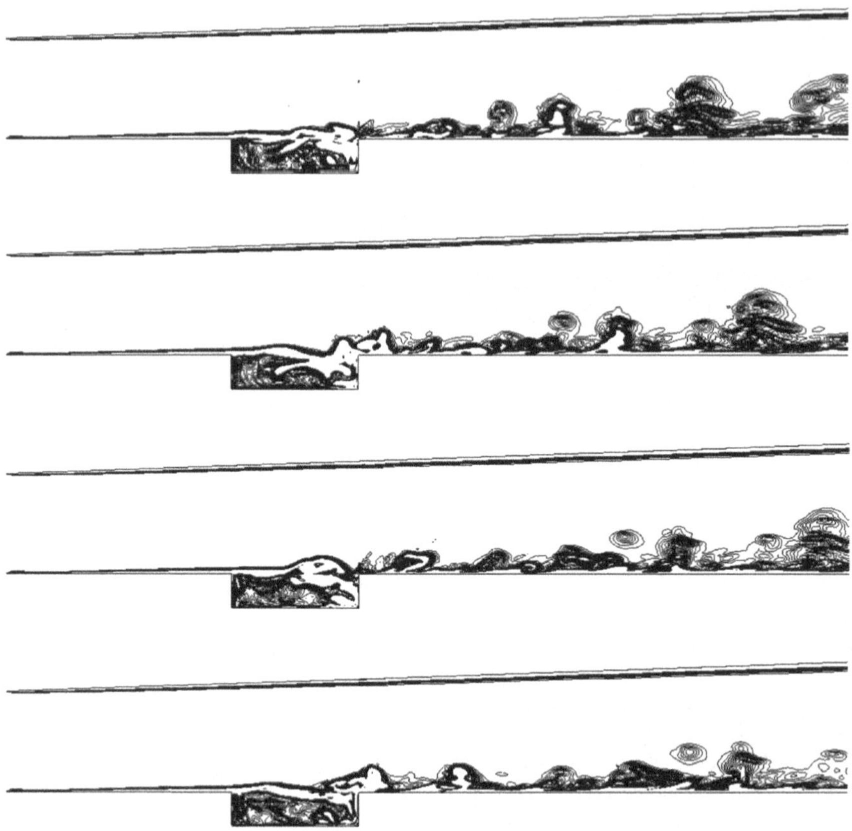

Fig. 2.6 Instantaneous results in stage 'S', $\Delta t = D/u_{\infty}$ [3]

into the cavity, impinging on the cavity base, as observed by Rowley et al. [8]. Furthermore, the vortex has almost become saturated around the aft wall and there is no growing of its size downstream of the cavity.

Based on the analyses above, it is believed that the present supersonic cavity flow does experience a mixed shear-layer/wake oscillation mode. These two modes occur alternately, and stages like 'S' correspond to the shear-layer mode and those like 'W' are dominated by the wake mode. Firstly, the Kelvin-Helmholtz instability develops in the cavity shear layer, inducing shear-layer oscillation mode. As the Kelvin-Helmholtz instability grows to sufficient strength, a strong recirculating flow is induced within the cavity. The resulting flow becomes absolutely unstable later, leading to wake mode. As analyzed above, much energy is needed to sustain the wake oscillation mode. Under the present condition, however, there is not so much energy supplied by the freestream constantly. Accordingly, the wake mode cannot stand for quite a long time but recovers to the shear-layer mode after several periodic cycles. Again, as the Kelvin-Helmholtz instability grows to sufficient strength during

Fig. 2.7 Instantaneous isosurfaces of vorticity [3]

shear-layer mode, a second wake mode cycle occurs, and so on. It seems that the lowest frequency 1034 Hz may be the frequency associated with wake mode or the alternation between shear-layer mode and wake mode.

2.1.1.2 Feedback-Loop Mechanisms

Huerre and Monkewitz [9] have shown that for hyperbolic tangent profiles with greater than 13.6% backflow, the flow is absolutely unstable. Based on the present flow conditions, the free-stream velocity is about 1000 m/s. For point U_2 (Fig. 2.3b), one can see that the backflow in stage 'W' reaches a maximum of 33% and a mean of about 20% of the free-stream velocity. The backflow in stage 'S' is much smaller, reaching a mean of about 8% of the free-stream velocity, and never exceeding 16%. The cavity flow is not strictly parallel and the mean profiles are not well described by hyperbolic tangent profiles, but the main point is that there is significant backflow in the cavity during the wake mode. It is possible, so, that absolute instability may provide a feedback mechanism for transition to wake mode [8].

As has been previously pointed out, there are currently several interpretations for the feedback mechanism of the cavity oscillations under shear-layer mode but none of them has been agreed upon. In particular, Tam [10] and Li et al. [11, 12] suggest a wave-reflecting mechanism for supersonic cavity oscillations. The results based on case 1 appear to support this mechanism and a detailed analysis will be given below.

Figure 2.8 shows the instantaneous results of vorticity and pressure contours around the cavity in stage 'S'. One can see apparently the vortex shedding, developing in the shear layer and impinging on the aft wall. Neglecting the expansion fans and resonant shocklets around the shear layer, one can see four waves in the cavity. The downstream-traveling waves WD1 and WD2 are two low-pressure waves associated with vortices V1 and V2, and the upstream-traveling waves WU1 and WU2 are two compression waves generated near the aft wall. It is notable that the downstream-traveling wave associated with the shed vortex usually appears in the downstream portion of the cavity and grows stronger and stronger when traveling downstream. The upstream-traveling wave, however, usually becomes apparent as it travels to the upstream portion of the cavity, which may result from the fact that the downstream-traveling wave is much stronger than the upstream-traveling one in the downstream portion of the cavity. As discussed by Li et al. [11, 12], there are not

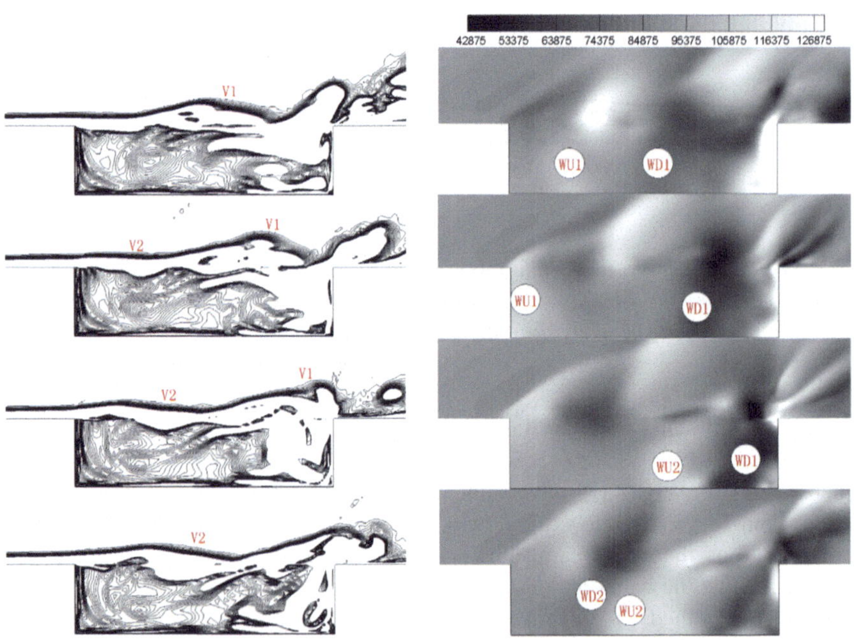

(a) Vorticity contours (b) Pressure contours

Fig. 2.8 Instantaneous results in stage 'S', $\Delta t = D/u_\infty$ [3]

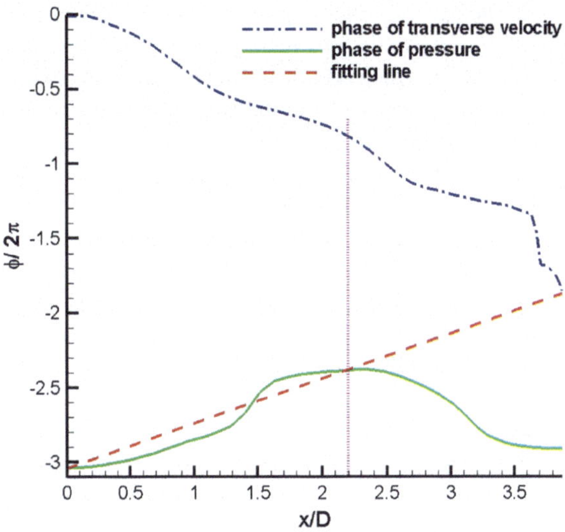

Fig. 2.9 Phase of transverse velocity and pressure at the frequency of the third Rossiter mode ($f = 22297$Hz). Transverse velocity in the shear layer along; pressure at cavity floor along [3]

only low-pressure regions associated with vortices but also high-pressure region in the connection of two adjacent vortices. These pressure wave-like regions alternatively/intermittently impinge on the trailing edge, which definitely results in pressure pulses. These pressure pulses are probably the generation mechanism of feedback acoustic waves.

A more quantitative analysis can be carried out based on the phase results presented in Fig. 2.9. The phase of the transverse velocity in the shear layer along $y/D = 0.0$ is plotted at the third resonant frequency, which represents the downstream-traveling vortex. Also plotted is the phase of the pressure at the cavity floor along $y/D = -1.0$, which represents the traveling pressure wave inside the cavity. These variables were chosen to better separate hydrodynamics from acoustic disturbances as discussed by Rowley et al. [8]. As expected, the phase of the transverse velocity confirms that the downstream traveling of the vortex in the shear layer. However, the phase of the pressure appears to show that there is an upstream-traveling pressure wave in the upstream portion ($x/D < 2.2$) but a downstream-traveling one in the downstream portion ($x/D > 2.2$), which is very different from the previous understanding of the subsonic cavity flow. Also can be seen is that the phases of the transverse velocity and the pressure show a very similar distribution in the downstream portion ($x/D > 2.2$), indicating the traveling pressure wave in this portion is probably forced by the shed vortex in the shear layer. As has been mentioned above, the pressure wave associated with the shed vortex becomes very strong compared to the upstream-traveling one generated around the aft wall in the downstream portion ($x/D > 2.2$) of the cavity. Accordingly, the upstream cavity floor ($x/D < 2.2$) sees an upstream-traveling wave while the downstream floor ($x/D > 2.2$) sees a downstream-traveling one instead.

Furthermore, if one supposes that the upstream-traveling wave has the same mean phase velocity in the upstream portion ($x/D < 2.2$) and in the downstream portion ($x/D > 2.2$), which is believed to be a good approximation considering the relatively uniform fluid properties in the streamwise direction near the cavity floor, then one can plot a fitting line to extend the phase velocity of the upstream-traveling wave in the upstream portion ($x/D < 2.2$) to the downstream part($x/D > 2.2$), as shown in Fig. 2.9. Notice that the total phase variation from the shear-layer convection (downstream) and acoustic propagation (upstream) is almost exactly $2\pi n$, where $n = 3$ is the index of the Rossiter mode. Notably, when the phase of the pressure along $y/D = -0.5$ or $y/D = -0.8$ is used instead, very similar results were obtained. This confirms the vortex convection-acoustic feedback mechanism of the shear-layer oscillation mode.

2.1.1.3 Influences of Mach Number and Upstream Injection

Figure 2.10 shows the velocity and pressure oscillation results for Ma = 2.52. From the velocity histories of point U_1 at the cavity leading, one can see that the upstream boundary layer never separates, indicating the wake mode does not occur under this flow condition. Notably, this is very different from the subsonic cavity flows, as shown by Rowley et al. [8], where the increase of Mach number usually tends to promote the transition to wake mode. The increase of Mach number tends to increase the momentum of the shear-layer fluid, leading to a stronger interaction between the shear layer and the cavity trailing edge. However, the supersonic cavity shear layer also becomes more stable with increasing Mach number. Thus there exist a competition between the increase of shear-layer momentum and the decrease of the amplitude of the fluctuations in the shear layer, and the latter seems to be dominant in the supersonic cavity flows. As pointed out b Rowley et al. [8], any effect that

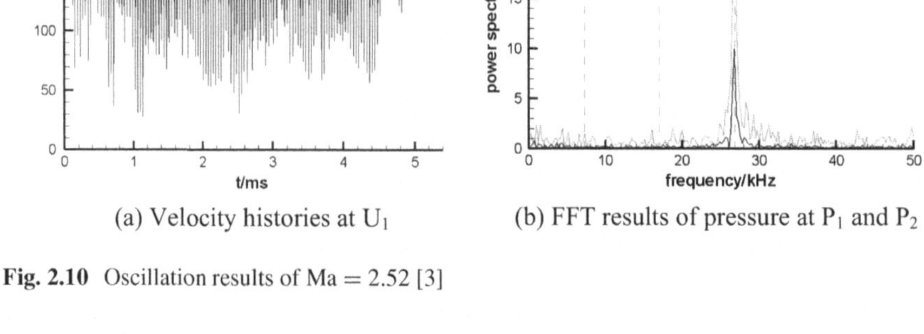

(a) Velocity histories at U_1 (b) FFT results of pressure at P_1 and P_2

Fig. 2.10 Oscillation results of Ma = 2.52 [3]

tends to decrease the amplitude of the fluctuations could inhibit the transition to wake mode.

The FFT results of the pressure indicate that the shear-layer mode oscillations still exist and agree will with the Rossiter relation. It is notable that all the oscillation energy appears to concentrate on the third mode, which is different from that for Ma = 1.75. So it seems that one can predict the possible dominant frequencies of the shear-layer mode but can not confirm which frequency or frequencies the oscillation energy will concentrate on beforehand.

Figure 2.11 shows that the upstream injection obviously weakens the oscillations within the cavity. However, there are still several dominant frequencies consistent with the shear-layer mode but shifting faintly, which indicates that the upstream injection does not destroy the vortex convection-acoustic feedback mechanism but only slightly changes the behavior of the cavity shear layer by moderate exchanging of mass and momentum between the jet and the shear layer.

Figure 2.12 shows the instantaneous isosurfaces of vorticity. It can be seen that the increase of Mach number makes the shear layer more stable and the rolling-up vortices become smaller. And the fluctuation intensity distributions of transverse velocity in the shear layer are consistent with these flow characteristics, as shown in Fig. 2.13. With upstream injection, the jet interacts with the cavity shear layer as it convects above the cavity and is gradually melted in the shear layer downstream of the cavity.

Figure 2.14 shows the sound pressure level (SPL) distributions along the cavity floor. The increase of Mach number does not change the SPL much while the upstream injection tends to weaken the pressure oscillations within the cavity. As has been pointed out earlier, the downstream-traveling pressure waves become stronger and stronger due to the developing of the associated vortices while the upstream-traveling ones usually become weaker and weaker due to the dissipation. So there is a trend that the oscillation energy is larger downstream. The periodic ups and downs of the

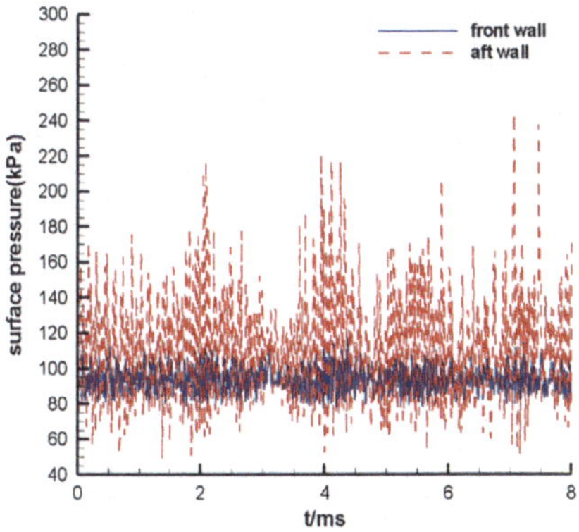

(a) Pressure histories

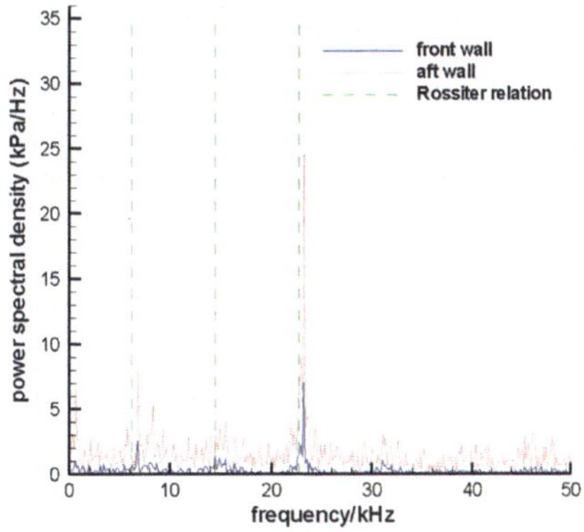

(b) FFT results

Fig. 2.11 Pressure histories and FFT results of points P1 and P2 (injecting, Ma = 1.75) [3]

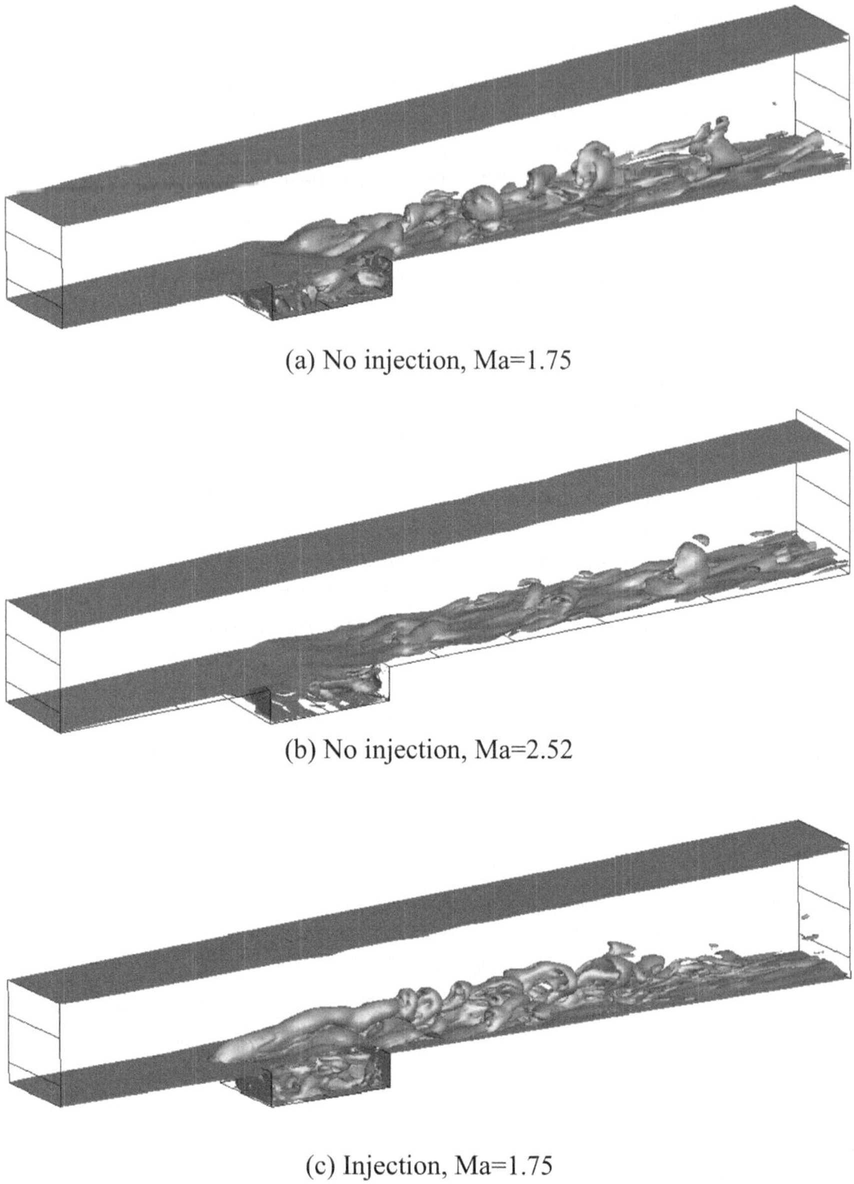

(a) No injection, Ma=1.75

(b) No injection, Ma=2.52

(c) Injection, Ma=1.75

Fig. 2.12 Instantaneous isosurfaces of vorticity [3]

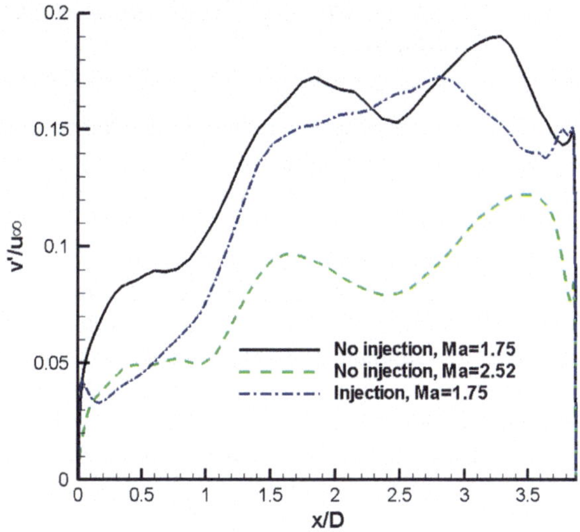

Fig. 2.13 Fluctuation intensity of transverse velocity along cavity mouth at $y/D = 0.0$ [3]

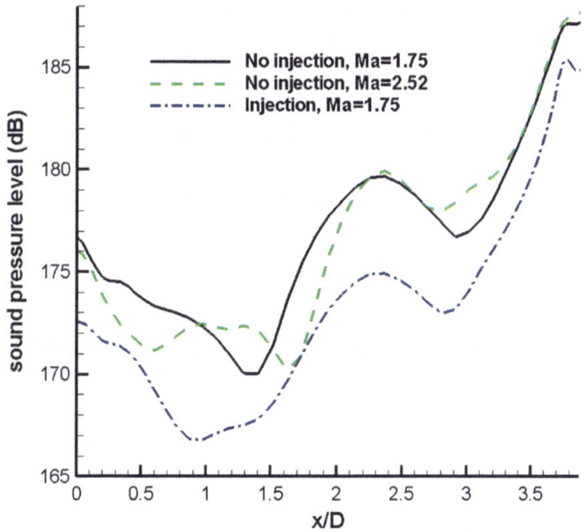

Fig. 2.14 Sound pressure level along cavity floor at $y/D = -1.0$ [3]

SPL suggest a possibility of existence of nodes and peaks of the static pressure in streamwise direction, which may be caused by the interactions of the streamwise-propagating pressure waves, as experimentally observed by Kegerise et al. [6].

2.1.2 Characteristics of Oscillations in Supersonic Cavity Combustion

The previous studies have clearly indicated the existence of combustion oscillations in the scramjet combustors. Wang et al. [13] investigated both the pressure and flame oscillations in a model scramjet combustor by experiments and numerical simulations.

The combustor that Wang et al. [13] investigated has a constant width of 50 mm and height of 40 mm, shown in Fig. 2.15. The model scramjet combustor is installed behind the nozzle of the air heater, which heats the air by means of air/O_2/alcohol combustion to simulate flight Mach 6 conditions, resulting in a Mach 2.52 and mass flow 1 kg/s stream in the combustor entrance. Detailed flow conditions at the nozzle exit and fuel jet exit are listed in Table 2.2, where the fuel is 99.5–99.8% pure H_2. The fuel can be injected transversely into the combustor from a 2 mm orifice, located 10 mm upstream of the cavity leading edge. Three injection conditions Pjet = 0.6, 1.2 and 1.8 MPa, resulting in equivalence ratios of 0.038, 0.076 and 0.114, are considered. Cavities of depth D = 8 mm, length to depth ratio L/D = 4 or 7, and aft angle A = 22.5°, 45° or 90° are adopted. For simplicity, we use LxAy to denote the cavity with L/D = x and A = y from now on.

2.1.2.1 Experimental Study of Combustion Oscillations

Figure 2.16 shows the frequency spectra of cavity oscillations without hydrogen injection for a typical cavity (L4A90). There seem to be three dominant frequencies around 7, 15.2 and 24.7 kHz, corresponding to the first three Rossiter modes

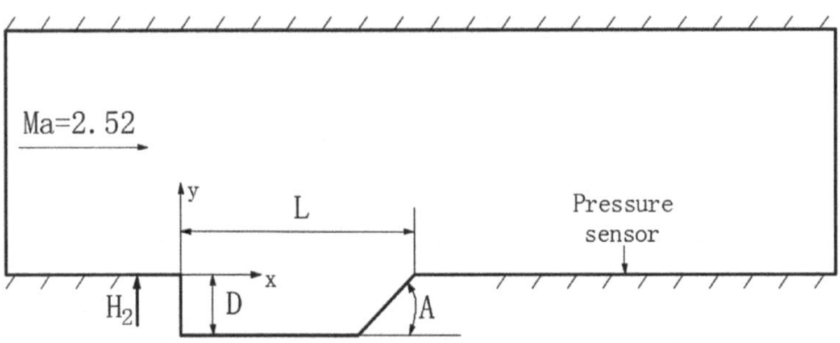

Fig. 2.15 Schematic of the scramjet combustor Wang et al. [13]

Table 2.2 Inflow conditions for cavity flows (Wang et al. [13])

Parameter	Air	Jet
T_0, K	1486	300
P_0, MPa	1.6	0.6/1.2/1.8
Ma	2.52	1.0

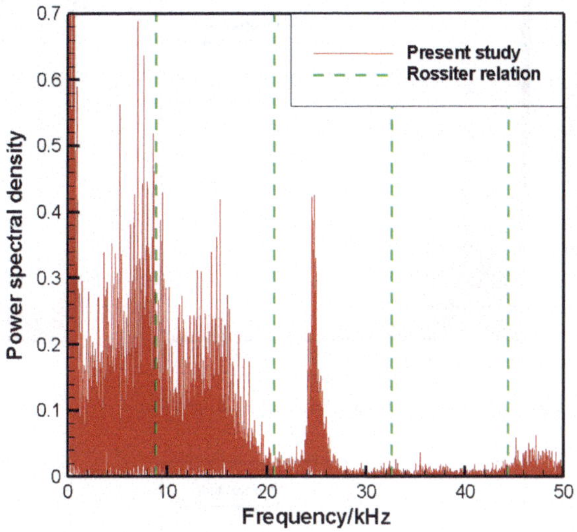

Fig. 2.16 Frequency spectra of pressure oscillations for cavity L4A90 without hydrogen injection [13]

[4, 5]. Rossiter mode suggests an acoustics-vortex resonant mechanism to explain the oscillations: the upstream traveling compression wave reaches the front wall, induces shed vortices or disturbances which are amplified subsequently via Kelvin-Helmholtz instability, the amplified structures in the shear layer impinges on the aft wall and generates another compression wave, closing the feedback loop. Although the present study appears to support the existence of Rossiter mode, there is a quantitative discrepancy between the experimental data and the predictions. Additionally, the first two modes are surrounded by a widely distributed spectra, similar to the observations of Chandra et al. [14] under some conditions. These results indicate that the cavity oscillations under the complicated conditions (turbulent incoming boundary layer, shock wave etc.) inside the scramjet combustor are very different from those under the ideal conditions as supposed in Rossiter mode.

Figure 2.17 shows the frequency spectra for cavity L7A90, where it is evident that the frequencies shift to higher levels due to the combustion and the increase of equivalence ratio. Figure 2.18 shows the frequency and intensity of pressure oscillations for both cold and combustion flows. Notably, it is found that flame can hardly be stabilized by shorter cavities with L/D = 4 under the present conditions, thus only the longer cavities with L/D = 7 are considered for the combustion cases.

In the present study, intensities of pressure and flamefront oscillations are defined as sound pressure level and rms (root mean square) results, respectively. One can see that cavities with larger aft angle tend to exhibit pressure oscillations of higher frequency and intensity. Additionally, the results indicate that the combustion of the hydrogen injected upstream of the cavity suppresses the lower frequencies but pushes oscillations to higher frequencies (basically located in 15–20 kHz). Furthermore, the frequency increases with increasing injection pressure. The change of the frequency may be attributed to the instability of the jet mixing layer developing above the cavity and the heat release from the combustion, which interact with the cavity

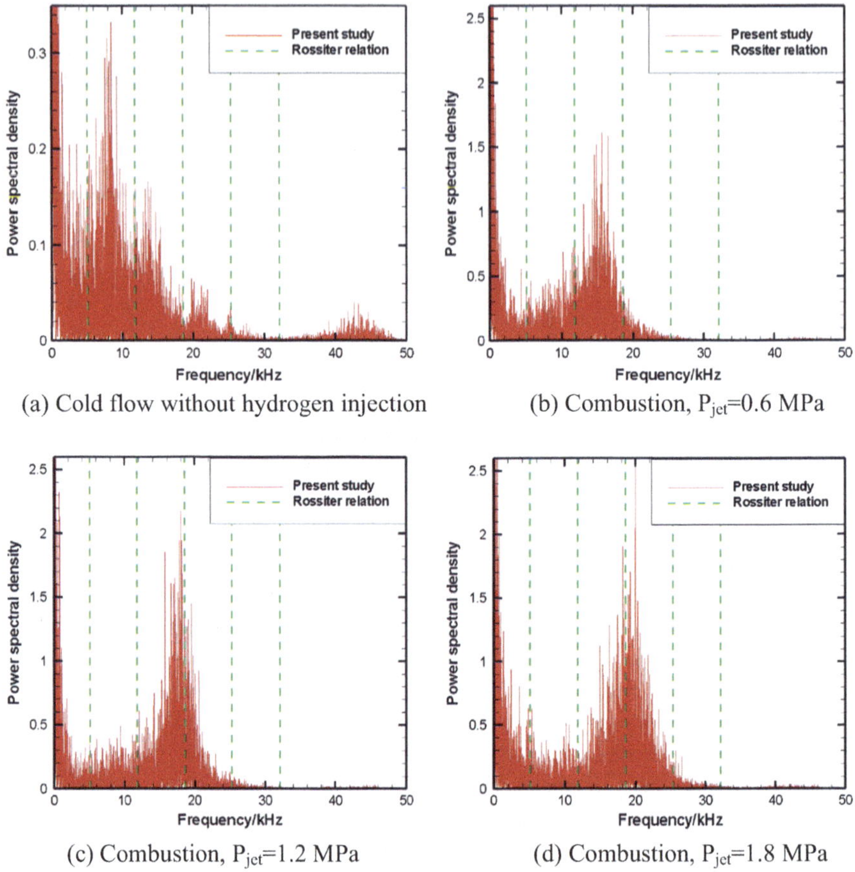

(a) Cold flow without hydrogen injection (b) Combustion, P_{jet}=0.6 MPa

(c) Combustion, P_{jet}=1.2 MPa (d) Combustion, P_{jet}=1.8 MPa

Fig. 2.17 Frequency spectra of pressure oscillations for cavity L7A90 [13]

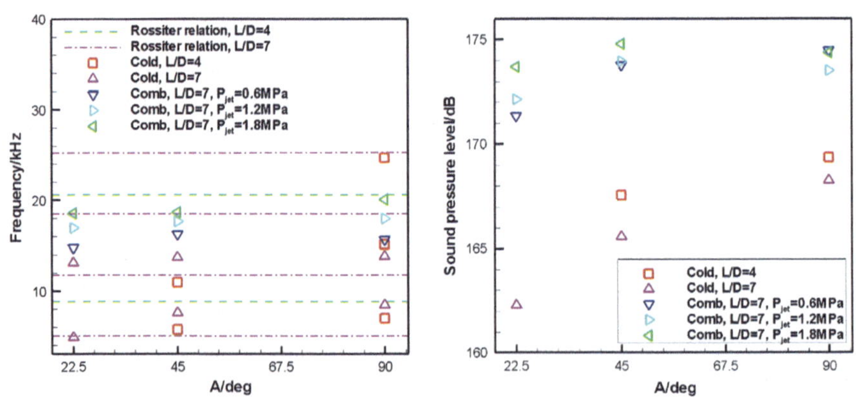

Fig. 2.18 Frequency and intensity of pressure oscillations [13]

shear layer and trigger high-frequency oscillations favored by the modified flowfield. Meanwhile, the oscillation intensity also markedly increases due to the combustion. One reason is the increase of the mean pressure caused by the heat release. Another reason may be the introduction of additional energy supplied by unsteady heat release to the oscillating system.

High-speed flame luminosity imaging is also adopted to investigate the flame oscillations. The camera images the combustion luminosity at 4000 frames per second through the combustor side window. Figure 2.19 shows the instantaneous flame structure of different cavities with moderate injection pressure Pjet = 1.2 MPa. Following the method of Micka [15], a combustion iso-luminosity contour of 10 counts is used to characterize the combustion zone outline. The streamwise location of the flamefront is then calculated for y/D = 0.5–1.5 above the cavity. Figure 2.20 shows the history and frequency spectra of the flamefront oscillations for cavity L7A90 with Pjet = 1.2 MPa. One can see that the amplitude of the flamefront oscillations in the streamwise is usually up to 10 mm. Also, there seems to be no dominant frequency less than 2 kHz for the flame oscillations, indicating the existence of no low-frequency combustion instabilities coupled with acoustics.

Figure 2.21 shows the average location and intensity of the flamefront oscillations for cavities with different aft angles and injection pressures. The cavity with larger aft angle shows stronger flame oscillations regardless of any injection pressure considered in the present study. Meanwhile, the cavity with larger aft angle also tends to stabilize the time-averaged flamefront more upstream, indicating a shorter ignition

Fig. 2.19 Instantaneous flame structure for P_{jet} = 1.2 MPa [13]

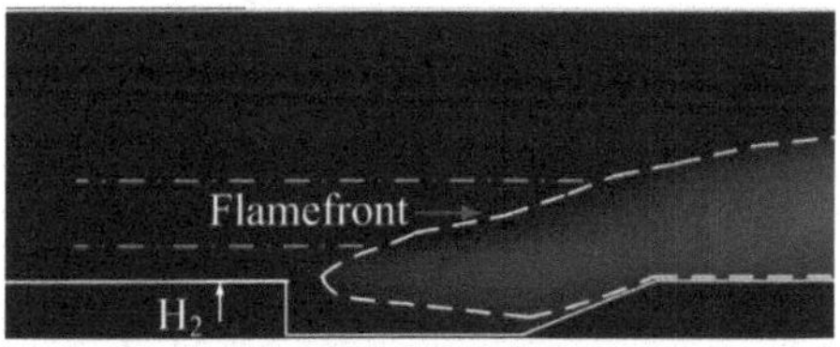

(a) Cavity L7A22.5

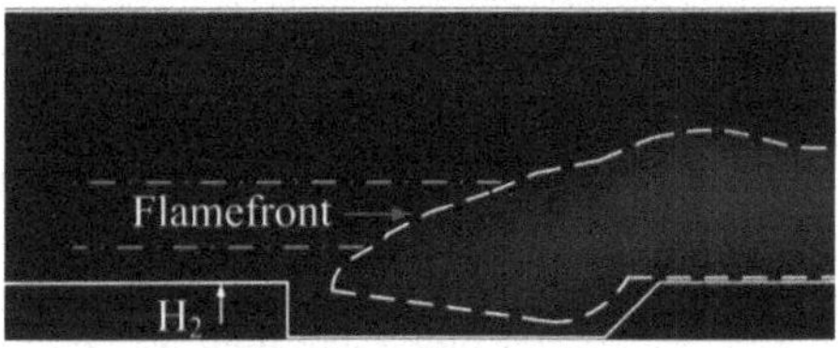

(b) Cavity L7A45

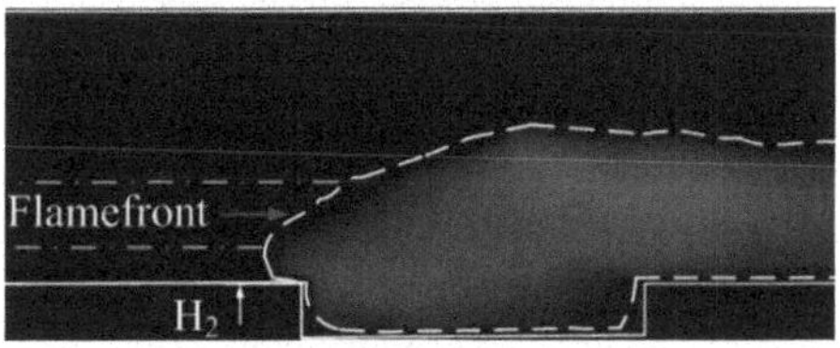

(c) Cavity L7A90

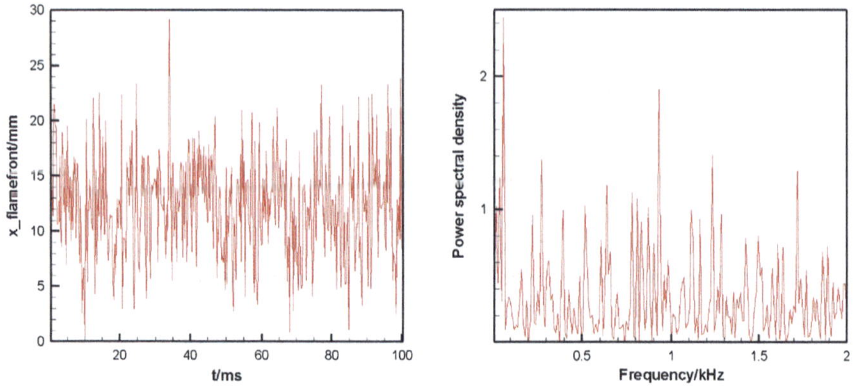

Fig. 2.20 History and frequency spectra of flamefront oscillations for cavity L7A90, $P_{jet} = 1.2$ MPa [13]

Fig. 2.21 Average location and rms of flamefront oscillations [13]

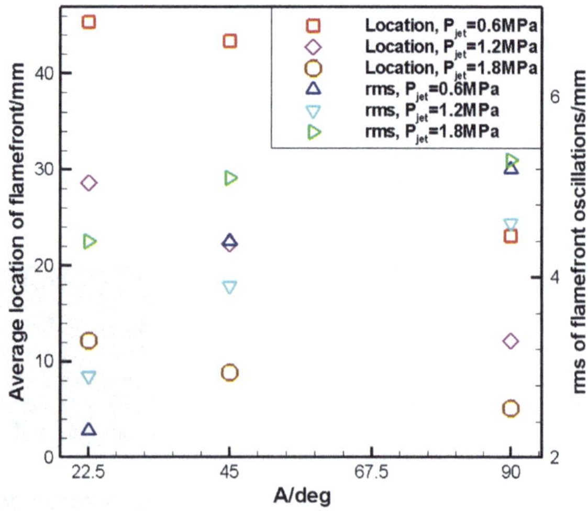

distance. Therefore, it is inferred that moderate oscillations may be beneficial to the ignition and combustion by enhancing the mixing process.

In summary, cavity oscillations inside the scramjet combustor appear to be dominated by Rossiter mode, but there is a quantitative discrepancy between the experimental data and the predictions due to the complicated incoming conditions. High-speed imaging results show that there exist intense flame oscillations though there is no dominant frequency less than 2 kHz. The cavity flameholder with larger aft angle tends to exhibit stronger flame oscillations as well as shorter ignition distance, suggesting that moderate oscillations may be beneficial to the ignition and combustion. Under the combustion conditions, both the frequency and intensity of the pressure oscillations shift to higher levels (15–20 kHz, >170 dB), indicating the existence of high-frequency, strong flow and combustion oscillations. The flame is expected to be influenced by these high-frequency, strong pressure oscillations.

2.1.2.2 Numerical Study of Combustion Oscillations

Due to the limitation of the spatial and temporal resolution of measurement, the experiments can not obtain enough flow field information to analyze the mechanism of high-frequency combustion oscillation. Therefore, Wang et al. [16] investigated combustion oscillations in a supersonic combustor with hydrogen injection upstream of a cavity flameholder using a hybrid RANS/LES method acting as a wall-modeled LES to analyze the mechanisms of high-frequency combustion oscillations in more details. The conditions of the simulation are shown in Table 2.2, except that the total pressure of the fuel jet is 1.8 Mpa.

Characteristics of jet-cavity combustion

In order to study the long-term behaviors of the combustion oscillations as well as to minimize the statistical error, the calculation is continued for more than twenty characteristic times after convergence, where the characteristic time is defined as L/U_∞.

Figure 2.22 shows the instantaneous flowfield around the reactive transverse jet with supersonic incoming turbulent boundary layer. Typical high-speed, low-speed streaks and turbulent structures in the boundary layer as well as the bow shock, separation shock and large-scale structures around the jet boundaries are clearly observed. Also, the interaction between the turbulent structures in the incoming boundary layer and the transverse jet can be obviously seen, which is expected to be important for the quick breakdown of the jet and for the generation of large-scale structures around the jet as will be seen below.

The three-dimensional features of vortical structures around the jet and the cavity shear layer are visualized by the instantaneous isosurfaces of the second invariant of

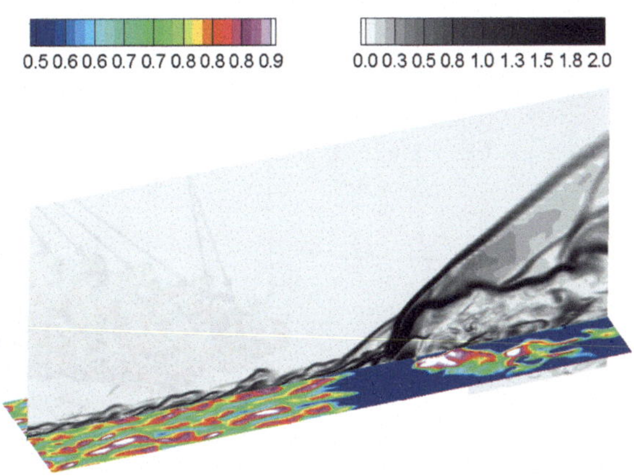

Fig. 2.22 Transverse jet with supersonic incoming turbulent boundary layer. Density gradient magnitude and streamwise velocity contours in centerplane and wall-parallel plane close to the wall ($y/\delta_{\text{inf}} = 0.2$) [16]

velocity gradient tensor Q in Fig. 2.23. The Q isosurfaces are colored by streamwise vorticity, which indicates that red and blue surfaces show clockwise and counter-clockwise rotating vortices with the axis in the streamwise direction while Q isosurfaces with green color show vortices without the axis in the streamwise direction. Two types of vortices around the fuel jet are observed. One is the counter-rotating vortex pair with the axis in the streamwise direction and the other is the hairpin-like vortices rolling up in the jet boundaries. As observed by Kawai and Lele [17], there should be two groups of hairpin-like vortices generated from the windward and leeward portion of the jet, respectively. In the present study, however, the hairpin-like vortices in the leeward portion of the jet are not evident, which may result from the interaction of the cavity shear layer and the leeward portion of the jet. The counter-rotating vortices may play an important role in the flame spreading as observed by Sun et al. [18], while the hairpin-like vortices have great influences on the possible auto-ignition process as will be seen below. Relatively fine vortical structures are also observed around the cavity shear layer, which are important to the mass exchange between the fluids in and out of the cavity as well as to the jet-cavity interaction, related to the flame holding and spreading.

Figure 2.24 shows the instantaneous H_2 and OH distributions together with sonic line in the axial slices. The stretching and distortion of the fuel jet can be clearly seen in the near field, which is beneficial to the subsequent mixing of the fuel and air. One portion of the jet mixing occurs in the subsonic region adjacent to the cavity shear layer and a small portion of the fuel can be even entrained into the cavity recirculation region, which indicates that a portion of combustion may occur within the cavity. Another portion of the jet mixing accompanied by the formation of large-scales in the windward jet boundary occurs in the supersonic region. The calculation shows

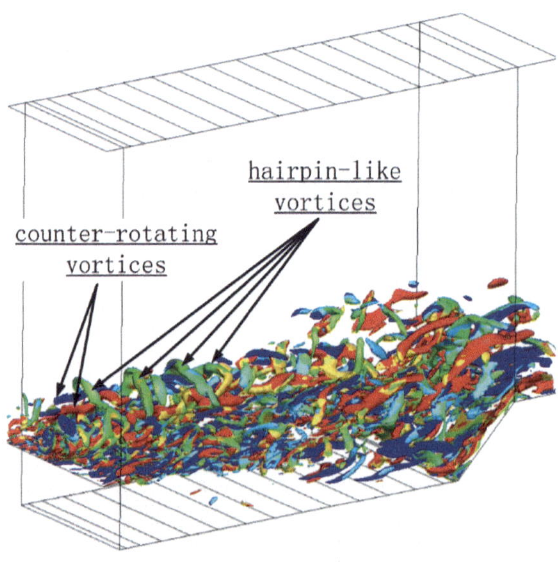

Fig. 2.23 Instantaneous isosurfaces of the second invariant of velocity gradient tensor Q colored by streamwise vorticity [16]

Fig. 2.24 Oblique views of axial slices for instantaneous contours together with sonic line [16]

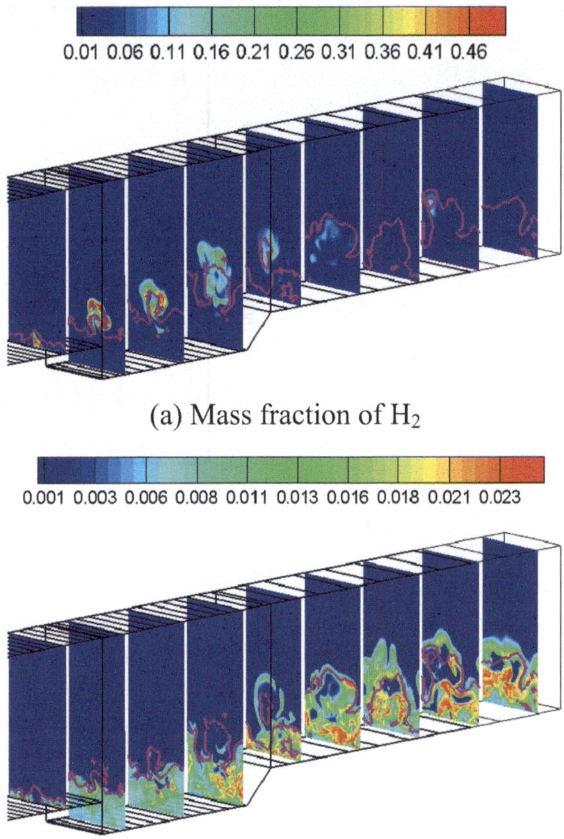

(a) Mass fraction of H_2

(b) Mass fraction of OH

that, however, the combustion is mainly confined in the subsonic regions though a small portion of reaction may occur in the supersonic regions far downstream of the cavity.

Although the fuel injection and combustion may mitigate the intensity and change the dominant frequencies of the cavity-shear oscillations, there still exist remarkable pressure oscillations in the flowfield around the cavity. Figure 2.25 shows the frequency spectra of pressure oscillations at the cavity floor. The acoustics-vortex resonant mechanism proposed by Rossiter [4] still seems to exist and the first two dominant frequencies, 6.7 and 17.4 kHz, appear to agree with the first and third mode of the Rossiter relation [4]. These two frequencies are found to couple with the combustion oscillations in the main stream as will be shown below. There are also several dominant frequencies ranged from 30 to 80 kHz, evident coupling of which between the combustion is not observed, so the analyses will not focus on them in the present study.

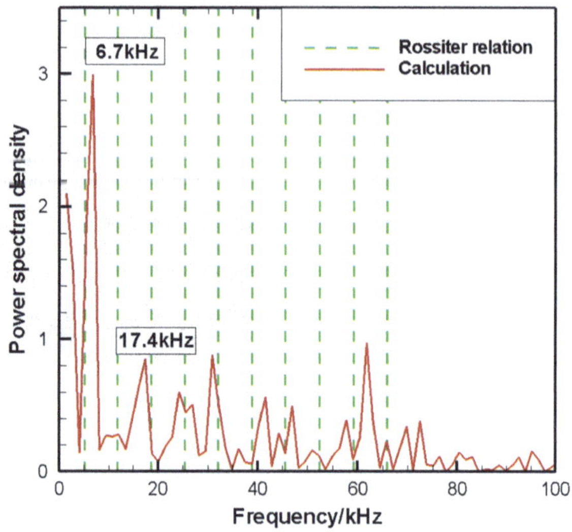

Fig. 2.25 Frequency spectra of pressure oscillations at cavity floor [16]

Combustion oscillations

It is found that the reaction zone in the cavity is relatively stable, making the cavity act as a flameholder. However, the combustion zones in the main stream oscillate remarkably. Since the temperature oscillations at one point can be influenced by too many factors and may be too chaotic, it is not appropriate to use them to demonstrate the combustion or flame oscillations. Here, we define a mass-averaged temperature to indicate the combustion oscillations. The mass-averaged temperature at the streamwise station x is defined as

$$T_{ma}(x) = \frac{\iint\limits_S \rho(x, y, z)u(x, y, z)T(x, y, z)dydz}{\iint\limits_S \rho(x, y, z)u(x, y, z)dydz} \tag{2.2}$$

where the integrating region S should be chosen to just include the possible combustion zone around the fuel jet in the main stream but exclude the influences of the combustion in the cavity shear layer as much as possible.

Figure 2.26 shows the time history and frequency spectra of the mass-averaged temperature oscillations at station $x/D = 4.5$. Based on the calculation observations, the integrating region S is chosen as $y/D = 1 \sim 2$ and $z/D = -0.6 \sim 0.6$ as marked by B in Fig. 2.27. As can be seen, there exist intense temperature oscillations though the temperature is a mass-averaged one in a relatively large region, indicating the existence of strong combustion or flame oscillations. The combustion oscillations show a widely distributed spectra ranged from several kHz to over one hundred kHz.

Recall that the two lower frequencies, 6.7 and 17.4 kHz, correspond to the cavity-shear oscillations as has been analyzed above. This means that the oscillations of the cavity shear layer have direct effects on the combustion oscillations in the main

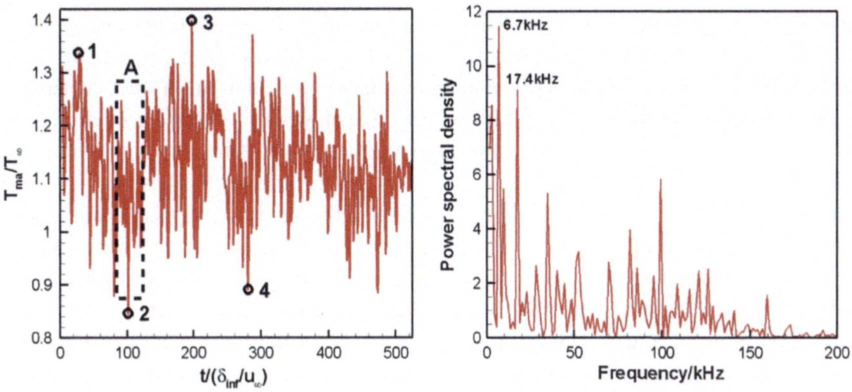

Fig. 2.26 Time history and frequency spectra of mass-averaged temperature oscillations at station $x/D = 4.5$, as marked by B in Fig. 2.27 [16]

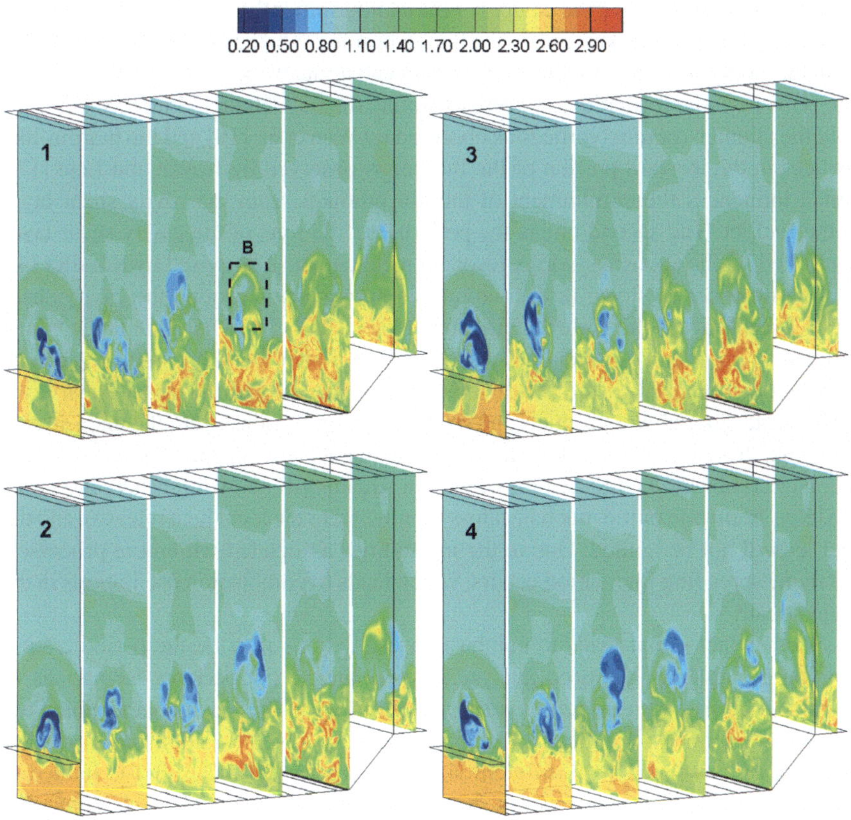

Fig. 2.27 Oblique views of streamwise slices around the cavity for instantaneous temperature contours [16]

stream. It can also be concluded from the experimental results in the last section, shown in Fig. 2.27d. The dominant frequency is 17.4 kHz, which agrees with the third mode of the Rossiter relation [4]. The main frequencies of calculation and experiment are inconsistent due to the sensitivity of the shear layer to the choice of dominant modes. However, it is shown that the high-frequency pressure oscillation may be caused by the self-excited oscillation of the cavity-shear layer. As has been pointed out above, a relatively stable reaction zone exists in the cavity, acting as a flameholder for the jet combustion. Once the flame is stabilized in the cavity shear layer or within the cavity, it needs to spread to the main flow and ignite the jet so that the flameholding process can be accomplished, where the interaction of the jet-with-cavity shear layer plays an important role. Representative snapshots of temperature contours in several streamwise slices are shown in Fig. 2.27. The mass-averaged temperature history data shown in Fig. 2.26 are obtained by integrating in region B as shown in the first snapshot. These four snapshots are taken at the time corresponding to the numbered markers in the temperature history. The simulated unsteady flowfield illustrates that the fuel jet passing above the cavity is ignited intermittently.

There may be three key processes that determine whether the flame can spread from the cavity shear layer to the main stream and ignite the jet successfully. The first one is the periodic formations of vortices around the fuel jet, including formations of both the counter-rotating vortices as observed by Won et al. [19] and the hairpin-like vortices in the leeward portion of the fuel jet as observed by Kawai and Lele [17], which influences the entrainment of the hot products from the cavity shear layer to the fuel jet. The second one is the periodic oscillations of the cavity shear layer controlled by the acoustics-vortex resonant mechanism as proposed by Rossiter [4], which influences the local flow condition and combustion around the cavity shear layer. The last one is the vortex interaction of the jet and the cavity shear layer, which is the most important one and finally determines whether the flame spreading can be successfully achieved. If the vortex interaction of the jet and the cavity shear layer is favorable, enough hot products and oxidant will be entrained into the main stream following the counter-rotating vortices, and the fuel jet will be ignited. Otherwise, the flame or combustion entrained into the main stream will be extinguished due to the too large strain rate or too much heat loss (compared to the local heat release), and the fuel jet will not be ignited. Due to the instabilities associated with these processes, the flame spreading fluctuation occurs, which leads to combustion oscillations in the main stream.

Another factor that leads to combustion oscillations may be the periodic auto-ignition process. As has been analyzed above, there exist periodic formations of hairpin-like vortices in the jet boundaries, accompanied by which combustible fluid packets are formed periodically. Free-jets generally have two dominate instability frequencies associated with different sizes of vortices [20]. The first is related to the instability of the shear layer at the jet orifice. The initial vortex shedding frequency, also called the most amplified frequency f_{θ_j}, scales with the initial shear layer momentum thickness θ_j and jet exit velocity U_j. The corresponding Strouhal number is

$$St_{\theta_j} = \frac{f_{\theta_j}\theta_j}{U_j} \tag{2.3}$$

where St_{θ_j} is found to be scattered from 0.01 to 0.18 [21]. The second dominant jet instability mode is related to larger structures present downstream of the jet potential core. The characteristic frequency is referred to as the preferred mode frequency f_j. The preferred mode frequency scales with the jet exit diameter d and velocity U_j to yield the corresponding Strouhal number

$$St_d = \frac{f_j d}{U_j} \tag{2.4}$$

Gutmark and Ho [21] found that the value of St_d varied between 0.24 and 0.64 based on the various initial conditions of different facilities. Fric [22] found that the value of St_d varied between 0.2 and 2 at different locations along the jet. The analyses from Ben-Yakar [20] indicated a value of $St_d \approx 1$. In the present study, the shear-layer thickness at the jet orifice is neglected, thus only the preferred mode can be considered. Based on the present flow conditions, the preferred mode frequency is estimated to vary between 65 kHz and 325 kHz, assuming a value of $St_d = 0.2 \sim 1$.

Representative snapshots of temperature contours and H_2 mass fraction distribution overlapped with vorticity magnitude contours in the centerplane are shown in Fig. 2.28 and Fig. 2.29, respectively. These eight snapshots are taken at the time corresponding to region A in the mass-averaged temperature history shown in Fig. 2.26. Within this stage, it is expected that the ignition probability of the fuel jet by the flame spreading from the cavity shear layer is small. So the auto-ignition process can be demonstrated more distinctly.

Here, we track three representative fluid packets formed in the windward portion of the fuel jet to analyze the auto-ignition. When the hairpin-like vortices are formed, hot air is entrained into the fuel jet boundaries and combustible fluid packets are formed accordingly, as illustrated in Figs. 2.28 and 2.29. It is observed that the combustible fluid packets are basically generated in the regions between the hairpin-like vortices and the main body of the fuel jet. Since the auto-ignition delay time and distance is controlled by the entire time history of the velocity, temperature and equivalence ratio of a fluid packet, the combustion zone resulting from the auto-ignition will not have a continuous structure as can be seen in Fig. 2.28. Because of the different initial condition (formation position, temperature, equivalence ratio, dimension, etc.) and passing route of these coherent structures, moreover, the resulting combustion zones oscillate in the main stream. For example, packet 1 seems to have a larger dimension than packets 2 and 3, so it results in a larger reaction zone, which is also beneficial to ignite the whole fuel jet earlier. Notably, the time interval between the formations of packets 1 and 2 is approximately 15 μs, and that between the formations of packets 2 and 3 is about 8 μs. These periodic events correspond to frequencies ranged from 67 to 125 kHz, which agree well with the relatively higher frequencies of the mass-averaged temperature oscillations shown in Fig. 2.29, fitly located in the previously estimated range of the preferred mode frequency of the jet. So it is expected that

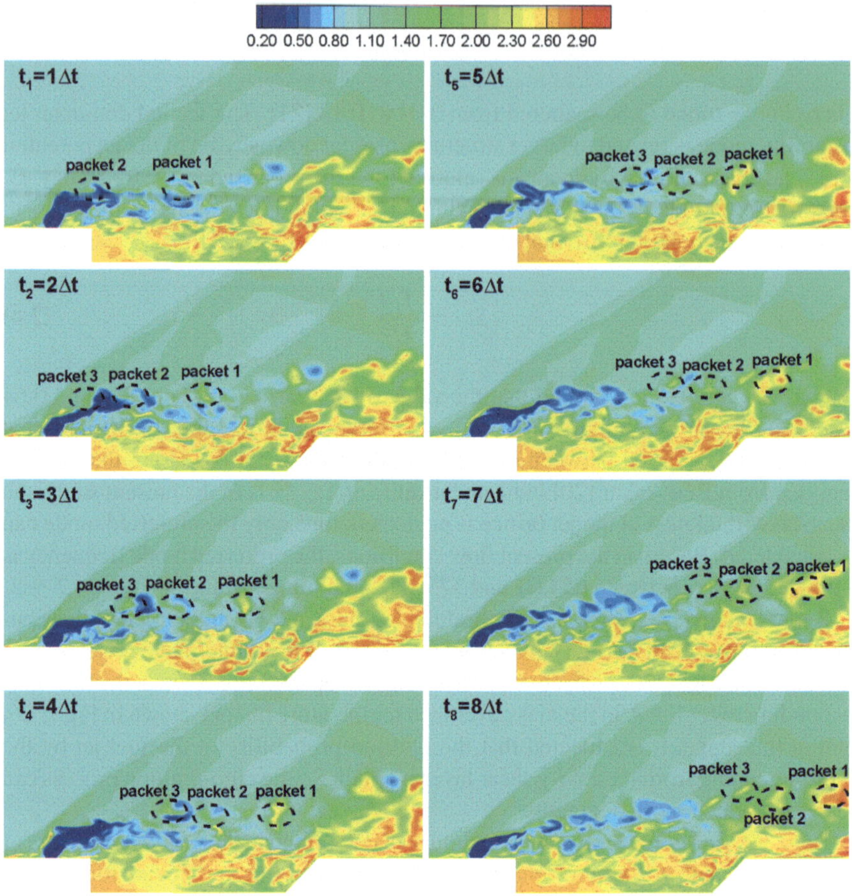

Fig. 2.28 Representative snapshots of temperature contours in centerplane, where $\Delta t = 4\delta_{\text{inf}}/u_\infty$ [16]

the auto-ignition process accompanied by the formation of the hairpin-like coherent vortices also has evident influences on the combustion oscillations in the main stream.

In sum, the combustion oscillations in the cavity-based combustor can be induced by two unsteady processes. One is the intermittent flame spreading from the cavity shear layer to the main stream and the other is the auto-ignition of the combustible fluid packets formed by the hairpin-like vortices in the jet boundaries. In fact, these two processes may interact with each other, making the oscillations more complicated. Notably, hydrogen is used as fuel in the present study, which has a relatively short ignition delay when compared to most hydrocarbon fuels. Thus, the oscillation mechanism related to the auto-ignition may become less evident when hydrocarbon fuels are used unless the flight Mach number or total temperature is well raised. It is inferred, however, that the fuel may have little effect on the other oscillation mechanism since the reactivity of the reactants (mainly determining the laminar burning

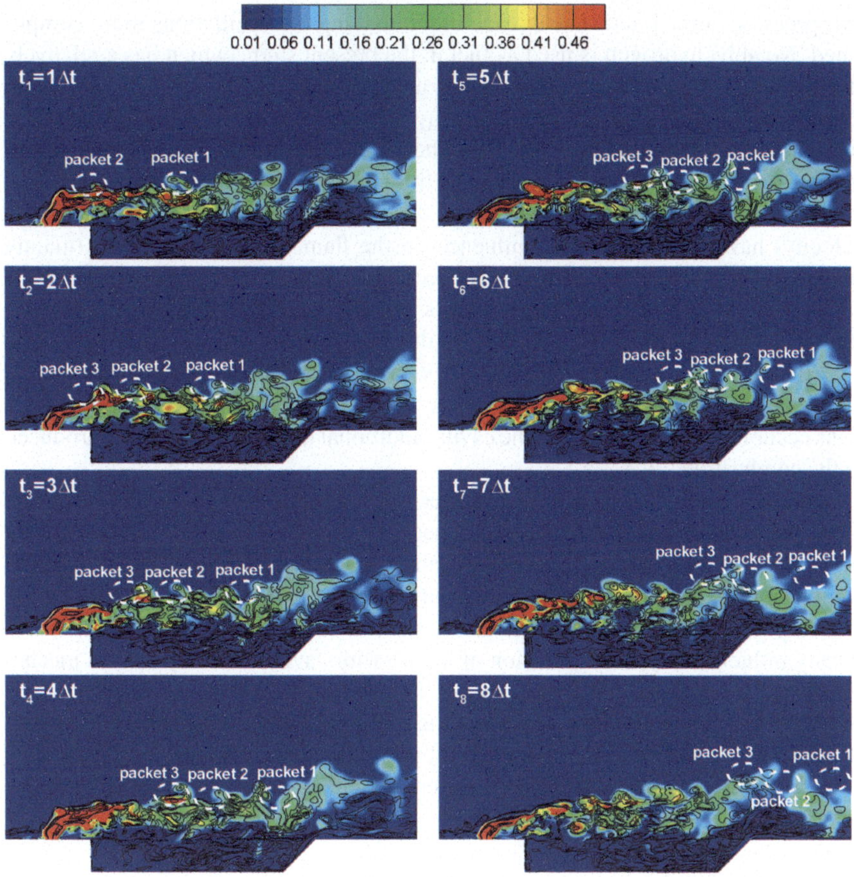

Fig. 2.29 Representative snapshots of H_2 mass fraction distribution overlapped with vorticity magnitude contours in centerplane, where $\Delta t = 4\delta_{\text{inf}}/u_\infty$ [16]

velocity) have relatively small influence on the flame spreading process (mainly determined by the turbulent transport) once the combustion is stabilized around the cavity flameholder.

In general, there exists a competition between the auto-ignition and flame propagation, which needs to be further explored in the future work. It is also notable that the shock/boundary-layer interaction near the upper wall is neglected in the present study, which weakens the reflection shock waves and delays the shock-combustion interactions. Once strong shock-combustion interactions occur in the regions around the cavity, additional instabilities may be introduced to the combustion.

In sum, the combustion oscillations in the cavity-based combustor can be induced by two unsteady processes. One is the intermittent flame spreading from the cavity shear layer to the main stream and the other is the auto-ignition of the combustible fluid packets formed by the hairpin-like vortices in the jet boundaries. In fact, these two processes may interact with each other, making the oscillations more complicated. Notably, hydrogen is used as fuel in the present study, which has a relatively short ignition delay when compared to most hydrocarbon fuels. Thus, the oscillation mechanism related to the auto-ignition may become less evident when hydrocarbon fuels are used unless the flight Mach number or total temperature is well raised. It is inferred, however, that the fuel may have little effect on the other oscillation mechanism since the reactivity of the reactants (mainly determining the laminar burning velocity) have relatively small influence on the flame spreading process (mainly determined by the turbulent transport) once the combustion is stabilized around the cavity flameholder. In general, there exists a competition between the auto-ignition and flame propagation. It is also notable that the shock/boundary-layer interaction near the upper wall is neglected, which weakens the reflection shock waves and delays the shock-combustion interactions. Once strong shock-combustion interactions occur in the regions around the cavity, additional instabilities may be introduced to the combustion.

Totally, it is believed a mixed shear-layer/wake oscillation mode in the flow, where these two modes occur alternately. The shear-layer mode and wake mode are driven by vortex convection-acoustic feedback and absolute instability, respectively. The combustion oscillations can mainly be attributed to two mechanisms. One is the unsteady flame spreading from the cavity shear layer to the main stream, which is greatly influenced by the interaction of the jet-with-cavity shear layer. This mechanism leads to relatively low-frequency oscillations that correspond to the cavity-shear layer oscillations. The other is the auto-ignition of the combustible fluid packets formed around the fuel jet accompanied by the generation of the hairpin-like vortices, which leads to relatively high-frequency oscillations that correspond to the jet instabilities.

2.2 Low Frequency Acoustic Oscillation

The combustion instabilities of scramjet can cause strong interaction among fuel jet, cavity flame stabilization zone, mainstream combustion zone and pre-combustion shock waves. And it leads to low-frequency acoustic oscillation in the combustion chamber flow field. Low-frequency acoustic oscillation can be propagated upstream through various subsonic zones such as boundary layer and flame-stabilized recirculation zone in scramjet. When the combustor operates in sub-combustible mode, the research results of Ma et al. [23] and Lin et al. [24] clearly showed the existence of low frequency acoustic oscillation. Li et al. [25] carried out three dimensional simulation of the ethylene-fueled scramjet combustor investigated by Ma et al. [23]. And the results displayed the oscillations of the flame and fuel distribution. Sun et al. investigated the effects of mixing status on the oscillation modes. However, when the combustor operates in super-combustion mode, the low frequency acoustic oscillation needs further study. Therefore, this paper mainly introduces the low frequency acoustic oscillation when the combustor works in sub-combustion mode.

2.2.1 *Effect of Cavity Parameters on the Acoustic Oscillation*

A series of experiments have been carried out by Ouyang et al. [26] to study the influence of cavity parameters on the combustion oscillation in scramjet combustor. The influence of flameholding cavity position, its length to depth ratio *L/D* and aft wall angle θ and amount on ethylene combustion oscillation characteristics in scramjet combustor has been researched. The obtained experimental results show that, as the premixing distance between ethylene injector and flameholding cavity varies, the ethylene combustion flame will take on two distinct forms, small-amplitude high frequency fluctuation, and large-amplitude low frequency oscillation. The dominant frequency of the large-amplitude combustion oscillation is in inverse proportion to the pre-mixing distance. Moreover, the influence of cavity length to depth ratio and the aft wall angle θ exists diversity when the flameholding cavity position is different and can be recognized as unnoticeable compared to the impact of the premixing distance. In addition, we also find that, when the premixing distance is identical and sufficient, increasing the amount of tandem flameholding cavity can change the dominant frequency of combustion oscillation hardly, let alone avoid the combustion oscillation. It is believed that the present investigation will provide a useful reference for the design of the scramjet combustor.

2.2.1.1 Experimental Description

A direct-connected test facility is used for the experiments. The facility is composed of air heater, super-sonic nozzle and scramjet combustor. The model scramjet combustor is directly mounted downstream the supersonic nozzle of the air heater which heats the air by means of air/ethanol/O_2 combustion. The repetitive ability of the experimental system has been validated in a large number of experiments [27]. The flow conditions of the supersonic nozzle exit, that is the scramjet combustor entry, are listed in Table 2.3. As Fig. 2.30 shows, the model scramjet combustor consists of a constant cross-section isolator and a single-side expansion combustor. The entry cross section of the combustor is 54.5 mm in height and 75 mm in width. The combustor has an expansion angle of 2.5° on the upside wall. The T1 cavity is arranged in the expansion-side wall (upside) as an ignition cavity. Its parameters is set to depth $D = 15$ mm, length to depth ratio $L/D = 7$, and the aft wall angle $\theta = 45°$. The forced spark ignition plug was installed in cavity T1. Its supply voltage was 220 V; however, the instantaneous discharge voltage was nearly 2000 V. The spark rate was 100 Hz. The pilot hydrogen is injected upstream T1 cavity. The ethylene injector (orifice number × diameter, 3 × 2.0 mm, the distance between the orifices is 15 mm)

Table 2.3 Flow conditions at the scramjet combustor entry [26]

Ma	*P*/KPA	*T*/K	P_0/MPa	T_0/K	Y_{o2}
2.1	71	528	0.65	947	23.3%

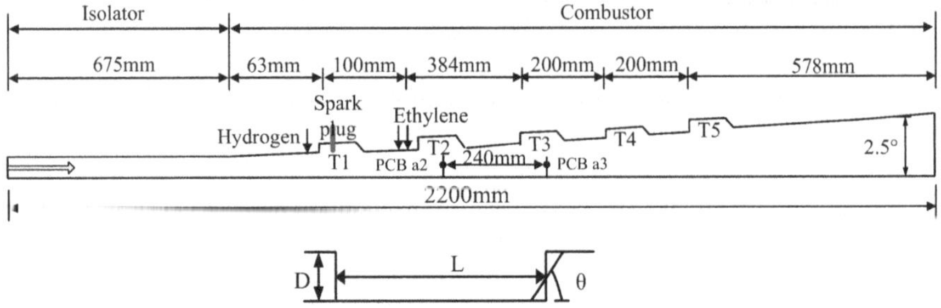

Fig. 2.30 Schematic diagram of the scramjet combustor model and cavity [26]

is set downstream T1 cavity, the global fuel equivalence ratio Φ is set 0.4. T2, T3, T4 and T5 cavity are designed as flameholding cavities in the experiments. Their depth D is set to 15 mm. Unlike T1 cavity, however, they will not keep open during every experiment. And their length L and the aft wall angle θ will also be adjusted to different experimental intention. The ethylene flame behavior is captured by a high speed movie camera through the quartz windows on the model, for which 6000 fps (frames per second) is chosen with a resolution of 1024 × 512 pixels and a shutter time of 1/8000 s. The high-frequency voltage signal corresponding to wall pressure change is acquired by a water-cooled high-frequency sensor (PCB model 112A05) at the points on the sidewall (shown in Fig. 2.30). The acquisition frequencies of the pressure signals are 50 kHz.

2.2.1.2 Effect of Flameholding Cavity Position

Firstly, the influence of flameholding cavity position, that is the premixing distance, has been studied by the comparative experiments listed in Table 2.4. It should be noted that, the ignition cavity T1 is not displayed in Table 2.4. The parameters of the flameholding cavity is unified to the length to depth ratio $L/D = 7$, and the aft wall angle $\theta = 45°$. As Fig. 2.31a shows, when the premixing distance between ethylene injector and flameholding cavity is 8 mm, the pressure oscillation intensity is in so low levels that the combustion can be regarded as stable. According to Fig. 2.32, the further finding indicates that, the combustion is in a state of small-amplitude high

Table 2.4 Comparative experiments of the influence of cavity position [26]

Run no.	Flameholding cavity	L/D	θ/°	Premixing distance/mm	Flame dynamics
01	T2	7	45	8	Stable combustion
02	T3			384	Periodic oscillation
03	T4			584	Periodic oscillation
04	T5			784	Periodic oscillation

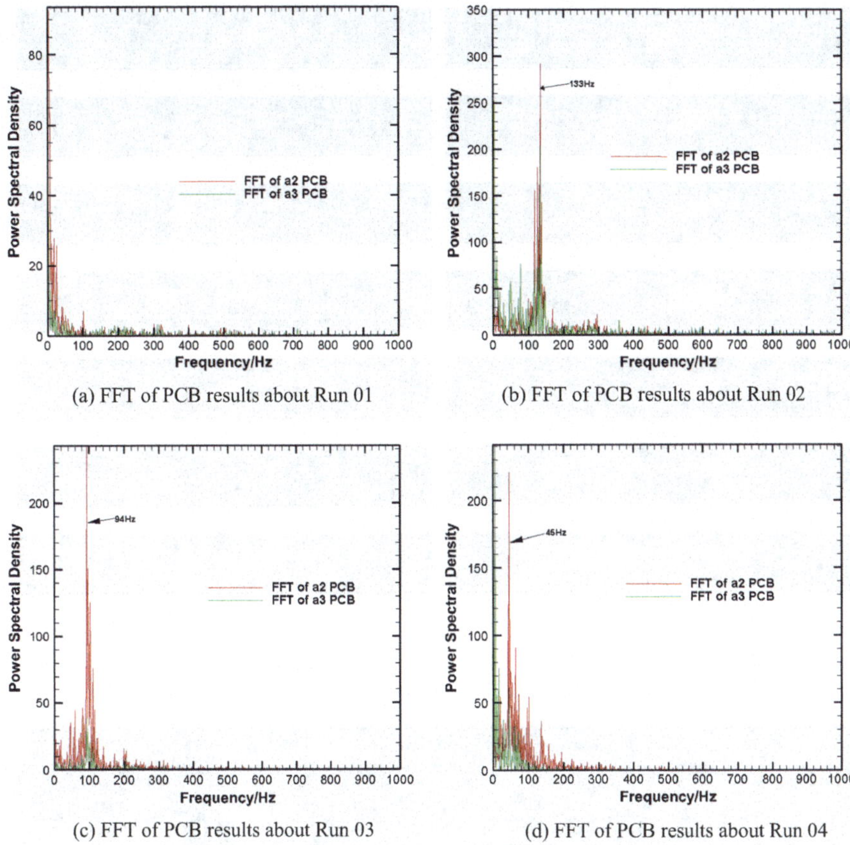

(a) FFT of PCB results about Run 01

(b) FFT of PCB results about Run 02

(c) FFT of PCB results about Run 03

(d) FFT of PCB results about Run 04

Fig. 2.31 FFT of PCB results about Run 01, 02, 03 and 04 [26]

frequency fluctuation in nature which will be attributed to the intrinsic instability of injection and high speed flow in scramjet combustor. As shown in Fig. 2.33, which is the average result of 100 flame frames, the combustion will keep stable overall. When the flameholding cavity is transferred to T3, T4 or T5, however, the noticeable distinction will take place as Fig. 2.31b–d shown. The pressure oscillation intensity will shift to higher levels and distinct dominant frequency can be obtained, which is about 133 Hz, 94 Hz, and 45 Hz, respectively. According to Fig. 2.34, the combustion is in a state of large-amplitude low frequency oscillation, the flame frames of run 03 and 04 are analogous to run 02 and can also flashback to T1 cavity, so they have been omitted to avoid repetition. Obviously, the large-amplitude combustion oscillation can be divided into three distinctive stages, which are flame re-holding, flame flashback and flame blown off, respectively.

As Fig. 2.35 shows, the dominant frequency of the large-amplitude combustion oscillation is in inverse proportion to the pre-mixing distance. This can be explained as follows. Although increasing the premixing distance between ethylene injector

Fig. 2.32 Instantaneous flame frame of run 01, $\Delta t = 1/6000$ s [26]

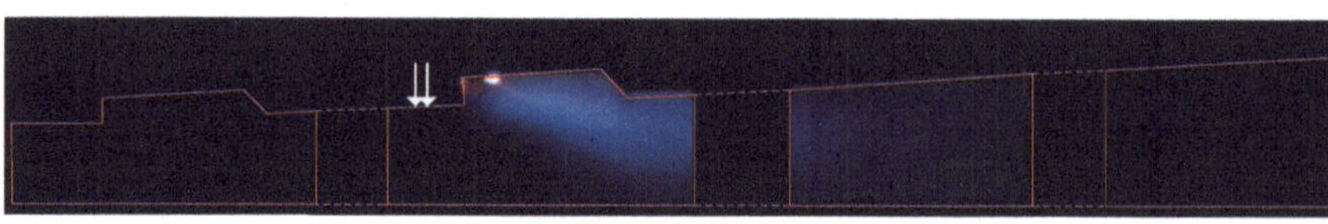

Fig. 2.33 Average result of 100 frames of run 01 [26]

and flameholding cavity can strengthen the premixing effect, which will shorten the flame re-holding time, it will lead to the higher local velocity and lower local pressure in the vicinity of flameholding cavity at the same time because of the single-side expansion scramjet combustor model, which will prolong the flame re-holding time. The increasing premixing distance between ethylene injector and flameholding cavity, moreover, will result in longer flame flashback and flame blown off time.

2.2.1.3 Effect of Cavity Length to Depth Ratio

In this section, the three group experiments listed in Table 2.5, corresponding to cavity T3, T4 and T5, respectively, will be performed to explore the influence of cavity *L/D* on the combustion oscillation. The cavity aft wall angle and depth will be hold as 45° and 15 mm, respectively. The cavity *L/D* can be changed to 4, 5 and 7 through the varying cavity length.

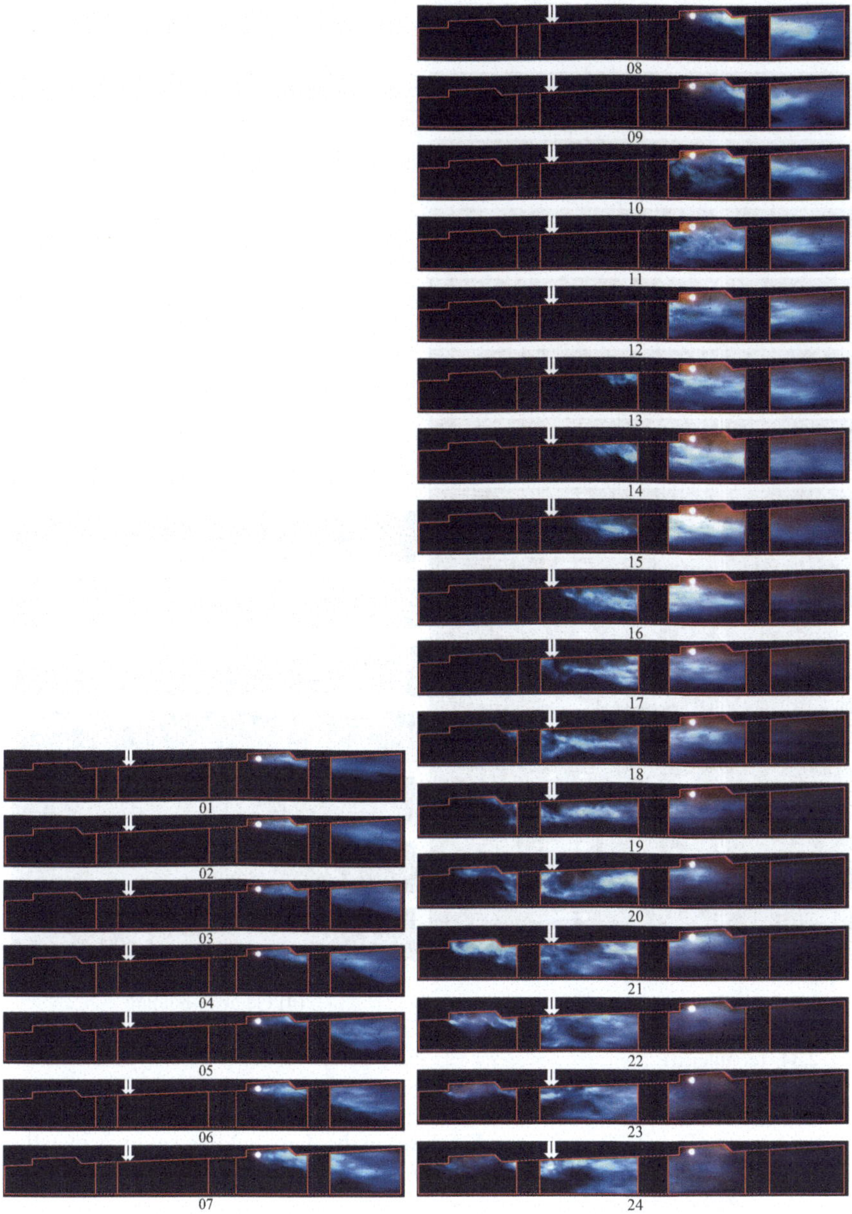

(a) Flame re-holding (b) Flame flashback

Fig. 2.34 Flame frames of one typical oscillation period in run 02, $\triangle t = 1/6000$ s [26]

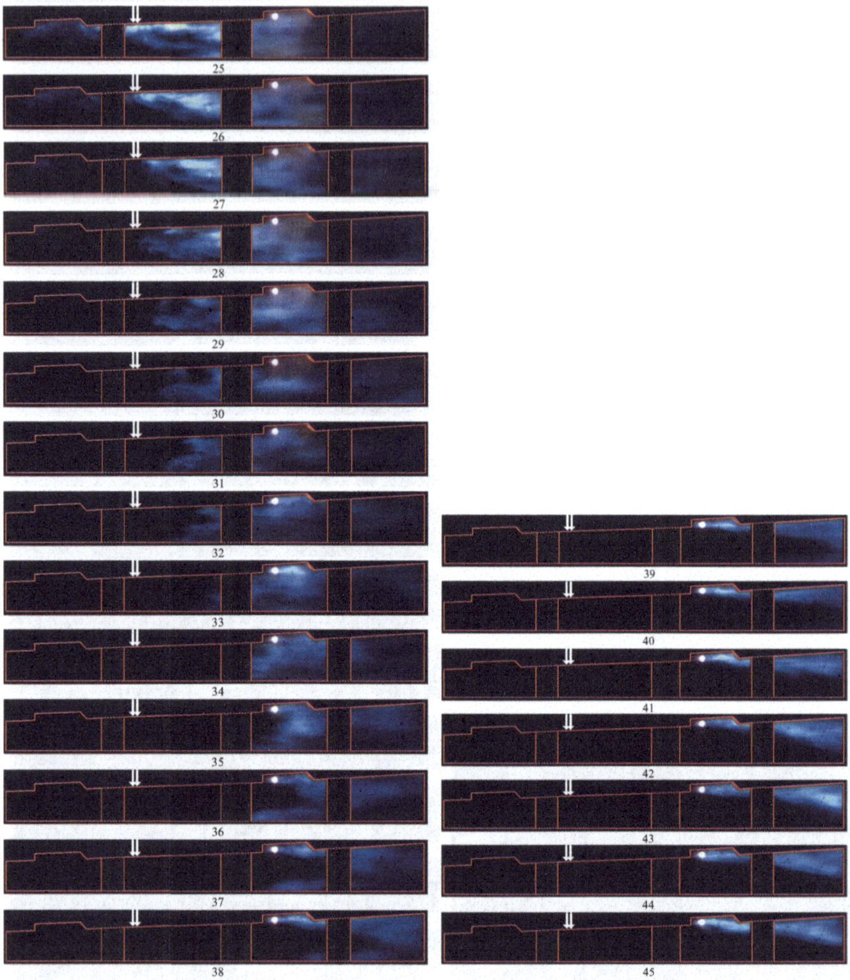

(c) Flame blown off (d) Flame re-holding

Fig. 2.34 (continued)

As shown in Fig. 2.36, on the one hand, when the flameholding cavity is installed in T3 or T4, the dominant frequency of combustion oscillation will become higher as cavity *L/D* increases. This is because the increasing cavity *L/D* will strengthen the mass and heat exchange between flameholding cavity shear layer and the core flow, which will shorten the flame re-holding time. When the flameholding cavity is installed in T5, however, the dominant frequency of cavity $L/D = 7$ will decrease, which maybe because that the higher local velocity flow in the vicinity of T5 cavity resulting from the single-side expansion scramjet combustor model will take more heat from flameholding cavity away to the combustor exit. We can find that, moreover,

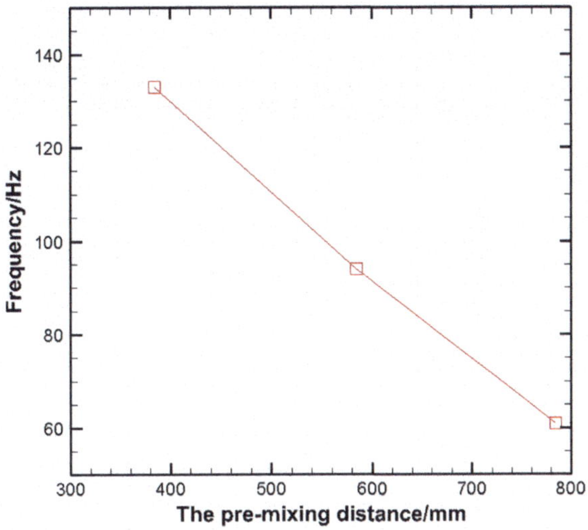

Fig. 2.35 Dominant frequency of different cavity positions [26]

Table 2.5 Comparative experiments of the influence of cavity length to depth ratio [26]

Group no.	Run no.	Flameholding cavity	L/D	θ/°	Dominant frequency/Hz
01	02	T3	7	45	133
	05		5		119
	06		4		118
02	03	T4	7		94
	07		5		87
	08		4		84
03	04	T5	7		45
	09		5		50
	10		4		40

the difference of dominant frequency results from different cavity position is more notable than that resulting from varying cavity *L/D*, which indicates that the influence of premixing distance is more important than that of cavity *L/D*.

2.2.1.4 Effect of Cavity Aft Wall Angle

In this part, the influence of cavity aft wall angle on the combustion oscillation will be studied. The cavity *L/D* is set to 4, and the cavity aft wall angle can be changed to 30°, 45°, 60°, 75° and 90°. There are also three group comparative experiments listed in Table 2.6, corresponding to cavity T3, T4 and T5, respectively.

The related results displayed in Table 2.6 and Fig. 2.37 indicate that, when the flameholding cavity is arranged in T4 or T5, the variation of combustion oscillation

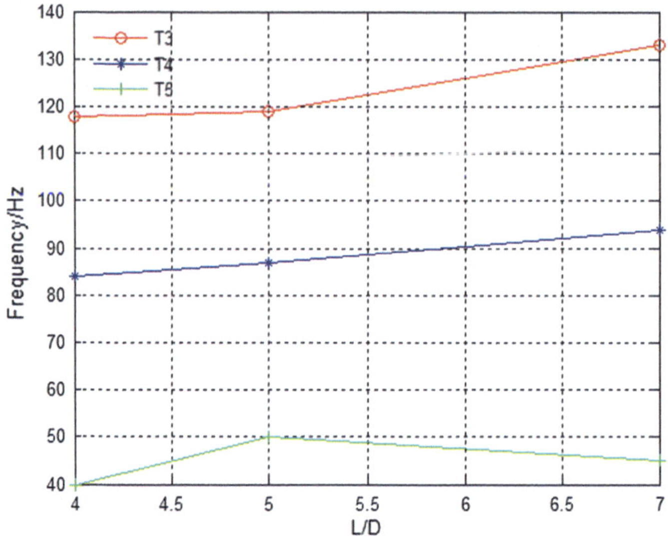

Fig. 2.36 Dominant frequency of different cavity L/D [26]

Table 2.6 Comparative experiments of the influence of cavity aft wall angle [26]

Group no.	Run no.	Flameholding cavity	L/D	θ/°	Dominant frequency/Hz
01	11	T3	4	30	114
	06			45	118
	12			60	124
	13			75	135
	14			90	116
02	15	T4		30	84
	08			45	84
	16			60	77
	17			75	77
	18			90	70
03	19	T5		30	35
	10			45	40
	20			60	35
	21			75	37
	22			90	41

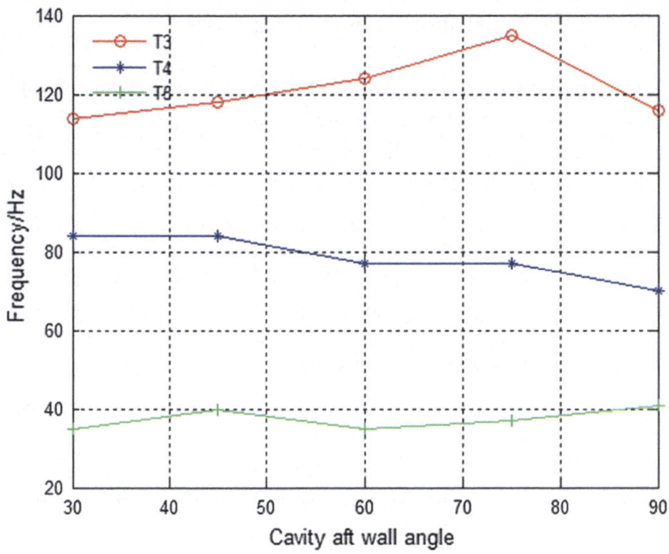

Fig. 2.37 Dominant frequency of different cavity aft wall angle [26]

dominant frequency caused by the change of aft wall angle is so small that can be ignored. When the flameholding cavity turn into T3, the combustion oscillation dominant frequency will increase as cavity aft wall angle enlarges excluding $\theta = 90°$. We suppose that cavity aft wall will act on the combustion oscillation mainly through the impinging shock wave in the cavity aft wall and the mass and heat exchange between cavity shear layer and the core flow. When the flameholding cavity is installed in T4 or T5, due to the increase of local flow velocity and the height of flow path, the action of the impinging shock wave will be weakened and the roll-up derived from the velocity deviation between cavity shear layer and the core flow will dominant the heat exchange, thus the influence of cavity aft wall will be weakened to be negligible. When the flameholding cavity turn into T3, the impinging shock wave will be strengthened as cavity aft wall angle shifts from 30° to 75°, to shorten the flame re-holding time. When the angle turn to 90°, however, it will maybe result in significant limitation on the mass and heat exchange between cavity shear layer and the core flow, thus the remarkable decrease of the combustion oscillation dominant frequency. In addition, as shown in Fig. 2.37, compared to the premixing distance, the influence of cavity aft wall angle θ, can be recognized as unnoticeable, which testifies to the significant impact of premixing distance on the large-amplitude low frequency combustion oscillation again.

2.2.1.5 Effect of the Amount of Flameholding Cavity

Unlike the aforementioned experiments, in which only one flameholding cavity is adopted. In this section, we will put multiple flameholding cavities into use to investigate the influence of the amount of flameholding cavity on the ethylene combustion oscillation characteristics in scramjet combustor. It should be noted that only the tandem cavities have been considered, and the parallel cavities have been omitted for the integration design of the whole aircraft. The *L/D* and aft wall angle of all cavities are set to 7 and 45°, respectively. The corresponding experimental condition and results are listed in Table 2.7. The results of Run 01 and 02 are also listed in Table 2.7 to enhance the comparability. Comparing Run 02, 24 and 26, it is revealed that, when the premixing distance is identical and sufficient, increasing the amount of tandem flameholding cavity is lost on changing the dominant frequency of combustion oscillation, let alone turns the oscillating combustion into stable combustion. It can be found that, the distance between ethylene injector and the first flameholding cavity, that is, the premixing distance, instead of the amount of flameholding cavity, will change the ethylene combustion oscillation characteristics in essence, which confirms the key of the large-amplitude low frequency combustion oscillation is premixing once again.

2.2.2 *Effect of Mixing Status on the Acoustic Oscillation*

Sun et al. [28] investigated combustion instabilities inside an ethylene-fueled scramjet combustor mounted on a Mach 2.1 direct-connect test facility with an inflow stagnation temperature of 846 K. Effects of fueling schemes on the combustion stability characteristics were examined. The experimental results suggest that the oscillation modes correlate with mixing status closely. For the cases with a quasi-steady thermal throat or stable shock trains, flame fluctuation exists in a mode of thermo-acoustic

Table 2.7 Comparative experiments of the influence of the amount of flameholding cavity [26]

Run No.	Flameholding cavities	L/D	θ	Flame dynamics	Dominant frequency
01	T2	7	45°	Stable combustion	—
02	T3			Periodic oscillation	133 Hz
23	T2 + T3			Stable combustion	—
24	T3 + T4			Periodic oscillation	135 Hz
25	T2 + T3 + T4			Stable combustion	—
26	T3 + T4 + T5			Periodic oscillation	136 Hz

type oscillation with a broad frequency range. For the cases with a transient thermal throat, if a fuel/air premixed region from the injection to the cavity flameholder exists, the cavity pilot flame could reignite the fuel/air mixture and undergo a process similar to deflagration–detonation transition (DDT). This process couples with the flame quenching upstream of the injection location, and a DDT-type low frequency oscillation can be formed.

A direct-connect test facility [29] was used for the experiments. The facility was composed of an air heater, a supersonic nozzle and a scramjet combustor. The air heater burned pure ethyl alcohol and oxygen continuously to heat air from the atmospheric temperature up to 846 K and increased the total pressure of vitiated air up to 0.76 MPa. The total mass flow rate of the vitiated air was 2.02 kg/s. The two dimensional converging–diverging M = 2.1 nozzle section, configured with a rectangular nozzle, was used to provide the designed inflow conditions. The model combustor shown in Fig. 2.38.

Table 2.8 shows the detailed injection schemes by combining different injector modules. Ethylene was injected at the condition of stagnation temperature T_{0i} = 300 K and stagnation pressure P_{0i} = 3.0 MPa. The fuel mass flow rate through the wall port was set to be 55 g/s for all schemes, which corresponded to a global equivalence ratio ϕ of 0.4. It is important to note that the local equivalence ratios in the fields near the upwall are higher than 0.4 since only the upwall injection is conducted.

Figure 2.39 shows the frequency power spectra of the pressure signals. For Scheme 1, there is no dominant frequency in the 0–1 kHz spectra. Although a low frequency ranging over 0–50 Hz exists, the intensity of the signal is at a very low level. For Schemes 2, 3 and 5 there are very sharp peaks in the FFT signal which corresponds to 150 Hz, 152 Hz and 183 Hz, respectively. For Schemes 4 and 6, the

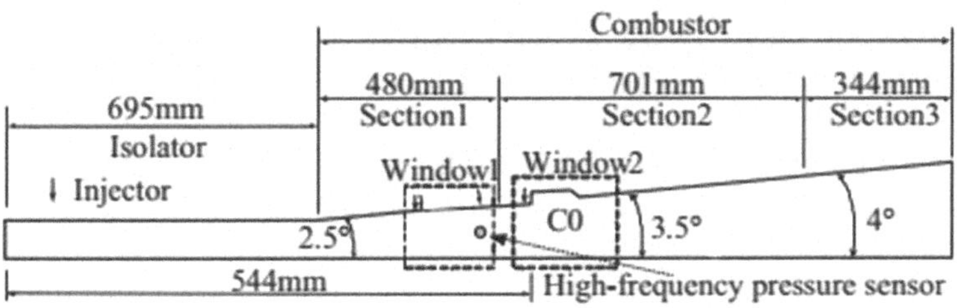

Fig. 2.38 Schematic of test section and cavity installation. Along the oblique direction of the C0 cavity in the combustion chamber, the distance from fuel injection positions I11, I12, I20, I31 and I32 to the C0 cavity leading edge are 10, 20, 120, 250 and 260 mm, respectively [28]

Table 2.8 Different injection schemes [28]

Scheme	1	2	3	4	5	6
Injector modules	No injection	I32 + I31	I31 + I20	I31 + I11	I20 + I11	I12 + I11

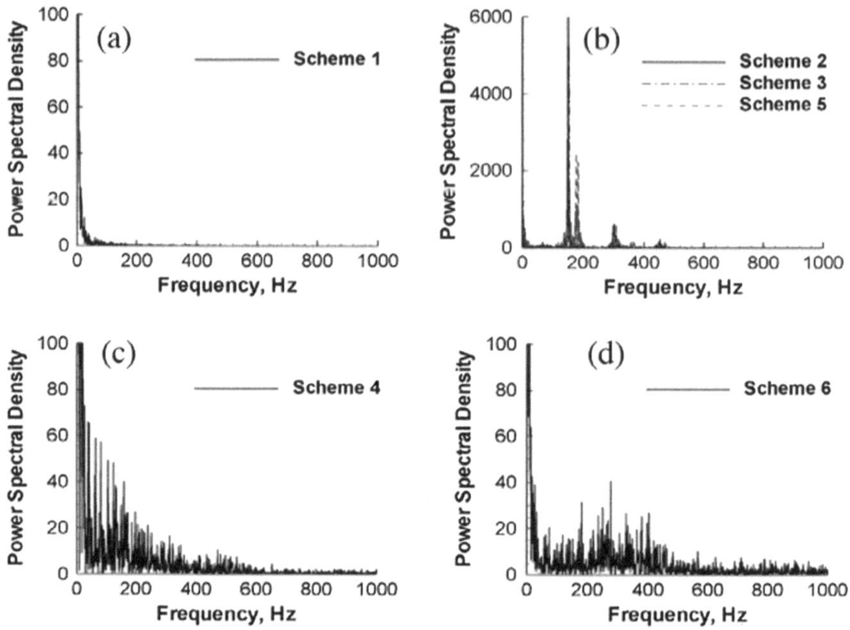

Fig. 2.39 Power spectra of pressure signals of various schemes [28]

FFT signals show the elevated fluctuations in a broad range of 50–450 Hz. The oscillation does not occur at a fixed frequency. According to traditional understanding, the oscillations of Schemes 2, 3 and 5 indicate a strong thermo-acoustic or fluid-dynamic instability mode. These instabilities are in a range of expected frequencies for thermo-acoustic modes such as those studied by Lin et al. [24], which have also been observed in a wide variety of devices utilizing premixed or partially premixed combustion [30]. These instabilities usually correspond to acoustic modes of the combustor.

Usually it is difficult to achieve flameholding for direct injection without a cavity under the conditions with a low air stagnation temperature. The insufficient fuel–air mixing and the high strain rate in the near field of injection, which was well described by Peters and Williams before [31], may result in significant difficulties for flameholding. Furthermore, it is observed that the separation region in front of the jet is unstable, which also tends to destroy the ability of the flame to self-sustain. The flame quickly goes into extinction and blows off gradually in the downstream direction until it reaches the C0 cavity location where the flame is self-sustained again in the shear layer. The C0 cavity flame is like a pilot flame to reignite the main flame after a certain time delay (about 1.9 ms on average). Since the flame movement is repeatable regularly and the imaging interval is 0.125 ms, the schlieren images (Fig. 2.40) were selected to compare the experimental results at the approximately same moment by luminosity with a maximum image matching error of 0.0625 ms. It is observed that the pre-combustion shock trains are pushed forward and the I31

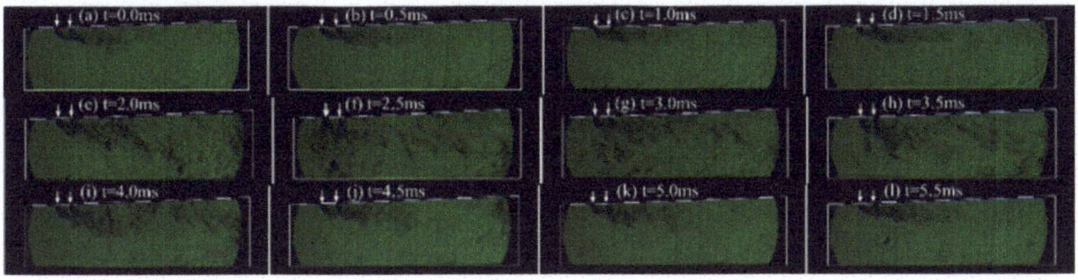

Fig. 2.40 Typical schlieren movie of flame flash-forward and blow-off event between cavity stabilized location and injection location for Scheme 2 [28]

injection has an increasingly higher penetration height during the process of the flame moving upstream to the jet. Meanwhile the combustion region transversely expands to the whole flow path around t = 3 ms, which indicates that a transient thermal throat occurs in the flow field at the moment. Obviously the flame movement changes fields of the total heat release rate and the local pressure. This explains why the pressure peak occurs at this moment. The luminosity and schlieren images show that there is no stable region for an acoustic feedback loop in the flow field since there does not exist any steady thermal throat, implying that the acoustic oscillation coupling with flame front movements is rare. It is inferred that the oscillation is a coupling of the mixing (not acoustic waves) with the heat release. For Scheme 6, Figs. 2.41 and 2.42 show that the flame extends into the main flow from the cavity shear layer, and forms a relatively stable cavity-stabilized combustion area. The shock train is also relatively stable and located upstream of the flame region, as shown in Fig. 2.42. For this scheme, a subsonic region is settled between the shock wave and the flame region. In this region a quasi-steady thermal throat exists, and it is possible for the heat release from the cavity stabilized combustion to couple with the acoustic waves. However, the shock-train location and the mixing efficiency change with the flame fluctuation. With the subsonic region fluctuating, the region boundary and the mixing level varies. It is impossible to form a fixed frequency oscillation under such

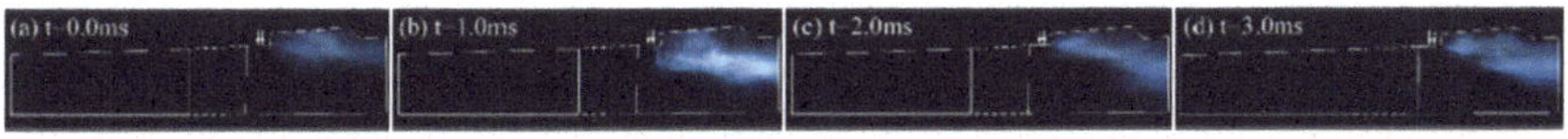

Fig. 2.41 Typical luminosity movie of flame fluctuation in cavity stabilized location for Scheme 6 [28]

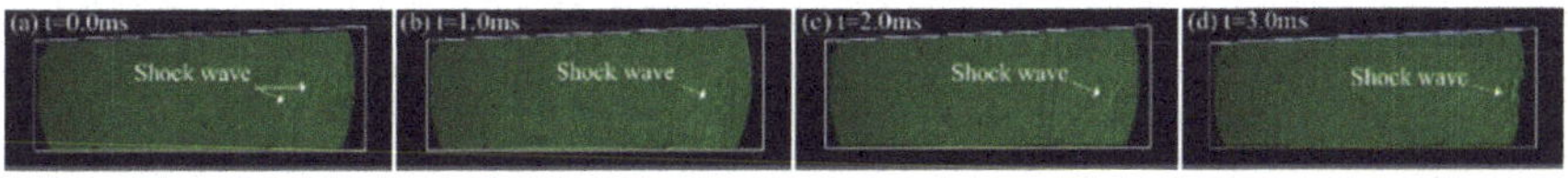

Fig. 2.42 Typical schlieren movie of flame fluctuation in cavity stabilized location for Scheme 6 [28]

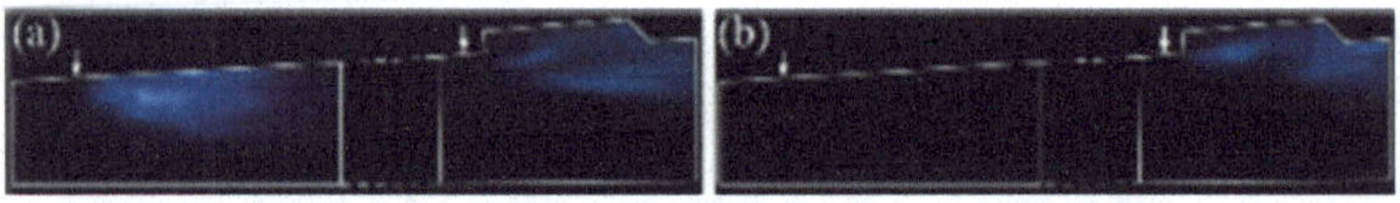

Fig. 2.43 Typical two flame stabilization locations for Scheme 4; **a** I31 jet-wake stabilized flame, and **b** C0 cavity stabilized flame [28]

a condition. This explains why the FFT spectra of Scheme 6 (shown in Fig. 2.41d) have a broad range of elevated fluctuations.

For Scheme 4, an intermittent flame oscillates between the C0 cavity stabilized location and I31 jet wake stabilized location, as shown in Fig. 2.43. The phenomenon is similar to that for Scheme 2 except that the oscillation is not periodic. From a careful observation, the leading flame is stabilized in the I31 jet wake at most times and there is a self-sustaining flame in the C0 cavity all the time. Occasionally the leading flame quenches and blows downstream to the C0 cavity location. However, the reignition occurs quickly and the flame propagates forward to the I31 position with a high pressure peak. For this scheme, i.e. Scheme 4, the leading flame zone is generally stable in the I31 jet-wake, or very quickly moving from the C0 cavity to I31 injection. The oscillation mode does not occur at a fixed frequency.

For Scheme 6, the measured low frequency has a wide range over 50–450 Hz. According to the analyses of Figs. 2.41 and 2.42, acoustic combustion instabilities may occur in this case while it is difficult to realize a fixed frequency oscillation. For Schemes 2, 3 and 5, the relatively low rate of oscillation with fixed frequency shows that the oscillation is not caused by auto-ignition in jet wakes, neither by thermo-acoustic nor periodic fluid dynamic instabilities. The results suggest that there is a new oscillation mode here.

The description of a typical DDT-type flame oscillation cycle is given here. At the cycle initiation, the fuel from the injected jet mixes with the air sufficiently with a long enough distance from the I31 injection location to the C0 cavity. Then a premixed condition with an appropriate equivalence ratio is formed near the C0 cavity. The flame in the C0 cavity shear layer heats the premixed mixture as a pilot flame. After a certain time delay, the mixture in the main flow is reignited. The generated flame propagates forward and accelerates into a phase of DDT. The flame velocity in the process is higher than the critical speed of the detonation onset, and even approaches the detonation speed. When the flame exceeds the I31 location, the jet penetration is greatly enhanced and the flame could not be sustained and quenches. Then the flame blows off and goes backward to the C0 cavity. The fuel mixes with air along the distance and after a period, reignition occurs again. Thus a low frequency combustion oscillation cycle is formed. It is mentioned that this type of oscillation is only for the flow field without fixed precombustion shock trains and any quasi-steady thermal throat.

Scheme 2 gives a fixed frequency oscillation using a concentrated injection. Scheme 3 changes the mixing status using a distributed injection, but the distance between these two injection locations L_{if} changes little. And the distance for mixing is still enough to form a premixed region. Therefore, the oscillation frequency

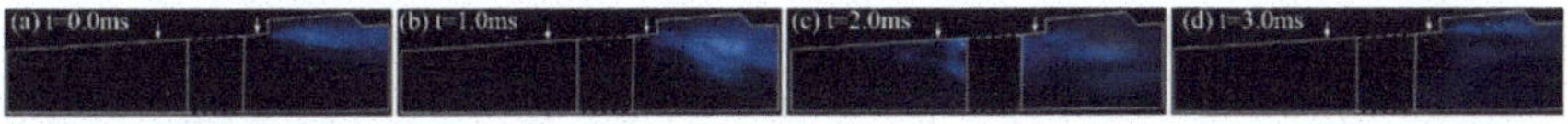

Fig. 2.44 Typical luminosity movie of flame fluctuation in cavity stabilized location for Scheme 5 [28]

does not change distinctly. Scheme 5 has a shorter mixing length L_{if} compared to Scheme 2, which results in an increment in the radical frequency (from 150 to 183 Hz). For Schemes 4 and 5, the I11 injector is adjacent to the C0 cavity, which results in a continuously self-sustaining flame in the C0 cavity (shown in Figs. 2.43 and 2.44). This flame fluctuation is similar to that of Scheme 6. For Scheme 4, I31 injection acquires a jet-wake stabilized flame at most times while I20 could not for Scheme 5. This might be due to the combustor configuration. The divergence angle of the upwall at I31 injection port is 2.5, which is smaller than that of I20 (3.5). The smaller divergence angle leads to a quicker pressure rise and a larger separation zone. The separation zone is pushed by the C0 flame zone and the continuous I31 flame self-sustenance becomes possible. For Scheme 5, the local velocity upstream of I20 injection is greater and the heat release leads to a smaller separation, which destroys the I20 jet-wake flame stabilization and finally a periodic extinction occurs. Schemes 2–5 indicate the sensitivity of the mixing process on the DDT type oscillation. Note that the pressure data obtained from experiments are highly limited since only one pressure sensor has been used in the current work. A group of pressure sensors should be applied in the future work.

2.2.3 *Numerical Analysis on Acoustic Oscillation*

To understand the underlying mechanisms for the observed flow oscillations, Cui [32] established a correlation model between flame position and pressure disturbance, and estimated the pressure oscillation frequency in the super-combustion combustion chamber by solving the quasi-one-dimensional control equations. Ma et al. [23] and Lin et al. [24] considered various feedback loops in the subsonic region bounded by the precombustion shock in the isolator and the thermal throat in the downstream region of the flame zone. Three different mechanisms were identified, involving the coupling of the flame zone with the precombustion shock and fuel injection.

2.2.3.1 Zero-Dimensional Numerical Analysis on Acoustic Oscillation with a Quasi-steady Thermal Throat

In order to explain these observed frequencies characteristic, Ma et al. [23] first identified two prospective mechanisms: interactions between the isolator shock and flame, and interactions between fuel injection/mixing and flame. Then Lin et al. [24] considered various feedback loops in the subsonic region bounded by the thermal throat in the downstream region of the flame zone and the precombustion shock in the isolator. They identified three prospective mechanisms, which involves the coupling of the flame zone with the precombustion shock and fuel injection, respectively, as illustrated in Fig. 2.45.

The first and second mechanisms involves the response of the shock wave to flow disturbances arising from the flame zone. Any acoustic wave generated by the heat release fluctuation in the flame zone can propagate upstream to interact with the shock wave in the isolator. Then the resultant flow oscillations in the isolator travels downstream in two possible forms: they either propagate downstream in the form of acoustic waves, or are convected downstream with the local flow velocity in the form of vorticity and entropy waves, which then reinforce the unsteady motions in the flame zone. Such phenomenon can be best characterized by the acoustic admittance function (at the acoustic reflection coefficient) of a shock wave [33–35], expressed as follows:

$$A_d = \left(\frac{u'/\bar{a}}{p'/\gamma \bar{p}} \right)_s \tag{2.5}$$

where u′ and p′ are respectively the velocity and pressure fluctuations immediately behind the shock wave, and $\bar{a}$ is the mean speed of sound, $\bar{p}$ is the mean pressure. The subscript s denotes the shock location. The admittance function can be expressed in terms of a dimensionless frequency defined as:

$$\Omega = \frac{2\pi f}{\bar{a}} \left(\frac{1}{A} \frac{dA}{dx} \right)_s^{-1} \tag{2.6}$$

where f is the oscillation frequency, A is the cross-sectional area of the flow passage. The acoustic reflection coefficient is related to the admittance function as:

$$\Omega = \frac{2\pi f}{\bar{a}} \left(\frac{1}{A} \frac{dA}{dx} \right)_s^{-1} \tag{2.7}$$

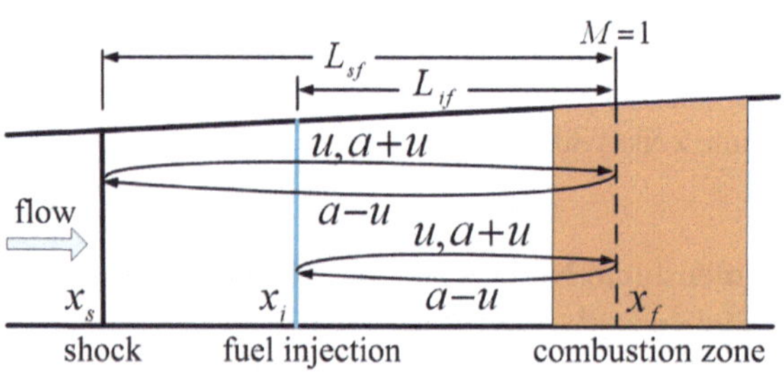

Fig. 2.45 Acoustic-convective feedback loops and associated characteristic velocities in a scramjet combustor (first: shock–flame acoustic feedback, second: shock–flame acoustic-convective feedback, and third: injector–flame feedback) [24]

The above analysis was established based on the assumption of inviscid flow. The shock acts as an effective acoustic damper absorbing disturbances arising from the downstream region [33–36]. However, due to the presence of boundary layers and their interactions with the shock wave, the acoustic reflection coefficient of a shock may reach a considerable value. Furthermore, the oscillating shock in the boundary layers may generate strong vortical waves. Shock-induced entropy waves and airflow fluctuation may also occur [34, 35]. So purely acoustic and acoustic-convective feedback loops between flame zone and the shock are established. And characteristic times that can be estimated respectively as:

$$\tau_{sf1} = \int_{x_s}^{x_f} \frac{dx}{a-u} + \int_{x_s}^{x_f} \frac{dx}{a+u} \approx \frac{L_{sf}}{\underline{a}-\underline{u}} + \frac{L_{sf}}{\underline{a}+\underline{u}} = \frac{2L_{sf}}{\underline{a}(1-\underline{M}^2)} \tag{2.8}$$

$$\tau_{sf1} = \int_{x_s}^{x_f} \frac{dx}{a-u} + \int_{x_s}^{x_f} \frac{dx}{u} \approx \frac{L_{sf}}{\underline{a}-\underline{u}} + \frac{L_{sf}}{\underline{u}} = \frac{L_{sf}}{\underline{a}\underline{M}(1-\underline{M})} \tag{2.9}$$

$\underline{a}$ and $\underline{M}$ are the speed of sound and Mach number longitudinally averaged between the shock and flame respectively, and x_s and x_f are the locations of the shock and flame, respectively. The corresponding oscillation frequencies f_{sf1} and f_{sf2} are:

$$f_{sfi} = 1/\tau_{sfi} \quad i = 1, 2 \tag{2.10}$$

The third mechanism is associated with the acoustic-convective interactions in the region between the fuel injection and the flame zone. The acoustic wave generated in the flame zone propagates upstream and causes an air mass flow-rate oscillation in the fuel injection/mixing region. The fuel/air mixture ratio then fluctuates according to the local air flow rate if the fuel injection rate is fixed (e.g., through a choked nozzle). The resultant oscillation is convected downstream to modify the stoichiometry in the flame zone [33]. The ensuing fluctuation in the overall heat-release rate $\dot{Q}$ can be determined by the fuel consumption rate $\dot{m}_f$. For longitudinal oscillations in a quasi-one-dimensional flow, the specific heat-release q can be written as:

$$q = \dot{Q}/\dot{m} = (\dot{m}_f/\dot{m})h_c \tag{2.11}$$

where $\dot{m}$ is the mass flow rate of the fuel/air mixture at a given axial location, and h_c is the heat of combustion per unit mass of fuel. The mass flow-rate fluctuation arising from the acoustic-pressure fluctuation is:

$$\frac{m'}{\bar{\dot{m}}} = \frac{\rho'}{\rho} + \frac{u'}{u} = \frac{1}{\gamma}\frac{p'}{p} - \frac{p'/\rho a}{Ma} = \left(1 - \frac{1}{M}\right)\frac{1}{\gamma}\frac{p'}{p} \tag{2.12}$$

The fluctuating heat release can be related to the acoustic oscillation as:

$$\left.\frac{q'}{q}\right|_{x_f,t} = \left.\frac{\dot{m}'_f}{\bar{\dot{m}}_f}\right|_{x_f,t} - \left.\frac{\dot{m}'}{\bar{\dot{m}}}\right|_{x_f,t} = -\left.\frac{\dot{m}'}{\bar{\dot{m}}}\right|_{x_i,t-\tau_c} = \left(\frac{1}{\bar{M}}-1\right)\frac{1}{\bar{\gamma}}\left.\frac{p'}{\bar{p}}\right|_{x_i,t-\tau_c} = \left(\frac{1}{\bar{M}}-1\right)\frac{1}{\gamma}\left.\frac{p'}{\bar{p}}\right|_{x_f,t-\tau_c-\tau_a} \quad (2.13)$$

where x_i is the location of the fuel injector, τ_a is the time for the acoustic wave propagating from the flame zone to the injector, and τ_c is the time for the fluctuating mixture convected from the fuel injector to the flame zone. Note that the fuel injection rate is assumed to be fixed in the above formulation. These two time scales are determined by the local acoustic and convective velocities as follows:

$$\tau_a = \int_{x_i}^{x_f} \frac{dx}{a-u} \approx \frac{L_{if}}{a-u} \quad (2.14)$$

$$\tau_c = \int_{x_i}^{x_f} \frac{dx}{u} \approx \frac{L_{if}}{u} \quad (2.15)$$

The characteristic frequency for the acoustic-convective feedback loop between the fuel injector and flame zone is:

$$f_{if} = 1/\tau_{if} = 1/(\tau_a + \tau_c) \approx \left[\frac{L_{if}}{aM(1-M)}\right]^{-1} \quad (2.16)$$

2.2.3.2 Quasi-one-Dimensional Numerical Analysis on Acoustic Oscillation

When the scramjet engine operates in the sub-combustion mode, the combustion is intensified due to an increase in the fuel-air equivalence ratio, and the increase in the combustion chamber area is insufficient to alleviate the thermal blocking effect caused by the heating in the supersonic flow, and the gas stream forms a thermodynamically blocked throat at a location downstream of the combustion chamber flame zone. A large back pressure is generated in the combustion chamber, and a pre-combustion shock train is generated in the isolation section. At this time, there is a large range of subsonic velocity in the combustion chamber, and low-frequency acoustic instability of acoustic coupling excitation may occur. The disturbance caused by the unsteady heat release of the downstream combustion zone can effectively affect the upstream pre-combustion shock or fuel injection. The mixing process forms a closed loop feedback. For this type of working condition, Cui [32] established a correlation model between flame position and pressure disturbance, and estimated the pressure oscillation frequency in the super-combustion combustion chamber by solving the quasi-one-dimensional control equations.

For a flow process with a combustion reaction, the schematic diagram of the quasi-one-dimensional theoretical model used in this paper is shown below.

The quasi-one-dimensional governing equations considering fuel charging, channel cross-section change, chemical reaction exothermic and inviscid are as follows (Fig. 2.46):

$$\frac{\partial U}{\partial t} + \frac{\partial F}{\partial x} = H \quad (2.17)$$

And:

$$U = \begin{bmatrix} \rho \\ \rho u \\ \rho e \\ \rho Y_k \end{bmatrix} \; F = \begin{bmatrix} \rho u \\ \rho u^2 + p \\ (\rho e + p)u \\ \rho u Y_k \end{bmatrix} \; H = \begin{bmatrix} -\frac{\rho u}{A} \cdot \frac{dA}{dx} + \frac{1}{A} \cdot \frac{\partial \dot{m}_f}{\partial x} \\ -\frac{\rho u^2}{A} \cdot \frac{dA}{dx} + \frac{u_j \cos\theta}{A} \cdot \frac{\partial \dot{m}_f}{\partial x} \\ -\frac{(\rho e+p)u}{A} \cdot \frac{dA}{dx} - \sum_{k=1}^{Ns} \dot{\omega}_k h^0_{fk} + \frac{h_{tf}}{A} \cdot \frac{\partial \dot{m}_f}{\partial x} \\ -\frac{\rho u Y_k}{A} \cdot \frac{dA}{dx} + \dot{\omega}_k + \frac{1}{A} \cdot \frac{\partial \dot{m}_{fk}}{\partial x} \end{bmatrix}$$

$k = 1, 2, \cdots, N_s - 1$, N_s is the total number of components.

U is a conservation variable, F is the non-viscous flux term, H is the source item; p, ρ, u, e are the pressure, density, velocity and specific energy of the mixed gas respectively; Y_k is the mass fraction of the component; h^0_{fk} is the standard enthalpy of production for component k (25 °C,1 atm), which can be checked; $\dot{\omega}_k$ is the net mass production rate of component k in the unit volume due to chemical reaction; $\dot{m}'''_f$ is the rate of change in mass volume caused by fuel injection; $h_{t,f}$ is total enthalpy of fuel; θ is the angle between fuel jet and axial direction; u_j is the fuel jet velocity; t is the time, x is axial distance, A is the cross-sectional area of the flow channel.

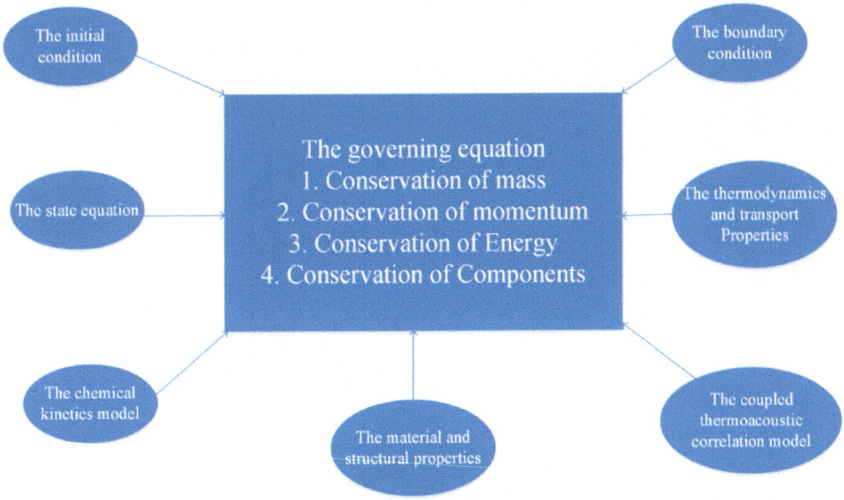

Fig. 2.46 The schematic diagram of the theoretical model

For chemical reactions, the one-dimensional program adopts the equilibrium chemical reaction hypothesis [37] By ignoring intermediate products and using simplified one-step or multi-step (mostly two-step) total package reactions, the exothermic reaction of chemical reaction is added to the right end of the energy equation as a source term, and the component mass fraction can be obtained under the condition of element conservation. This method is applicable when there are fewer components to be considered, but when there are more components to be considered, the method of equilibrium analysis must also be used.

In this paper, by introducing the concept of flame, a chemical reaction model [23] is established by correlating the mass generation rate w of chemical reaction with the flame velocity of the fuel in unit volume.

$$\dot{\omega}_f = -\left[\rho u_f W / \cos\theta\right] \cdot \left[f / (1 + f)\right] / A \tag{2.18}$$

W is the width of the engine runner, f is the combustion-air mass ratio of the flame front, u_f is the effective flame propagation speed, θ is the flame propagation angle. Figure 2.47 shows the geometric relationship between turbulent flame velocity s_t, inflow velocity u and flame propagation angle. Obviously, flame propagation angle can be obtained from the following formula:

$$\theta = \sin^{-1}\left(u_f / u\right) \tag{2.19}$$

The effective flame propagation speed u_f can be set in advance or calculated by multidimensional numerical analysis. In this paper, $\theta = 30°$, so $u_f = \sin\theta \cdot u = u/2$.

The derivation process of the chemical reaction model is given below:

The fuel mass density of the flame front is

$$\rho_f = \rho\frac{m_f}{m} = \rho\frac{m_f}{m_f + m_{air}} = \rho\frac{f}{1 + f} \tag{2.20}$$

So the fuel mass flow rate of the flame front is

$$\rho_f u_f = \rho u_f \frac{f}{1 + f} \tag{2.21}$$

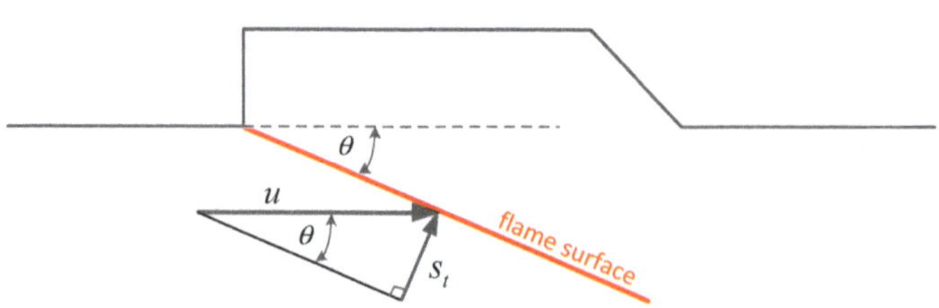

Fig. 2.47 The geometric relationship between turbulent flame velocity, inflow velocity and flame propagation angle [32]

Assuming that this part of the fuel is consumed within the flame propagation distance L, then:

$$\dot{\omega}_f = \frac{d\rho_f}{dt} = -\frac{\rho_f}{t} = -\frac{\rho_f}{L/u_f} \tag{2.22}$$

Considering the geometric relationship between flame propagation distance and runner width and runner cross-sectional area, so

$$L = \frac{A}{W}\cos\theta \tag{2.23}$$

In order to further verify the practical calculation ability of the one-dimensional program adopted in this paper, one-dimensional calculation of pressure distribution along a certain working condition is carried out and compared with the experimental results [38]. The one-dimensional calculation of pressure distribution in the concave chamber of the combustor is basically consistent with the experimental results, which shows that the chemical reaction model adopted in this paper can be applied to the evaluation of the heating effect of the actual combustor combustion, and the results obtained by quasi-one-dimensional calculation are reasonable. The calculated pressure and test value are quite different between the expansion section and the isometric isolation section of the combustor. The reason is that in actual conditions, a pre-combustion shock train is generated between the combustion chamber and the isolation section due to the combustion heating effect. The length can be evaluated by the formula in the literature. The velocity of the incoming flow decreases and the pressure rises after the pre-ignition shock train passes through. However, the quasi-one-dimensional calculation program in this paper has not been established. The model of pre-ignited shock train gives a positive shock wave (as shown in the one-dimensional calculation of pressure step), such as the end shock of pre-ignited shock train. In the expansion section at the end of the combustion chamber, the pressure decreases with the increase of cross-section area, but in actual conditions, the pressure decreases slowly because the incomplete combustion mixture continues to burn and heat at the tail of the engine, which is different from the one-dimensional calculation results.

Cui [32] found through analysis that it is workable to correlate changes in pressure with changes in the position of the flame zone. For the oscillation mechanism proposed by Lin et al. [24, 39] and the relationship between heat release fluctuation and acoustic pressure disturbance, a correction coefficient is used to correlate flame position oscillation with heat release.

$$\left.\frac{x'}{x_{reac} - x_{shock}}\right|_{x_{f,t}} = C_s \left.\frac{q'}{q}\right|_{x_{f,t}} = C_s\left(\frac{1}{M} - 1\right)\frac{1}{\gamma}\left.\frac{p'}{p}\right|_{x_{f,t-\Delta t}} \tag{2.24}$$

x', x_{shock} and x_{reac} respectively represent the flame position pulsation, shock position and flame position (reaction position),and the correction factor C_S is affected by various factors, such as local flow Mach number, cross-sectional area, area change rate, etc., the forms are as follows:

$$C_S = f(M, A, \frac{dA}{dx}, \ldots) \tag{2.25}$$

Before the quasi-one-dimensional numerical calculation using the above analytical model, the reaction location needs to be determined first. There are two ways to determine the reaction position. One is that the cavity is usually used as flame stabilizer in the experiment, and the high static temperature and long residence time in the recirculation zone inside the cavity are used to maintain or stabilize the flame, or as a necessary duty flame to ignite the upstream inflow premix. Usually the reaction position is chosen according to the position of flame stabilizer or the average position of flame front. Another is that according to Lin's point of view [24, 39], the reaction position is chosen at the turning point where the combustion chamber pressure begins to decrease, which represents the average position area of the combustion flame. Cui [32] adopted the latter point of view. The pressure distribution along the combustion of the model engine is given in Fig. 2.48. The selection of relevant parameters such as reaction position in this model is shown in Fig. 2.49.

According to the inflow conditions and the configuration of unilateral expanding combustor in Sun's experiment, the pressure oscillation process and its spectrum results are obtained by quasi-one-dimensional numerical calculation under different correction coefficients, respectively, as shown in Figs. 2.50, 2.51, 2.52 and 2.53.

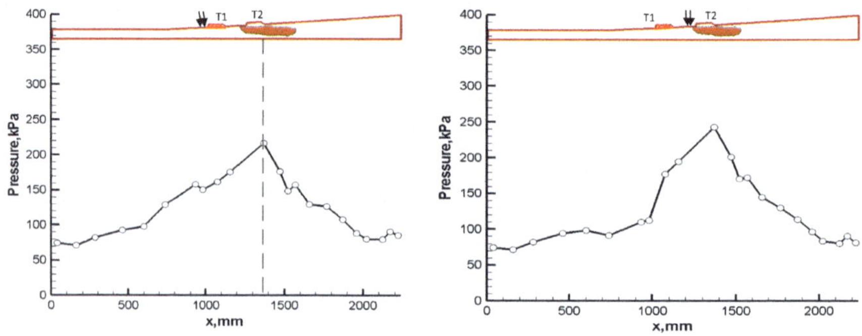

Fig. 2.48 Distributions of pressure under different injection schemes [32]

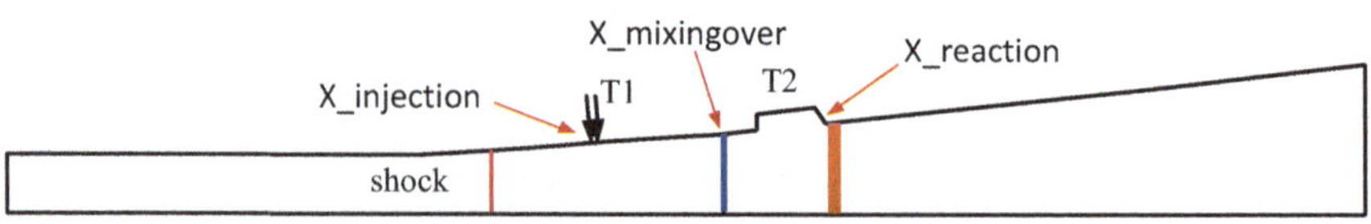

Fig. 2.49 The selection of relevant parameters (such as reaction position) [32]

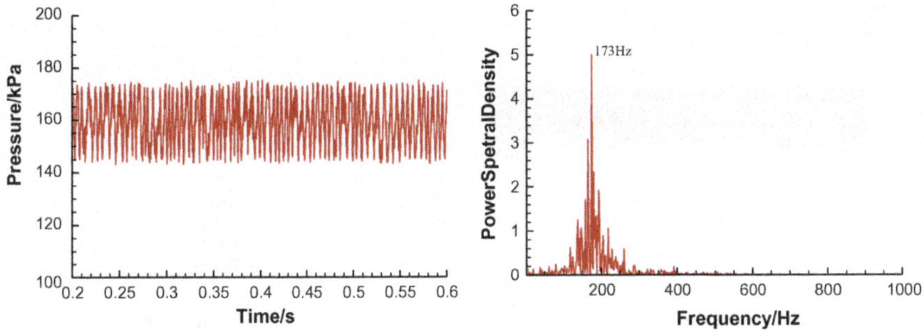

(a) the pressure oscillation process (b) the results of FFT

Fig. 2.50 Time history and frequency spectrum of pressure oscillation obtained by quasi-one-dimensional numerical calculation under condition $C_s = 1.8$ [32]

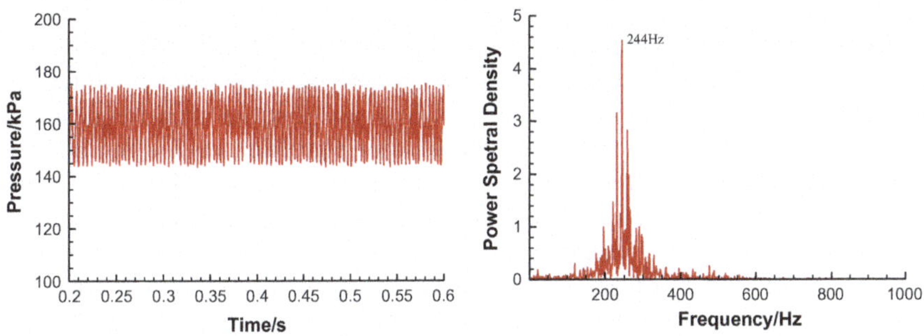

(a) the pressure oscillation process (b) the results of FFT

Fig. 2.51 Time history and frequency spectrum of pressure oscillation obtained by quasi-one-dimensional numerical calculation under condition $C_s = 1.9$ [32]

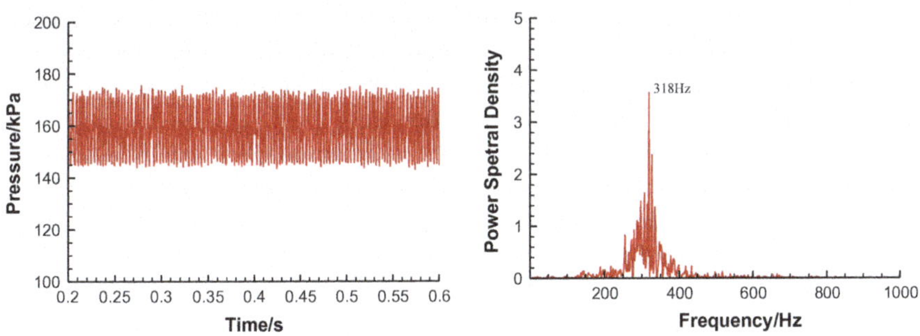

(a) the pressure oscillation process (b) the results of FFT

Fig. 2.52 Time history and frequency spectrum of pressure oscillation obtained by quasi-one-dimensional numerical calculation under condition $C_s = 2.0$ [32]

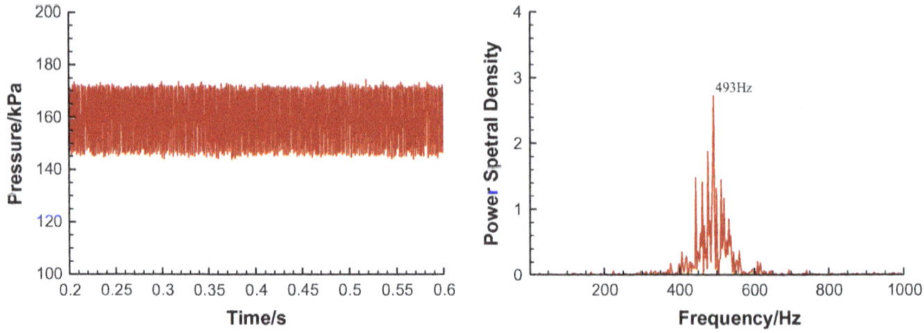

(a) the pressure oscillation process (b) the results of FFT

Fig. 2.53 Time history and frequency spectrum of pressure oscillation obtained by quasi-one-dimensional numerical calculation under condition $C_s = 2.1$ [32]

The spectrum analysis of pressure oscillation data shows that the main oscillation frequencies of 173, 244 318 and 493 Hz are obtained respectively when the correction coefficients are 1.8, 1.9, 2.0 and 2.1. It can be found that the main frequency of pressure oscillation increases with the increase of correction coefficient, and the main frequency of pressure oscillation corresponding to correction coefficient 1.9 and 2.0 happens to be in the range of low frequency oscillation frequency distribution (180–400 Hz) of acoustic coupling excitation in Sun's experimental study. This further indicates that the feedback loop between the pre-combustion shock wave and the flame zone may be the dominant mechanism of low-frequency oscillation, and the quasi-one-dimensional numerical calculation based on the above analysis model is reasonable to study the thermoacoustic instability in the scramjet combustor. Time history and frequency spectrum of pressure oscillation obtained by quasi-one-dimensional numerical calculation under condition.

In order to further explore the regularity relationship between the correction coefficient and the main frequency of the oscillation, Cui [32] expands the range of the correction coefficient, as shown in Fig. 2.54. It can be seen that the oscillation frequency has both high frequency and low frequency.

Figures 2.55 and 2.56 show the pressure distribution obtained by quasi-one-dimensional numerical calculation under $C_s = 1.9$ and $C_s = 2.0$, respectively. The pressure is non-dimensionalized by $\rho_0 u_0^2$. And the axial length of the engine is measured by the inlet height H of the isolation section. It can be seen from Figs. 2.55 and 2.56 that due to the chemical heat release, a large back pressure is generated in the combustion chamber, and a positive shock wave is generated between the isolation section and the combustion. At the same time, there is a large pressure fluctuation between the zones between the positive shock wave and the downstream combustion reaction in the combustion chamber, which further indicates that the low-frequency combustion oscillation is caused by the feedback loop between the pre-combustion shock and the flame zone. And the pressure is gradually increases to the maximum value (at the combustion reaction position) from the injection position; then the decompression effect plays a leading role and the pressure gradually decreases due

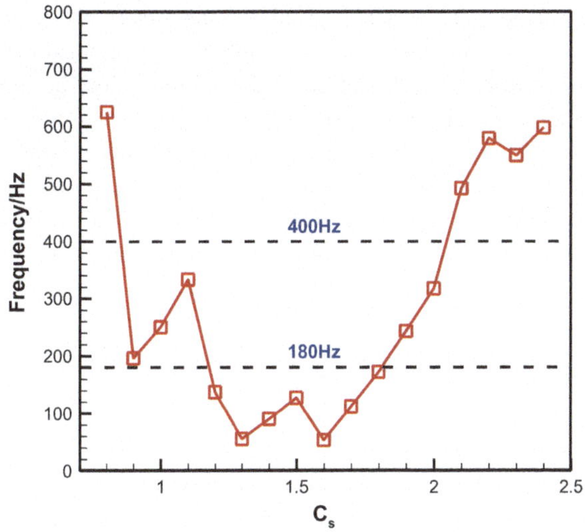

Fig. 2.54 The regularity relationship between the correction coefficient and the main frequency of the oscillation [32]

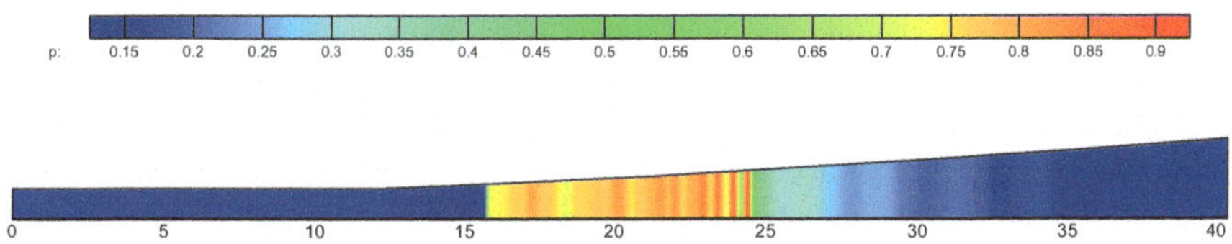

Fig. 2.55 Pressure distribution cloud by quasi-one-dimensional numerical calculation under $C_s = 1.9$ [32]

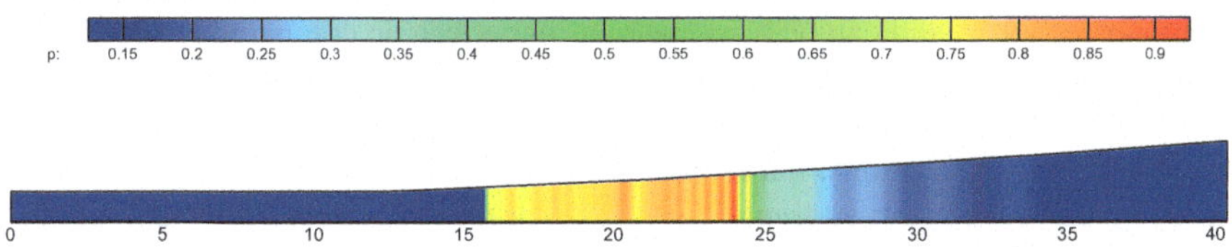

Fig. 2.56 Pressure distribution obtained by quasi-one-dimensional numerical calculation under $C_s = 2.0$ [32]

to the expansion of the combustion chamber. The calculated pressure distribution is basically consistent with the actual situation.

In summary, the model established in this section has certain research value in the analysis of thermoacoustic instability. Quasi-one-dimensional numerical method can be used to preliminarily evaluate combustion oscillation in combustor. The results show that the acoustic-convective feedback cycle between the pre-ignition shock wave coupled with the acoustic wave and the flame zone is the source of pressure oscillation. It should be pointed out that although the one-dimensional calculation is small and economical, the simplification of the complex flow field of the scramjet will inevitably lead to the absence of some influencing factors due to the neglect of the radial parameter changes. Therefore, the analysis results obtained from the one-dimensional calculation can only be used as a reference, and more complex experiments, more precise two-dimensional and three-dimensional numerical calculation are needed to explore the detailed combustion oscillation in the scramjet.

2.3 Summary

In this chapter, the acoustic oscillations in scramjet have been introduced. Based on the related researches, a few conclusions can be drawn as follows:

Totally, it is believed a mixed shear-layer/wake oscillation mode in the flow, where these two modes occur alternately. The shear-layer mode and wake mode are driven by vortex convection-acoustic feedback and absolute instability, respectively. The combustion oscillations can mainly be attributed to two mechanisms. One is the unsteady flame spreading from the cavity shear layer to the main stream, which is greatly influenced by the interaction of the jet-with-cavity shear layer. This mechanism leads to relatively low-frequency oscillations that correspond to the cavity-shear layer oscillations. The other is the auto-ignition of the combustible fluid packets formed around the fuel jet accompanied by the generation of the hairpin-like vortices, which leads to relatively high-frequency oscillations that correspond to the jet instabilities.

The low-frequency oscillations can be influenced by cavity parameters, mixing status of fuel and air, and so on. It can be found that the influence of cavity length to depth ratio L/D and the aft wall angle θ exists diversity when the flameholding cavity position is different and can be recognized as unnoticeable compared to the impact of the premixing distance; And the experimental results suggest that the oscillation modes correlate with mixing status closely. For the cases with a quasi-steady thermal throat or stable shock trains, flame fluctuation exists in a mode of thermo-acoustic type oscillation with a broad frequency range. For the cases with a transient thermal throat, if a fuel/air premixed region from the injection to the cavity flame holder exists, a DDT-type low frequency oscillation can be formed.

To understand the underlying mechanisms for the observed flow oscillations, Various feedback loops in the subsonic region bounded by the precombustion shock in the isolator and the thermal throat in the downstream region of the flame zone were considered. And three different mechanisms were identified, involving the coupling of the flame zone with the precombustion shock and fuel injection. Except that, a correlation model between flame position and pressure disturbance was established, and the pressure oscillation frequency in the super-combustion combustion chamber was estimated by solving the quasi-one-dimensional control equations.

References

1. Rockwell, D., & Naudascher, E. (1978). Review-self-sustaining oscillations of flow past cavities. *Journal of Fluids Engineering, 100*(6), 152–165.
2. Sahoo, D., Annaswamy, A., Zhuang, N., & Alvi, F. (2005). Control of cavity tones in supersonic flow. In *43rd AIAA Aerospace Sciences Meeting and Exhibit* (p. 793). Reno, Nevada: AIAA.
3. Wang, H., Sun, M., Qin, N., Wu, H., & Wang, Z. (2013). Characteristics of oscillations in supersonic open cavity flows. *Flow, turbulence and combustion, 90*(1), 121–142.
4. Rossiter, J. E. (1964). Wind tunnel experiments on the flow over rectangular cavities at subsonic and transonic speeds. Ministry of Aviation; Royal Aircraft Establishment; RAE Farnborough.
5. Heller, H., & Bliss, D. (1975, March). The physical mechanism of flow-induced pressure fluctuations in cavities and concepts for their suppression. In *2nd Aeroacoustics Conference* (p. 491). Hampton, VA, U.S.A: AIAA.
6. Kegerise, M., Spina, E., & Cattafesta, III, L. (1999, November). An experimental investigation of flow-induced cavity oscillations. In *30th Fluid Dynamics Conference* (p. 3705). Norfolk, VA, U.S.A: AIAA.
7. Colonius, T., Basu, A., & Rowley, C. (1999). Numerical investigation of the flow past a cavity. In *5th AIAA/CEAS aeroacoustics conference and exhibit* (p. 1912). Bellevue, WA, U.S.A: AIAA.
8. Rowley, C. W., Colonius, T., & Basu, A. J. (2002). On self-sustained oscillations in two-dimensional compressible flow over rectangular cavities. *Journal of Fluid Mechanics, 455,* 315–346.
9. Huerre, P., & Monkewitz, P. A. (1985). Absolute and convective instabilities in free shear layers. *Journal of Fluid Mechanics, 159,* 151–168.
10. Tam, C. J. (1996). Algebraic turbulence model simulations of supersonic open-cavity flow physics. *AIAA Journal, 34*(11), 2255–2260.
11. Li, W., Nonomura, T., Oyama, A., & Fujii, K. (2010). LES study of feedback-loop mechanism of supersonic open cavity flows. In *40th Fluid Dynamics Conference and Exhibit* (p. 5112). Chicago, Illinois: AIAA.
12. Li, W., Nonomura, T., & Fujii, K. (2011). Effects of shear-layer characteristic on the Feedback-loop Mechanism in supersonic open cavity flows. In *49th AIAA Aerospace Sciences Meeting including the New Horizons Forum and Aerospace Exposition* (p. 1218). Orlando, Florida: AIAA.
13. Wang, H., Wang, Z., & Sun, M. (2013). Experimental study of oscillations in a scramjet combustor with cavity flameholders. *Experimental thermal and fluid science, 45*(2), 259–263.
14. Chandra, B., Bharathraj, R., & Chakravarthy, S. (2003). Fuel-air mixing by cavity-induced acoustic oscillations in confined supersonic flows. In *41st Aerospace Sciences Meeting and Exhibit* (p. 1206). Reno, Nevada: AIAA.
15. Micka, D. J. (2010). *Combustion stabilization, structure, and spreading in a laboratory dual-mode scramjet combustor* (Doctoral dissertation, University of Michigan).
16. Wang, H., Wang, Z., Sun, M., & Qin, N. (2013). Large-Eddy/Reynolds-averaged Navier–Stokes simulation of combustion oscillations in a cavity-based supersonic combustor. *international journal of hydrogen energy, 38*(14), 5918–5927.
17. Kawai, S., & Lele, S. (2009). Large-eddy simulation of jet mixing in a supersonic turbulent crossflow. In *19th AIAA Computational Fluid Dynamics* (p. 3795). San Antonio, Texas: AIAA.
18. Sun, M. B., Wang, Z. G., Liang, J. H., & Geng, H. (2008). Flame characteristics in a supersonic combustor with hydrogen injection upstream of a cavity flameholder. *Journal of Propulsion and Power, 24*(4), 688–696.
19. Won, S. H., Jeung, I. S., Parent, B., & Choi, J. Y. (2010). Numerical investigation of transverse hydrogen jet into supersonic crossflow using detached-eddy simulation. *AIAA journal, 48*(6), 1047–1058.
20. Ben-Yakar, A. (2000). *Experimental investigation of mixing and ignition of transverse jets in supersonic crossflows*. PhD thesis, Stanford university.

21. Gutmark, E., & Ho, C. (1983). Preferred modes and the spreading rate of jets. *Physics of Fluids, 26*(10), 2932–2938.
22. Fric, T. F. (1990). *Structure in the near field of the transverse jet*. PhD thesis, California Institute of Technology.
23. Ma, F., Li, J., Yang, V., Lin, K. C., & Jackson, T. (2005). Thermoacoustic flow instability in a scramjet combustor. In *41st AIAA/ASME/SAE/ASEE Joint Propulsion Conference & Exhibit* (p. 3824). Tucson, Arizona: AIAA.
24. Lin, K. C., Jackson, K., Behdadnia, R., Jackson, T., Ma, F., Li, J., et al. (2007). Acoustic characterization of an ethylene-fueled scramjet combustor with a rectangular cavity flameholder. In *43rd AIAA/ASME/SAE/ASEE Joint Propulsion Conference & Exhibit* (p. 5382). Cincinnati, OH: AIAA.
25. Li, J., Ma, F., Yang, V., Lin, K. C., & Jackson, T. (2007). A comprehensive study of combustion oscillations in a hydrocarbon-fueled scramjet engine. In *45th AIAA Aerospace Sciences Meeting and Exhibit* (p. 836). Reno, Nevada: AIAA.
26. Ouyang, H., Liu, W., & Sun, M. (2017). The influence of cavity parameters on the combustion oscillation in a single-side expansion scramjet combustor. *Acta Astronautica, 137,* 52–59.
27. Sun, M. B., Zhong, Z., Liang, J. H., & Wang, Z. G. (2014). Experimental investigation of supersonic model combustor with distributed injection of supercritical kerosene. *Journal of Propulsion and Power, 30*(6), 1537–1542.
28. Wang, Z. G., Sun, M. B., Wang, H. B., Yu, J. F., Liang, J. H., & Zhuang, F. C. (2015). Mixing-related low frequency oscillation of combustion in an ethylene-fueled supersonic combustor. *Proceedings of the Combustion Institute, 35*(2), 2137–2144.
29. Sun, M. B., Gong, C., Zhang, S. P., Liang, J. H., Liu, W. D., & Wang, Z. G. (2012). Spark ignition process in a scramjet combustor fueled by hydrogen and equipped with multi-cavities at Mach 4 flight condition. *Experimental Thermal and Fluid Science, 43,* 90–96.
30. Konnov, A. A. (2009). Implementation of the NCN pathway of prompt-NO formation in the detailed reaction mechanism. *Combustion and Flame, 156*(11), 2093–2105.
31. Xu, C., & Konnov, A. A. (2012). Validation and analysis of detailed kinetic models for ethylene combustion. *Energy, 43*(1), 19–29.
32. Cui, X. D. (2014). *Investigations on low frequency combustion oscillations in supersonic flow* (Master thesis, National University of Defense Technology).
33. Culick, F. E. C., & Rogers, T. (1983). The response of normal shocks in diffusers. *AIAA Journal, 21*(10), 1382–1390.
34. Oh, J. Y., Ma, F., Hsieh, S. Y., & Yang, V. (2005). Interactions between shock and acoustic waves in a supersonic inlet diffuser. *Journal of Propulsion and Power, 21*(3), 486–495.
35. Yang, V., & Culick, F. E. C. (1985). Analysis of unsteady inviscid diffuser flow with a shock wave. *Journal of Propulsion and Power, 1*(3), 222–228.
36. Wong, H. Y. W. (2006). Overview of flow oscillations in transonic and supersonic nozzles. *Journal of Propulsion and Power, 22*(4), 705–720.
37. Liu, J. H., Ling, W. H., & Liu, L. (1998). A quasi-one dimensional unsteady numerical analysis of supersonic combustor performance. *Journal of Propulsion Technology, 19*(1), 1–6.
38. Ouyang, H. (2013). *Investigations on the influence of injection scheme on the supersonic combustion performance of hydrocarbon fuel*. National University of Defence Technology.
39. Lin, K. C., Jackson, K., Behdadnia, R., Jackson, T. A., Ma, F., & Yang, V. (2010). Acoustic characterization of an ethylene-fueled scramjet combustor with a cavity flameholder. *Journal of Propulsion and Power, 26*(6), 1161–1170.

Chapter 3
Flow Dominating Instability in Supersonic Flows

The shock-induced separation is a common phenomenon in supersonic combustion, especially when the scramjet combustor works under high equivalence ratios. The low-speed recirculation zone and strong heat release enable the separated region to be an ideal place for flame stabilization. Asymmetric separation usually occurs in the combustor with high backpressure [1, 2], which generates asymmetric combustion combined with different flame stabilization modes. Oscillations of separated regions induced by the shock will lead to fluctuations of backpressure and heat release, which is a certain cause of unsteady combustion [3]. Following a series of systematic studies in this chapter, it is found that this kind of unsteady combustion is dominated by flow instabilities.

3.1 Asymmetric and Dynamic Combustion Behaviors in Strong Separated Flows

For a rectangular supersonic combustor with dual parallel cavities and near-cavity fuel injections, flowfield structures generally transform from symmetry into asymmetry as the increment of equivalence ratio. The asymmetric combustion flowfield is accompanied by strong separation. Under an intermediate equivalence ratio with threshold heat release, intermittent dynamic combustion occurs [3]. When this unsteady combustion happens, the axial movement of shock train undergoes a high-amplitude, low-frequency oscillation.

M. Sun et al., *Unsteady Supersonic Combustion*,
https://doi.org/10.1007/978-981-15-3595-6_3

3.1.1 Experimental Setup and Numerical Methodology for High-Temperature Cases

3.1.1.1 Direct-Connected Test Facility and Supersonic Combustor

A direct-connected test facility [4] is used for the experiments. The facility is composed of an air heater, a supersonic nozzle, and a scramjet combustor. The air heater burns pure ethylalcohol and oxygen continuously to heat air from room temperature up to 1480 K and increases the total pressure of vitiated air up to 3.6 MPa. The total mass flow rate of vitiated air is 1.73 kg/s. The two-dimensional converging-diverging Mach 3.46 nozzle, configured with a rectangular section, is adopted to provide the designed inflow conditions. The repeatability of the experimental system has been validated by a large number of experiments [5, 6].

The model combustor as shown in Fig. 3.1 has a total length of 2200 mm and consists of one constant area section and three divergent sections with the expansion angles of 2.5°, 3.5°, and 4°, respectively. The entry cross-section of the combustor is 54.5 mm in height and 75 mm in width. There are two cavity installations in the test section. Here for brevity, we denote the upper cavity by 'T_1' and the lower cavity by 'B_1'. The distance is 978 mm and 962 mm respectively from T_1 and B_1 cavity leading edge to isolator inlet. The parameters of the cavity are set to depth $D = 15$ mm, length to depth ratio $L/D = 7$, and the aft wall angle $A = 45°$. Figure 3.1 also shows fuel injection locations. Injectors I_1 and I_2 are mounted upstream of two parallel cavities. Injector configuration (orifice number × diameter, 3 × 2.0 mm, the distance between the orifices is 15 mm) is conducted. The distance is 10 mm from the injector to cavity leading edge. The flowfield is visualized through the quartz window (shown in Fig. 3.1) with a high-speed camera to capture flame chemiluminescence and schlieren images. Static pressure distribution in the axial direction is measured along the centerline of the combustor upper wall.

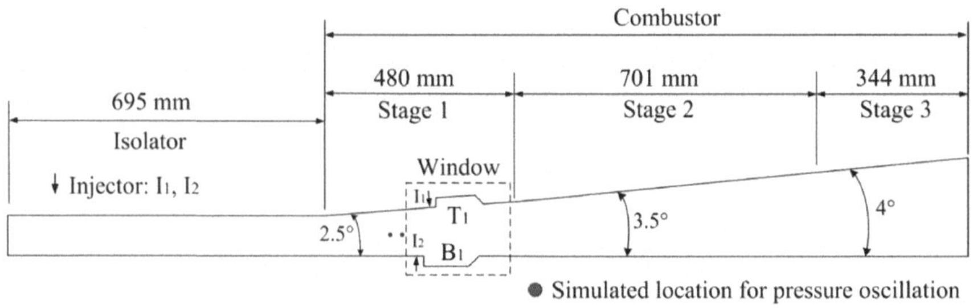

Fig. 3.1 Schematic of the supersonic combustor and cavity-injection installation [3]

3.1.1.2 Unsteady RANS Method and Computational Grid

The computational domain includes the entire experimental flowfield without the first section of the isolator, the axial length of which is 1.9 m. The configuration and distribution of cavities are exactly the same as in the experiment. Furthermore, the Mach number and the temperature of the gas in isolator entrance, along with the injection pressure of the injectors are settled to be the same as their values under experimental conditions. As 3D mixing effect cannot be simulated in a 2D simulation accurately, premixing injection is applied to compensate for the difference between 2D and 3D simulation. The premixing injection requires that the overall equivalence ratio is kept as a constant parameter. The fuel and the air are premixed with overall equivalence ratio and injected through the injector. Thus the mass fraction of ethylene is less than 1.0 in the jet, and the air added into I1 and I2 is less than 2% of total mass flow rate of the vitiated air. The overall equivalence ratios of ethylene are 0.502, 0.375 and 0.261. No-slip and adiabatic boundary conditions are used for the top and bottom walls.

Numerical simulations are implemented by using a RANS code with k-ω SST model and an uncoupled non-equilibrium reaction solver [7, 8]. The finite-rate reaction mechanism for ethylene–oxygen combustion consists of five reactive species ($C_2H_4 + 2O_2 \rightarrow 2CO + 2H_2O$, $2CO + O_2 \rightarrow 2CO_2$) and an overall reaction model [9]. There are no expansions along the sidewalls of the entire text facility. A two-dimensional structural grid is utilized for simulations, and the quasi-two-dimensional phenomenon of combustion flowfield will be discussed in Sect. 3.1.2. The total nodes of the grid are 3×10^5, and are refined in important areas such as the wall, shear layer, etc., in order to improve local numerical precision (shown in Fig. 3.2). In unsteady simulations, judging by the time scale of the chemical reaction (between 10^{-6} and 10^{-7} s approximately) and computational expense, the time step is set to

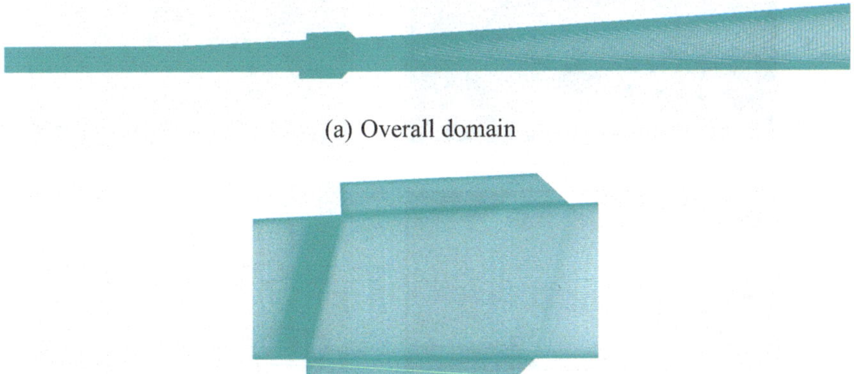

(a) Overall domain

(b) The grid refinement around the injection and dual cavities

Fig. 3.2 Computational mesh of model supersonic combustor [3]

9×10^{-7} s. Two characteristic locations (shown in Fig. 3.1) are selected to capture the static pressure oscillation history of the flow field. The axial locations of the simulated points are settled at 10 mm and 50 mm upstream of I2, respectively.

3.1.2 Combustion Characteristics Under Different Equivalence Ratios

The combustion flowfield shows apparently various structures under different equivalence ratios. Ethylene is injected at the condition of stagnation temperature $T_{0i} = 300$ K, stagnation pressure P_{0i} from 1.06 to 2.14 MPa, and the combustion mode gradually transfers from scramjet mode to ramjet mode. Figure 3.3a–c shows averaged flame chemiluminescence images and instantaneous schlieren images at pure scramjet mode, transition mode and pure ramjet mode, respectively. As depicted in Fig. 3.3, oblique shock waves in the shock train are very thin, so the combustion in these cases is a quasi-two-dimensional phenomenon. As expected, the flame brightness is raised and the flame distribution zone is enlarged as the equivalence ratio

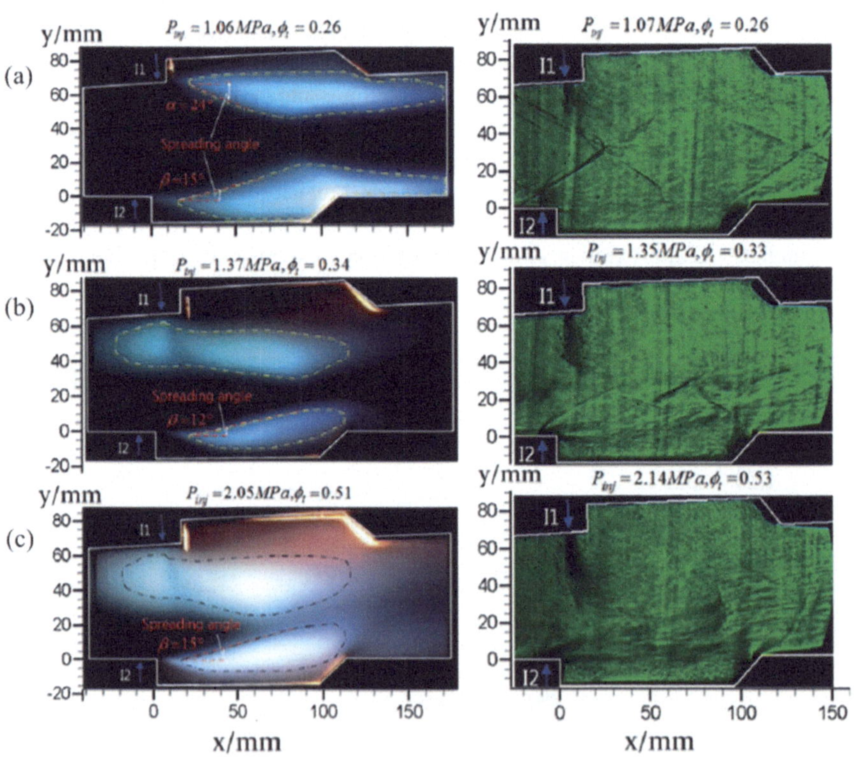

Fig. 3.3 Averaged flame chemiluminescence and instantaneous schlieren images at different equivalence ratios: **a** symmetric flame in scramjet mode, **b** asymmetric flame in transition mode, **c** asymmetric flame in ramjet mode [10]

increases. For the scramjet mode ($\Phi = 0.26$), flames in T1 and B1 are both stabilized in shear-layer, and the flame structure is symmetric. The limited heat release is not sufficient to choke the flow and the mainstream could still maintain supersonic. In this mode, the limited heat release near parallel cavities has a weak influence on the flow. There is no obvious separation on walls upstream of the injectors I_1 and I_2. As the equivalence ratio rises, the flame structure becomes asymmetric under transition mode ($\Phi = 0.34$). The flame in T_1 is stabilized in the jet-wake mode, while the flame in B1 is stabilized in the cavity-assistance shear layer mode. The asymmetric heat release region (separation zone) generates an oblique shock train. When the operating mode in combustor reaches the pure ramjet mode ($\Phi = 0.51$), the main flow is decelerated to subsonic due to the strong heat release. The pre-combustion shock train is pushed to an upstream location beyond the window. For this chock mode, there is a thermal throat in the subsonic region. The most noticeable character is that the strong pressure rise induces a large-scale recirculation zone on the upper wall upstream of the injector I_1. The flame T_1 spreads into the region upstream of the injector I_1, indicating that some fuel has been transported to the upstream region through separated flow. Though the injection stagnation pressure is the same, the penetration heights of injectors I_1 and I_2 are significantly different. The fuel injected through the injector I_1 has a high penetration height and is approximately vertically injected into the main flow.

The simulated asymmetric combustion is illustrated in Fig. 3.4b, and the symmetric case under low ϕ is presented as a contrast. It is observed that the calculated flame characteristics fit well with the flame chemiluminescence images in Fig. 3.3a, c.

As depicted in Fig. 3.4b, the asymmetric characteristics of the flow structure are quite obvious. The fuel from I1 penetrates deeply into the main flow and forms a fully developed separated region around T1. A large separated region enables the flame to travel upstream and stabilize in the jet wake. On the other hand, the supersonic region

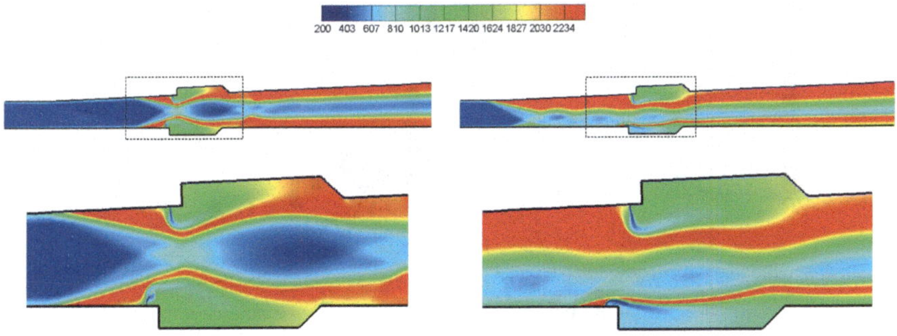

(a) Symmetric combustion, $\phi = 0.261$ (b) Asymmetric combustion, $\phi = 0.502$

Fig. 3.4 Temperature contours of the symmetric and asymmetric combustion based on numerical simulations [11] at the bottom wall bends the fuel jet to limited penetration depth and forms a shear-layer flame. The cause of the asymmetric combustion can be explained as follows. There are competitions between the two separated regions around the injectors at the top and the bottom. When counteracting the backpressure, the boundary layer becomes more easily separated at the expansion wall than that at the straight side. The separated region induced by the pressure rise at the expansion wall forms a high-temperature, low-momentum area around the top injector, which enables the flame to stabilize in the jet-wake mode. As the large separated region forms at the expansion wall, it suppresses the development of the separated region at the straight wall. As a result, the supersonic flow attaches to the straight wall, which leads to a shear-layer flame. Finally, the 'top jet-wake combustion' appears. The formation of asymmetric combustion will be further discussed in Sect. 3.3.

Figure 3.5 displays the time-averaged static pressure distributions along the centerline of the combustor upper wall under different operating modes (ramjet: $\phi = 0.502$, transition: $\phi = 0.375$, scramjet $\phi = 0.261$). It is seen that the numerical results correlate closely with the experiment results and precisely predict the peak of static pressure. The average shock location at $\phi = 0.261$ is better captured owing to the weak unsteady behaviors at low ϕ. The overall reaction simplification contributes to a higher reaction speed that forms a higher pressure peak around the cavities (clearly shown in operating conditions of $\phi = 0.502$ and 0.375). Due to the side wall effects and intermittent combustion that occurs downstream, the experimental data tends

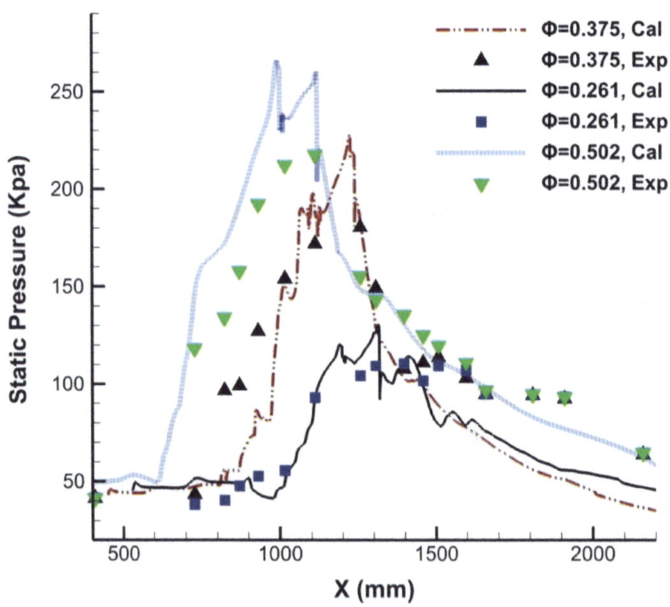

Fig. 3.5 Time-averaged static pressure along the upper wall centerline under different operating modes/equivalence ratios [3]

to be higher than the simulated results in the downstream expansion section of the combustor.

3.1.3 Dynamic Combustion Under Intermediate Heat Release

When the combustion condition stays in the transition mode (equivalence ratio 0.375), the flame stabilized mode transfers between cavity shear layer stabilized mode and jet wake stabilized mode randomly. The flow behaviors in combustor are more complicated. Sometimes, the asymmetric flame stabilizes in one state, while at other times, unsteady combustion occurs. The flame behaviors of the steady and unsteady processes are exemplified in Fig. 3.6.

Images of the unsteady process are chosen from the experimental flame chemiluminescence images. The method of Wang et al. [12] is adopted to characterize the flame zone, and a combustion iso-luminosity contour (shown in Fig. 3.7a) is captured. The streamwise location of the flame front is then obtained. The leading edge of the cavity is selected as the reference location X. If the flame travels upstream of X, this means that it stabilizes in the jet wake. Otherwise, the flame only appears in the shear layer. The reaction zones of the dual parallel cavities exchange between the jet wake and the shear layer during the dynamic process. These are depicted in Fig. 3.7b. Figure 3.8a–f illustrates the experimental chemiluminescence images in the flame switching process. The time of the first image is defined as T_0, and the interval between the two images is 0.5 ms. The reverse transition is depicted in Fig. 3.8g–j.

As depicted in Fig. 3.9, during the transition of the flame stabilized mode, the shock train travels along the streamwise direction. When the shock train travels further upstream of the cavities, it stays on one side of the flow path. When the shock train is located around the cavities, it is pushed up and down in the vertical direction, which causes the switch of operating conditions. After the flame switching process, one of the separated regions becomes fully developed and forms the 'jet-wake flame'. Meanwhile, the flame on the other side is suppressed in the shear layer and forms the 'shear-layer flame'. The entire switching process is illustrated in Fig. 3.9.

As shown in Fig. 3.9, the simulated flow fields Fig. 3.9a–j correspond to the experimental images (Fig. 3.8a–j) in flow time. The numerical results fit the experimental images very well in terms of both the shock train and the flame switching process. A detailed analysis is introduced in several steps, as indicated below. We refer to the back pressure of the shock train as 'back pressure' for short.

Figure 3.9a shows that the flame stays in the jet wake of I1 and the shear layer of I2 at first (the flame can appear at the jet wake of I2 from the beginning as well). A disturbance leads to a rise or fall of the back pressure. If the back pressure falls, the shock train travels downstream. As the shock train moves downstream, it becomes increasingly stronger and forms a more symmetric structure (Fig. 3.9b, c). A more symmetric shock causes the large separated region at the top to be pushed downstream, and the small separated region at the bottom to grow larger, as indicated

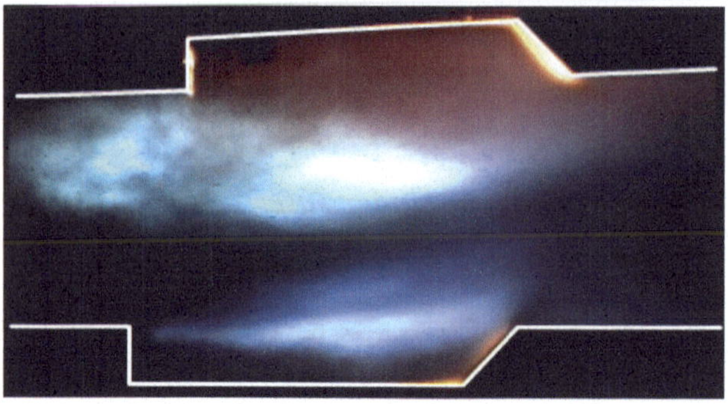

(a) Steady process

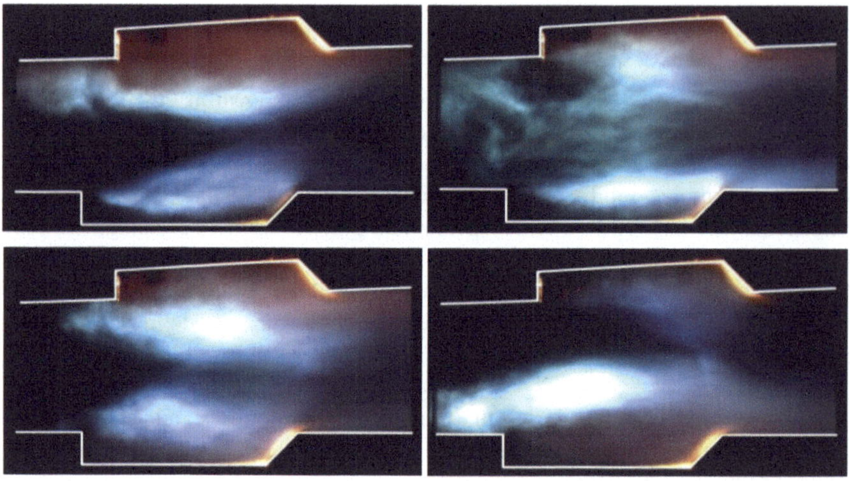

(b) Unsteady process

Fig. 3.6 Characteristic operating conditions during the steady and unsteady process for $\phi = 0.375$ [3]

in Fig. 3.9d. The larger separated region at the bottom promotes local combustion intensity, and the pressure rise induced by the stronger heat release generates an even larger separated region, which forms positive feedback. Finally, the separation at the bottom is strong enough to drive the shock train to the top wall, and the penetration of the injection at the bottom reaches its maximum value (Fig. 3.9e). The back pressure increases owing to the strong heat release around the bottom cavity, which enables the shock train to travel upstream. As the shock train moves upstream, it becomes gradually weaker, which weakens the separation at the bottom (Fig. 3.9f). The smaller separated region reduces the local fuel penetration depth and heat release, which results in the fall of the back pressure. Finally, the shock train stops at a certain point and travels downstream again (Fig. 3.9f, g). Figure 3.9g–j

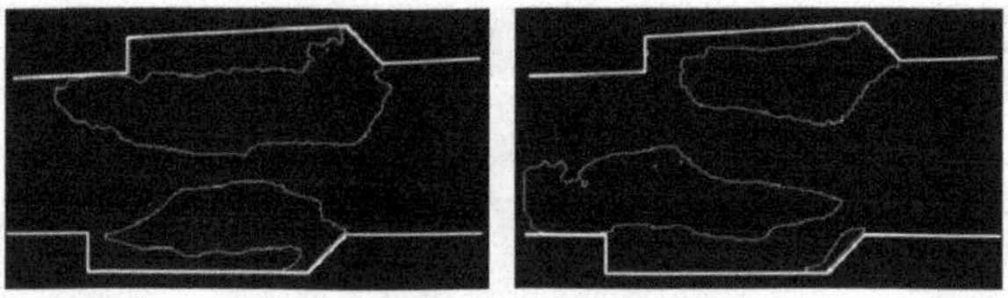

(a) Contours of the flame zone at T_0 and T_0+2.0ms

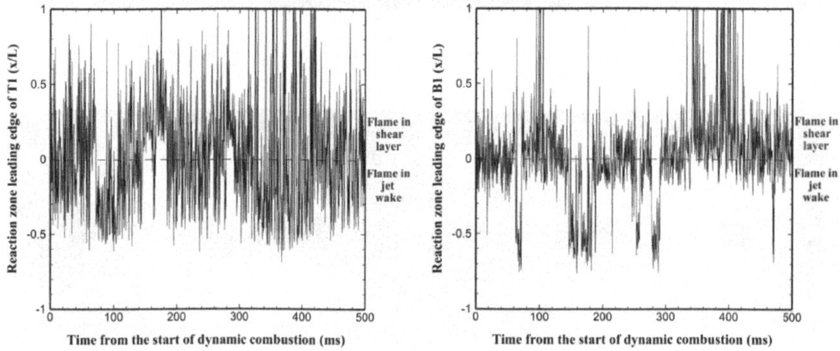

(b) Oscillating histories of the leading edge of the flame zone for T1 and B1

Fig. 3.7 Contours of flame zones and their oscillation histories [3]

depicts the switching process of the jet-wake flame from the bottom to the top, whose mechanism is similar to the previous description of Fig. 3.9a–e.

The following analysis is implemented based on the flame zone images obtained in Sect. 3.1.2. The Fast Fourier Transform (FFT) was adopted in order to analyze the oscillation history of the reaction zone (Fig. 3.7b), and the average frequency of the flame oscillation from both cavities was calculated. The final results are illustrated in Fig. 3.10. The peak frequency of the flame at the top is 156.750 and 180.961 Hz at the bottom.

Figure 3.11 depicts the static pressure oscillations from two characteristic locations (shown in Fig. 3.1) of the flowfield during the simulation of the dynamic process. It is observed that the oscillations have periodic characteristics. The static pressure oscillation histories of the different locations are similar due to the periodic wide range movement of the shock train. The peak value of the pressure curve indicates that a shock wave in the shock train has moved upstream of the characteristic location, causing a pressure rise at that point, thereby indicating the occurrence of the flame exchange. FFT analysis is adopted again in order to calculate the frequency of the static pressure oscillation.

As demonstrated in Fig. 3.11, the peak frequencies of the two locations occur at 469.882 Hz and 445.605 Hz. During one cycle of unsteady combustion, the flame from dual parallel cavities exchanges twice and forms two pressure peaks. Thus, the frequency of the flame is only half of the static pressure. Based on Fig. 3.11, the simulated average frequency of unsteady combustion is 228.871 Hz. Compared with

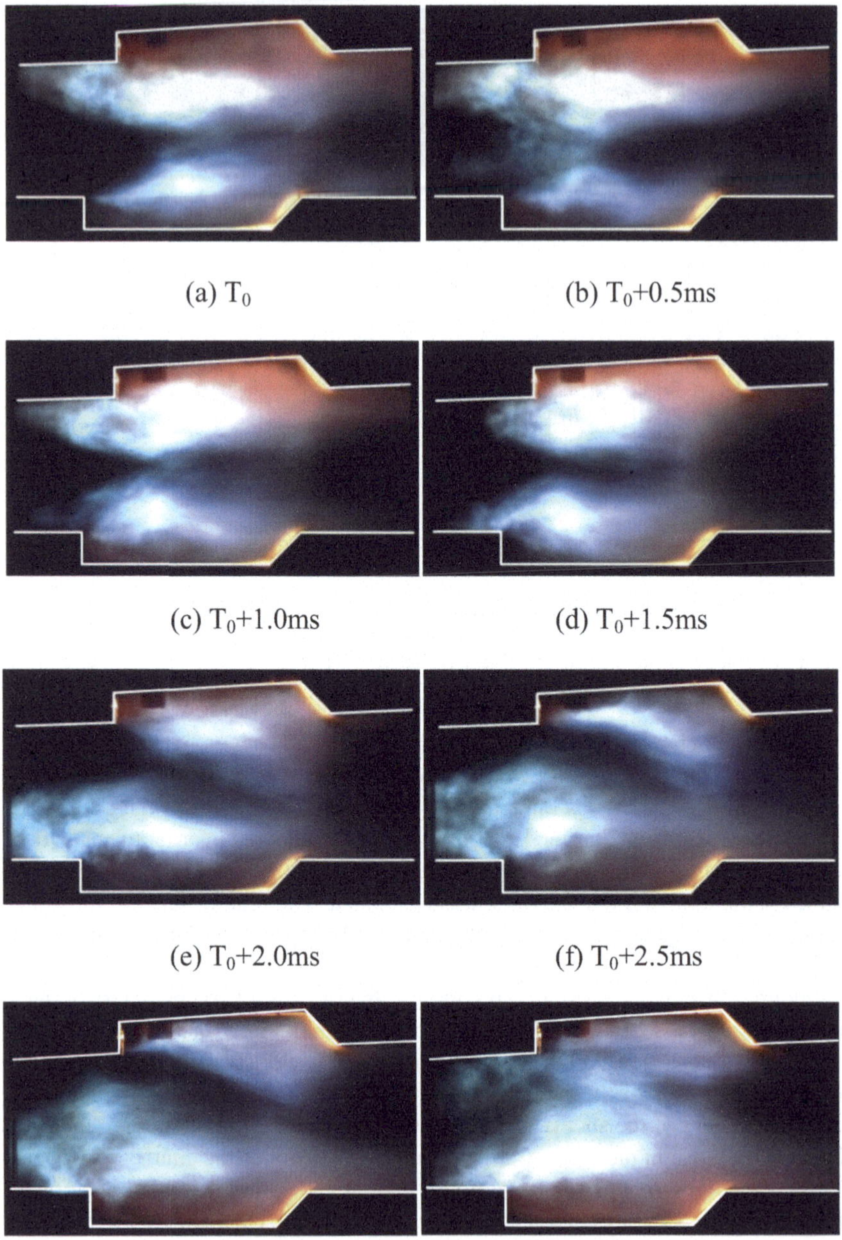

(a) T_0 (b) T_0+0.5ms

(c) T_0+1.0ms (d) T_0+1.5ms

(e) T_0+2.0ms (f) T_0+2.5ms

(g) T_1 (h) T_1+0.5ms

Fig. 3.8 Experimental luminosity images in the flame switching process [3]

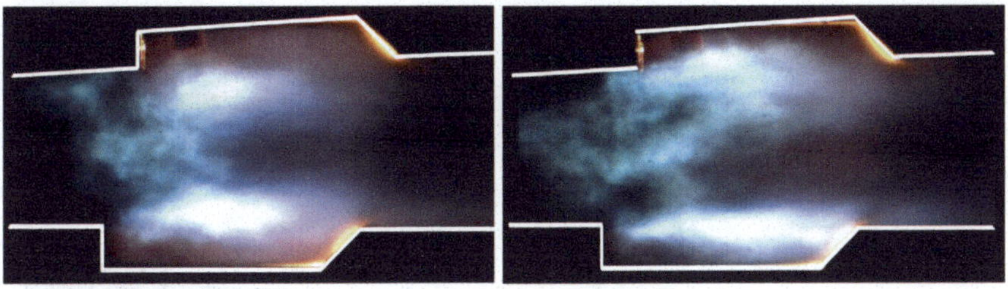

(i) T_1+1.0ms (j) T_1+1.5ms

Fig. 3.8 (continued)

the experimental frequencies, the relative errors are 31.5% (T1) and 20.9% (B1). The cause of the errors is attributed to several factors. Firstly, the overall reaction simplification (for saving computational cost) may speed up the reaction and heat release, which results in a higher simulated frequency. The premixing injection assumption may shorten the mixing time, which contributes to shorter numerical cycle time. Moreover, the intermittent unsteady process makes it even harder to attain a close fit for the numerical and experimental results. In the experiments, the jet wake stabilized flame shifts between T1 and B1 intermittently but not periodically. However, the simulation attains a quasi-periodic oscillating process, while the intermittent process could still be captured (e.g. t = 60–80 ms). This reveals that the URANS simulation captures the key unsteady characteristics of the dynamic combustion procedure.

The analyses in the current section used RANS in turbulence simulations, so the influence of eddies is eliminated. As a result, the cause of flame oscillation cannot be the acoustic-eddy mechanism [13]. An intrinsic low-frequency oscillation (unsteady behavior of the separation point) lies in the shockwave boundary layer interaction [14] which is a quasi-two-dimensional phenomenon. Such a low-frequency oscillation is greatly influenced by the scale of the separated region and downstream disturbances. The low-frequency oscillation of SWBLI is amplified by the large separated region induced by heat release. When it is influenced by downstream pressure fluctuations, these unsteady behaviors mentioned above appear.

3.2 Decoupling Analysis of the Unsteady Combustion

Strong coupling effects exist between flow and heat release in supersonic combustion, and various influencing factors make it difficult to uncover the mechanism of unsteady combustion. Main impact factors of unsteady combustion in the current chapter are summarized as four: flame, jet, back pressure, and cavity. In the current section, these possible influencing factors of combustion instability are decoupled step by step through controlling variable method. Characteristic phenomena of flowfields with different decoupling schemes are listed in Table 3.1 [15].

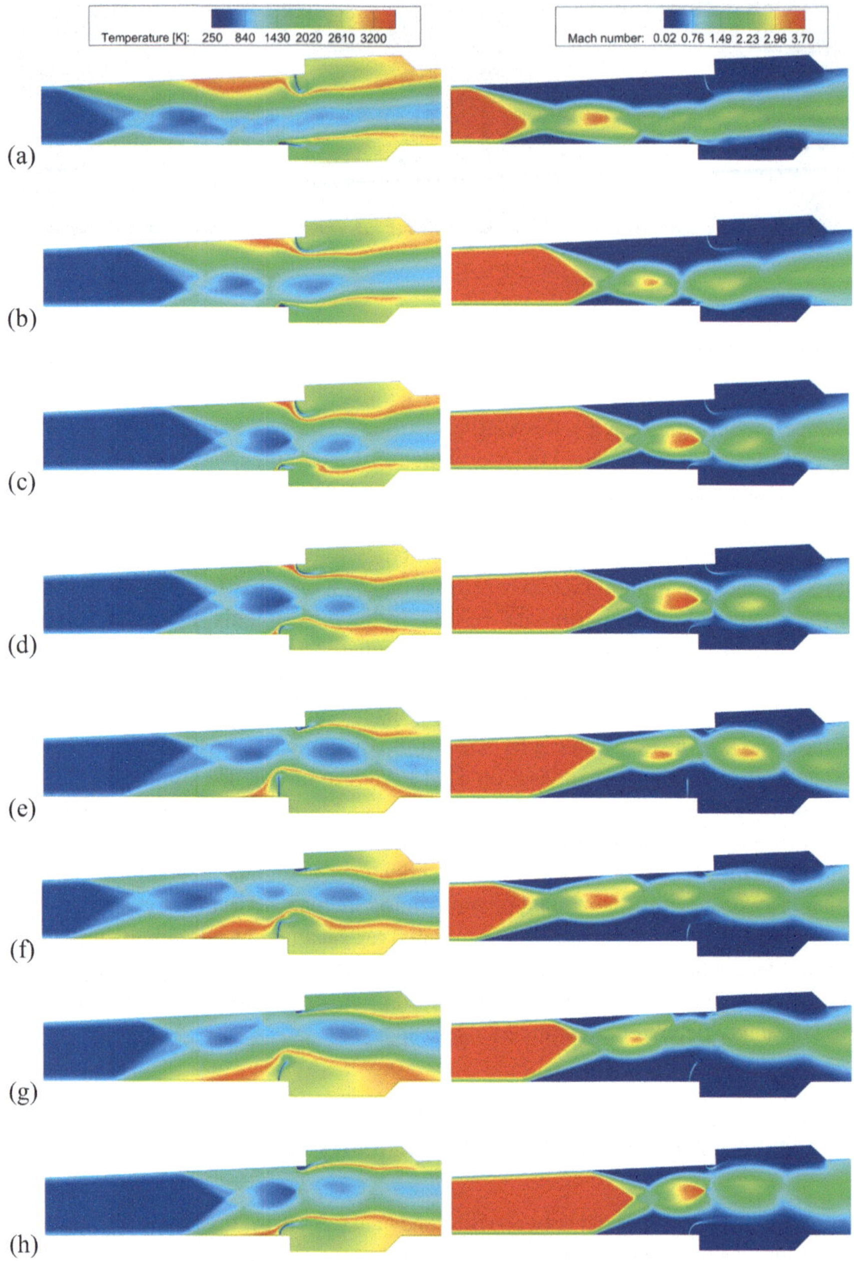

Fig. 3.9 Temperature and Mach number field distributions during the switching process of the operating condition [3]

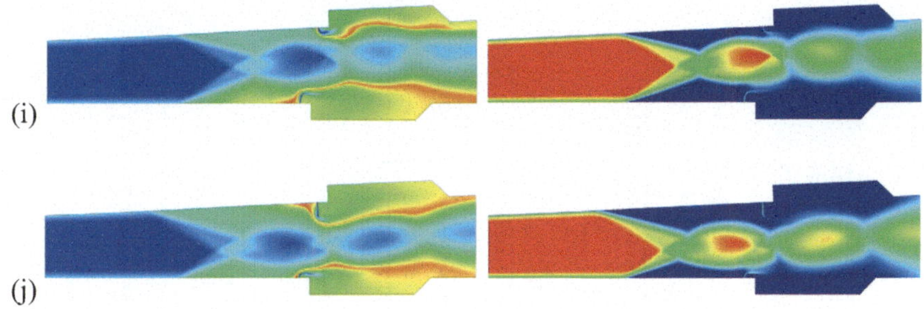

Fig. 3.9 (continued)

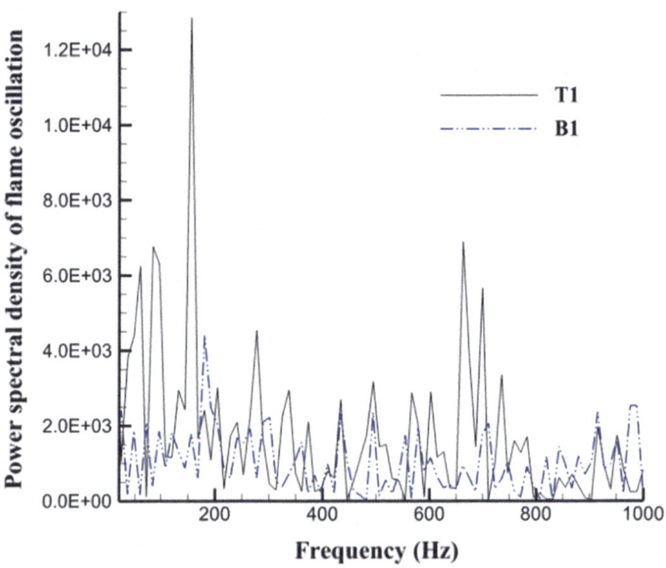

Fig. 3.10 Spectral energy distributions for the experimental oscillating history of flame leading edge for cavities T1 and B1 [3]

3.2.1 *Impact Factors of the Separation Dominating Unsteady Combustion*

3.2.1.1 The Effect of Flame on Unsteady Combustion

Keeping the original cavity configuration and position unchanged, fuel jet was replaced by air jet. A cylinder was located transversely at the average position of the thermal throat to substitute for the pressure rise caused by heat release in combustion. In the vertical direction, the cylinder was located in the channel center (equidistant from the upper and lower walls). The effect of flame on combustion instability was

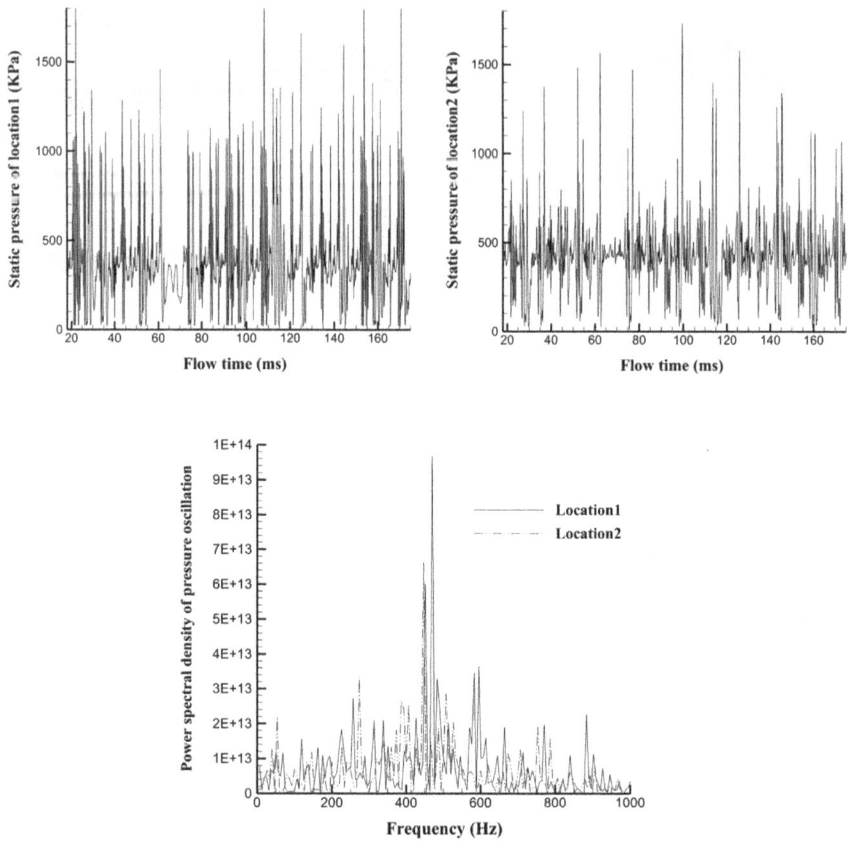

Fig. 3.11 Simulated static pressure oscillation histories from two locations, and power spectra calculated by FFT [3]

stripped. Figures 3.12 and 3.13 show that the flowfield without flame still presents the basic characteristics of unsteady combustion flowfield, such as that large-scale separation may occur on either side of the wall, accompanied by the large-scale axial oscillation of shock wave train. The simulation results demonstrate the competition between the two separation zones on the opposite side walls: when one separation zone develops sufficiently, the other will be inhibited. Under the same initial and boundary conditions, the two separation zones both have the possibility to develop sufficiently and become dominant.

3.2.1.2 The Effect of Jet on Unsteady Combustion

Keep the configuration of combustor unchanged, and the influence of back pressure was stripped by injecting air jet into the cold flow field with the mass flow rate of fuel

Table 3.1 Operating conditions of decoupling analysis [15]

No.	Effects included	Effects decoupled	Characteristic phenomena of the flow field
Case 1	Injection, back pressure, cavity	Flame	Unsteady, large separation exists on expansion wall, the complex oscillation of shock train
Case 2	Injection, back pressure, cavity	Flame	Unsteady, large separation exists on a straight wall, the complex oscillation of shock train
Case 3	Injection, cavity	Flame, backpressure	Unsteady, symmetric separation, periodical oscillation of shock train
Case 4	Back pressure, cavity	Flame, injection	Unsteady, asymmetric separation switches between two walls, complex oscillation of shock train
Case 5	Injection	Flame, back pressure, cavity	Steady
Case 6	Back pressure	Flame, injection, cavity	Unsteady, asymmetric separation switches between two walls, complex oscillation of shock train

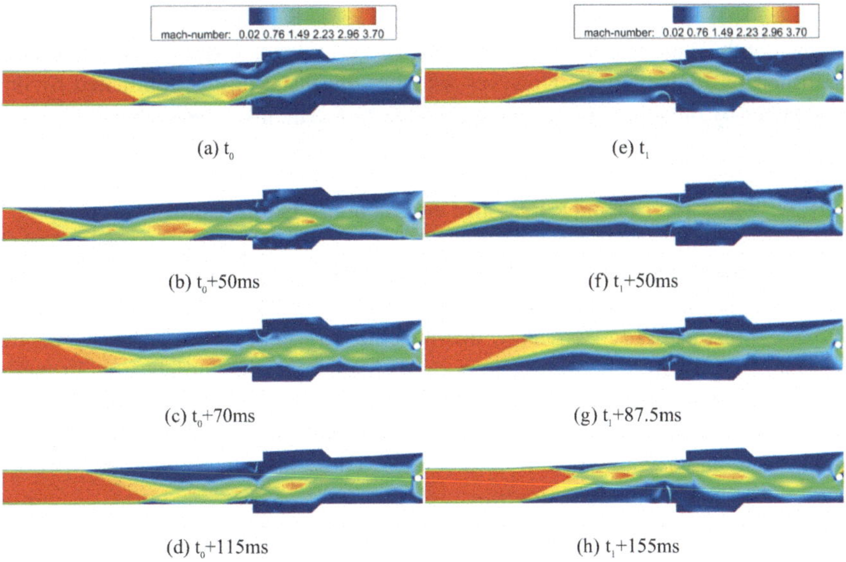

Fig. 3.12 Mach number contours of the unsteady process in cold case 1 (**a**–**d**), and case 2 (**e**–**h**) [15]

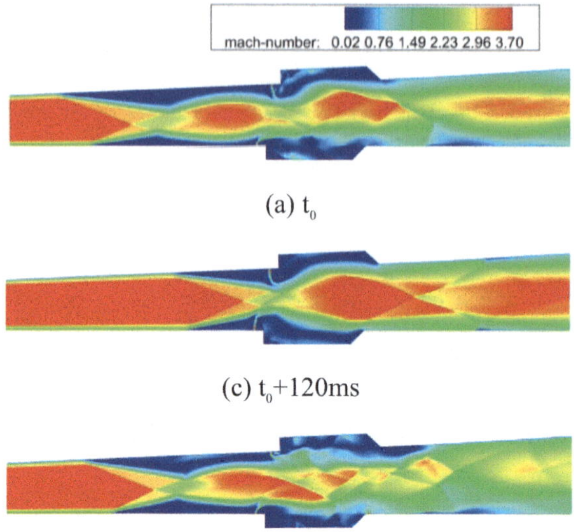

(a) t_0

(c) t_0+120ms

(e) t_0+196ms

Fig. 3.13 Mach number contours of the unsteady process in cold case 3 [15]

jet in the combustion condition. In this case, the flow field exhibits an obvious periodic unsteady characteristic, which is similar to the symmetric combustion case under low equivalence ratio. This typical process is shown in Fig. 3.13. A comparison of the dynamic characteristics of the cold flow field with air jet injected and the combustion flow field with an equivalent ratio of 0.261 is shown in Fig. 3.14. It can be seen from the figure that the pressure oscillation is very regular in both the cold flow field and the combustion flow field under symmetric combustion condition, which is also confirmed by FFT analysis. However, whether the instability is dominated by the jet alone or by the jet-cavity together, this question remains to be analyzed.

3.2.1.3 The Effect of Back Pressure on Unsteady Combustion

The geometrical configuration of the combustor remains the same, except that a cylinder is located at the average position of the thermal throat and there is no injection. Numerical results in Fig. 3.15 show that the cold flow field driven by back pressure can qualitatively reproduce all the important characteristics of combustion instability under medium equivalence ratios (large oscillation of shock train, switching of large-scale separation zones on upper and lower walls, the transformation between symmetrical and asymmetrical flow structure). Thus, back pressure is an important factor affecting combustion instability.

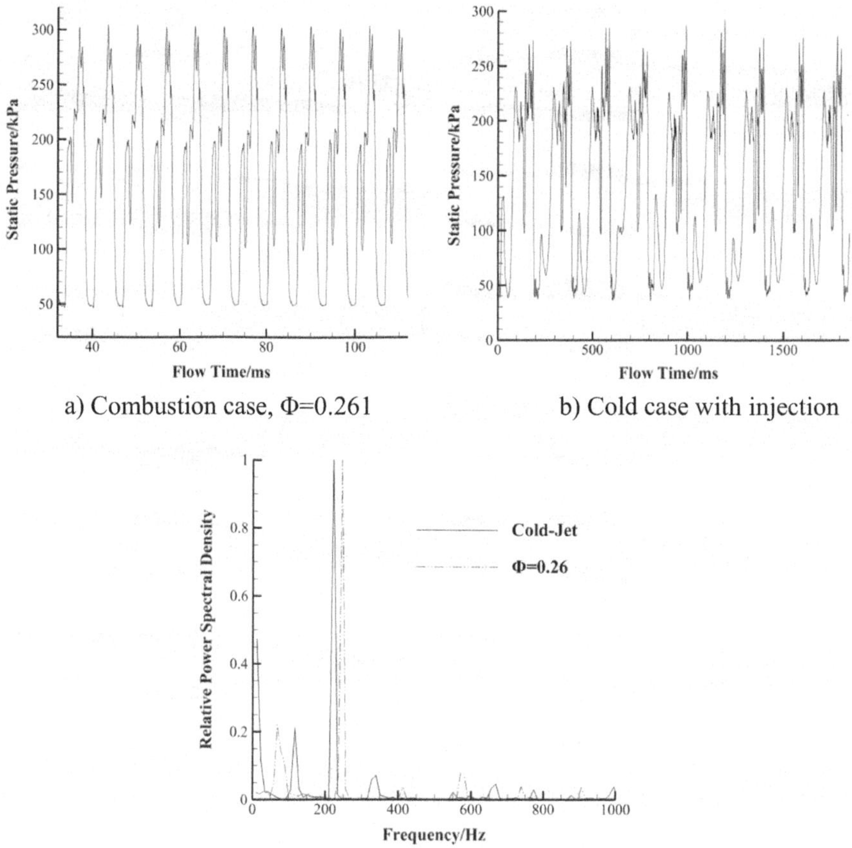

a) Combustion case, Φ=0.261

b) Cold case with injection

c) Power spectra calculated by FFT

Fig. 3.14 Comparison of the static pressure oscillation histories and power spectra, from combustion case $\Phi = 0.261$ and cold case with injection [15]

3.2.1.4 The Effect of the Cavity on Unsteady Combustion

The cavity is removed, and the other conditions are the same as those in Sect. 3.2.1.3. Figure 3.16 illustrates Mach number contours of the cold flow field with air jet injected in the cavity-free combustor configuration. The results show that the unsteady flow field does not appear in the cavity-free configuration, regardless of whether the two jets are oppositely aligned or misaligned. This indicates that the periodic unsteady phenomena in Sect. 3.2.1.2 are caused by the interaction between the jets and the cavity, instead of the inherent instability of the jet.

Figure 3.17 illustrates Mach number contours of the unsteady process of a cold flow field driven by back pressure (caused by cylinder) in the cavity-free combustor configuration. It can be seen that the supersonic flow in the expanding channel will undergo unsteady process characterized by the large axial oscillation of shock train

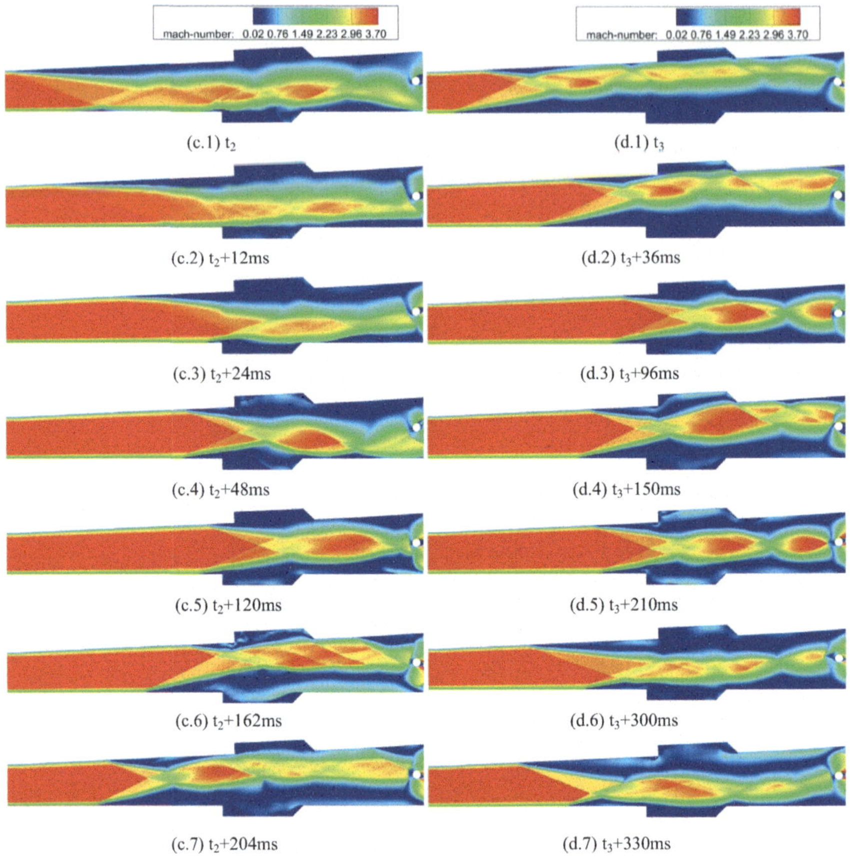

Fig. 3.15 Mach number contours of characteristic operating conditions in cold case 4 [15]

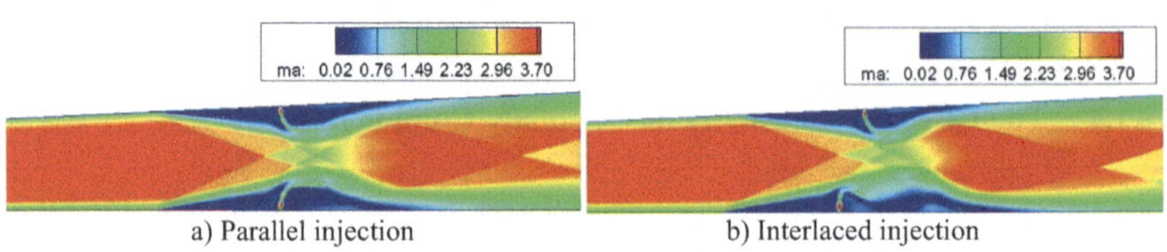

Fig. 3.16 Mach number contours of different injection schemes in cold case 5 [15]

and switching of asymmetric separation zones between two walls under the driving of back pressure. Based on the numerical simulation results and analysis in this section and Sect. 3.2.1.3, it can be concluded that back pressure is the determinant factor of combustion instability at medium equivalence ratios.

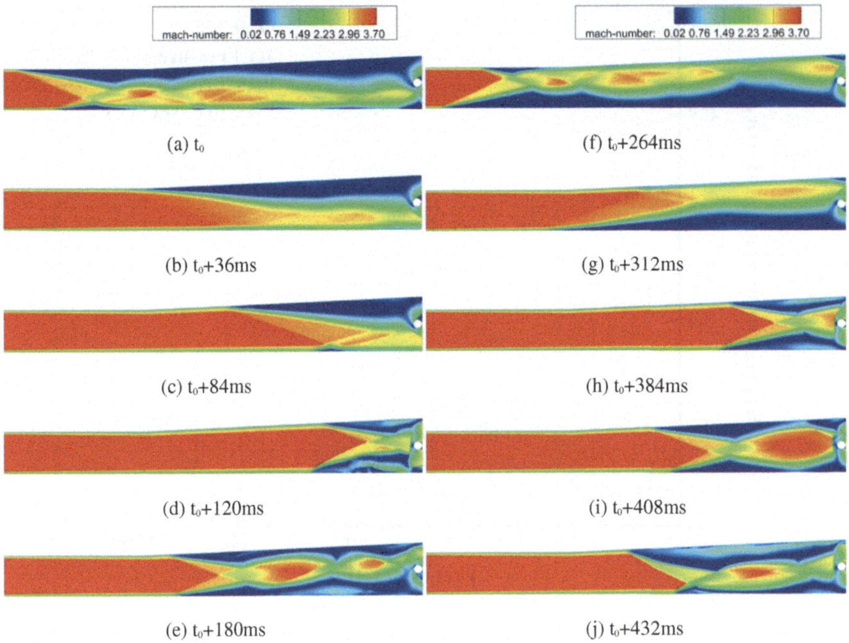

Fig. 3.17 Mach number contours of the unsteady process in cold case 6 [15]

3.2.1.5 The Influence of Decoupling Analysis on Frequency Behaviors

Figure 3.18 illustrates the effect of decoupling analysis on the frequency characteristics of the flow field, in which all the cases can represent the typical process and feature structures of the unsteady combustion flow field. The results show that with the simplification of the influence factors by decoupling analysis step by step, fewer and fewer frequency characteristics of unsteady combustion flow field can be captured. Although various factors can be separated from the combustion flow field by decoupling analysis, this method cannot represent the dynamic characteristics of the flowfield (period, amplitude, etc.). In order to predict the oscillation frequency of the combustion flow field more accurately, the decoupling analysis method should be used cautiously.

3.2.2 Dynamic Behaviors in High-Temperature Separated Flow Induced by Backpressure

In this section, typical cases of decoupling analysis in Sect. 3.2.1 are validated by experiments. The original configuration of the combustor is retained, that is, a rectangular single-side expanded combustor with parallel cavities for flame stabilization.

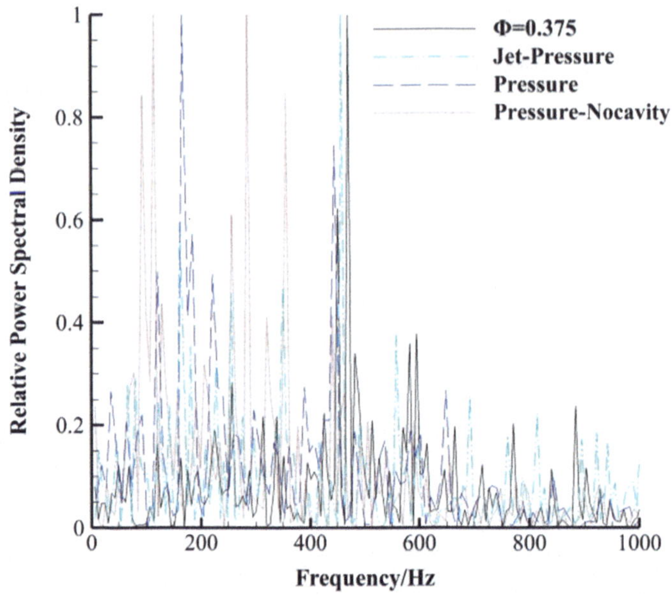

Fig. 3.18 The effects of decoupling analysis on dynamic features (frequency of static pressure oscillation) of the flow field [15]

A cylinder is located at the average position of the thermal throat in combustion experiments to generate the backpressure. Both effects of fuel jet and flame are excluded.

3.2.2.1 Symmetric Stable Separation Under Low Backpressure

Cylinders with different diameters were respectively located at the average position of the thermal throat during combustion experiments. When the diameter was 16 mm, the cold flowfield is similar to that of the symmetric combustion under low equivalence ratios. As shown in Fig. 3.19, the flow field structure is very similar to the symmetric separation flow field formed by symmetrical combustion with the low equivalent ratio of 0.261. The structure of the shockwave is stable, the shock train has no significant oscillation and the unsteady effect is very weak.

The comparison of time-averaged static pressure along the upper wall centerline is shown in Fig. 3.20. The peak pressure and the average position of the shockwave of the cold case fit well with the combustion case. This indicates that symmetrical combustion with low equivalence ratios can be regarded as a local choke dominated by low backpressure. In this case, the leading edge of the shock train is offset, and the flow field can be regarded as quasi-steady.

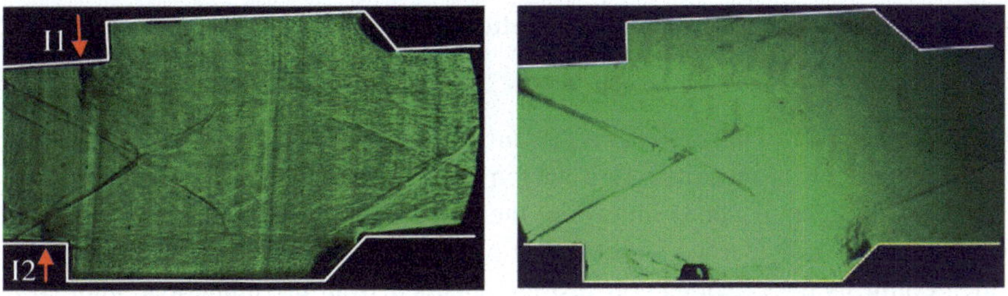

(a) Combustion case, Φ=0.261 (b) Cold case

Fig. 3.19 Comparison of the schlieren images from combustion and cold cases in symmetric separation mode [15]

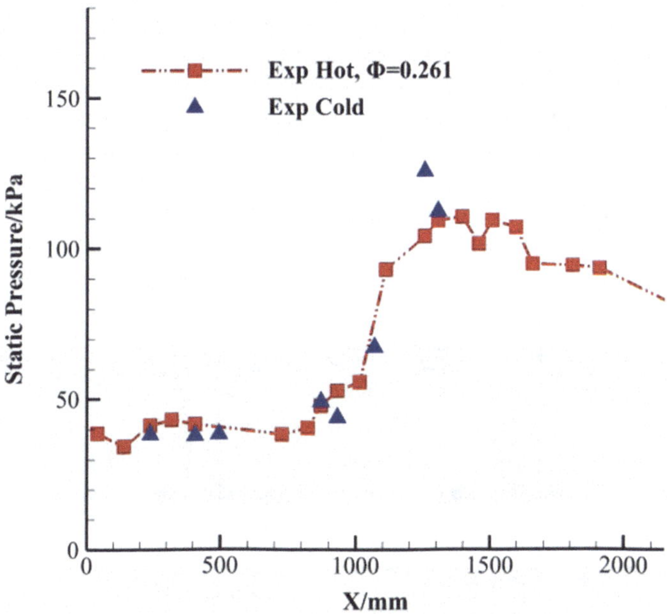

Fig. 3.20 Comparison of the time-averaged static pressure along the upper wall centerline, from combustion case $\Phi = 0.261$ and cold case [15]

The experimental results show that when the backpressure is low, it is not the dominant factor of unsteadiness. Thus, it is validated indirectly that the periodic oscillation of symmetric separation under low equivalent ratios is possibly caused by the interaction between the jet and the cavity.

3.2.2.2 Asymmetric Unsteady Separation Under Intermediate Backpressure

When the diameter of the cylinder is 20 mm, intermittent switching of the separation occurs in the cold flow experiment, which is the same as that in combustion case under moderate equivalence ratios and the numerical simulation results. It can be seen from Fig. 3.21 that the large separation zone switches between the two walls, which confirms the previous conclusions obtained from the numerical simulation analysis.

Figure 3.22 shows a comparison of the time-averaged static pressure along the upper wall centerline of the combustor under the switching condition of the cold and heat flow separation zone. It can be seen from the figure that the peak pressure under the cold flow condition agrees well with the combustion condition, but the average position of the shock wave is slightly lower than that under the combustion condition (this is due to the fact that no jet is introduced into the cold flow experiment). The peak pressure and shock location of the cold flow experiment are in good agreement with the numerical simulation results, which verifies that the numerical simulation results of cold flow under back pressure can reproduce the complex unsteady phenomena in the switching process of separation zone under combustion conditions.

Generally speaking, the transient structure, transition process and time-averaged pressure of the cold flow field merely driven by back pressure basically reflect the main characteristics of the switching conditions in the separation zone of the combus-

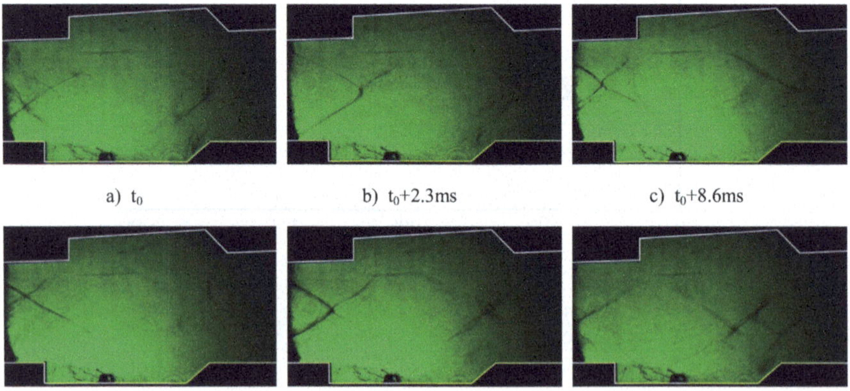

a) t_0 b) t_0+2.3ms c) t_0+8.6ms

d) t_0+16.6ms e) t_0+17.5ms f) t_0+20.1ms

Fig. 3.21 Schlieren images of the asymmetric separation switch process induced by backpressure [15]

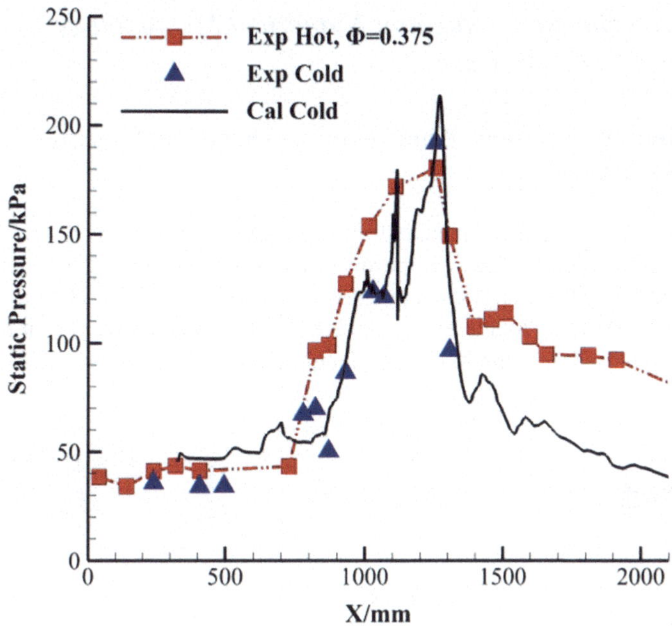

Fig. 3.22 Comparison of the time-averaged static pressure along the upper wall centerline, from combustion case $\Phi = 0.375$ and cold case [15]

tion flow field under medium back pressure. It can be concluded that the intermittent switching of asymmetric separated regions is driven by backpressure.

3.3 Cold Flow Analysis: Asymmetric Separation Induced by Boundary Layer Transformation

Cold flow analysis [16] is applied in a model scramjet combustor to obtain insights into the key physics of the symmetric/asymmetric combustion modes under different equivalence ratios. Systematic experiments have been implemented in a single-expanding duct with back pressure produced by a cylinder at Ma 3. Fine structures of separated flow have been gained by the NPLS system. Velocity profiles obtained from PIV are validated by numerical simulations. The formation mechanism of asymmetric separation is fully uncovered by a DNS study.

3.3.1 Experimental Setup and Numerical Methodology for Cold Flow Cases

3.3.1.1 Indraft Supersonic Wind Tunnel and Simplified Supersonic Combustor

Experiments were conducted in an indraft supersonic wind tunnel [17] at Mach 3.0. The upstream of the wind tunnel is the ambient atmosphere, and the downstream is a vacuum tank. A schematic of the apparatus is shown in Fig. 3.23. The incoming air of test section has 1 atm total pressure and 300 K total temperature. The air flows into the test section with a 5 mm thick turbulent boundary layer. Figure 3.24 illustrates

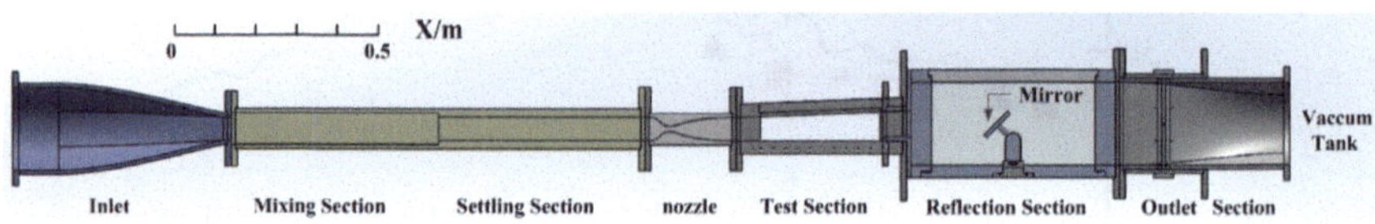

Fig. 3.23 Schematic of the experimental apparatus [16]

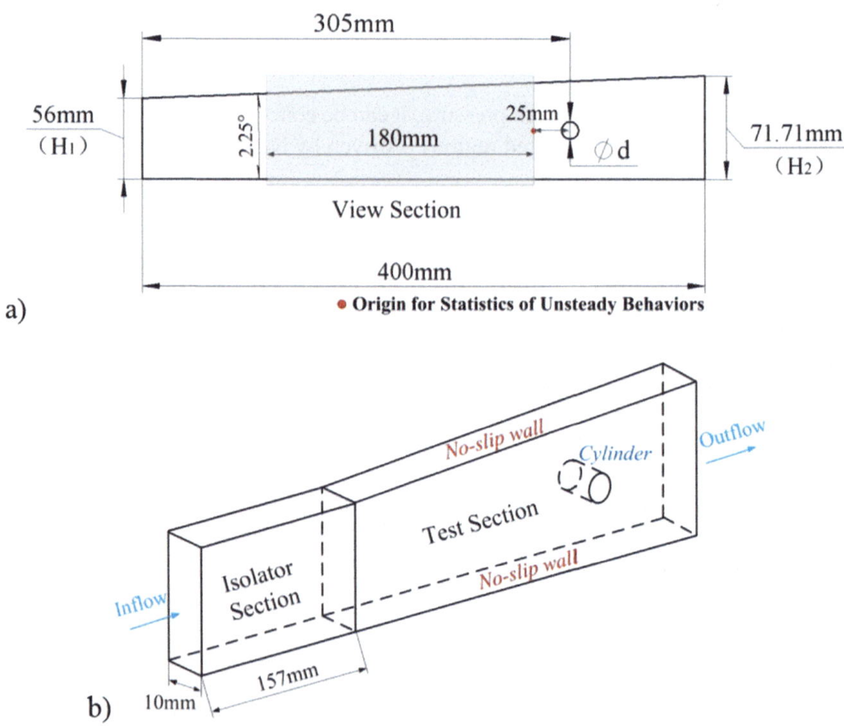

Fig. 3.24 Schematic of the flowfield: **a** experimental domain; **b** computational domain (DES) [23]

the single-expanding test section (120 mm in width) in detail, which has a cylinder located 305 mm downstream of the nozzle exit. The diameter and position of the cylinder are changeable for partial substitution of pressure rise induced by heat release in combustors (different diameters simulating the change in combustion intensity, variation of position mimicking the fluctuation of the thermal throat and asymmetric heat release). View section extends 25–205 mm upstream from the cylinder in the axial direction.

The fine structures of separated flow were visualized by nano-based planar laser scattering (NPLS) system [18, 19]. The mirror in the reflection section served as a medium to change the direction of the laser sheet, which illuminated the symmetry plane of the test section. Ten high-frequency pressure sensors were placed along the centerline of expansion wall, whose axial locations were 50, 80, 110, 140, 170, 230, 260, 290, 320 and 350 mm downstream of the nozzle exit, respectively. The acquisition frequency of the pressure signals was 20 kHz. Velocity profiles of the flow field were obtained by particle image velocimetry (PIV) method, which had been applied to a wide variety of supersonic experiments [20–22].

3.3.1.2 RANS Method for Characteristic Separated Flowfield Simulation

Characteristic separated flowfield simulation was implemented by a RANS code with k-ω SST model [8]. The Mach number, stagnation pressure and stagnation temperature of test section entrance were set the same as experimental conditions. Non-slip wall boundary conditions were used for the walls and cylinder. An isolator of 157 mm in length was settled upstream the test section to generate the 5 mm thick turbulent boundary layer from incoming flow. The flow field was quasi-two-dimensional, therefore the spanwise width (Z direction) of the computational domain was set as 5 mm. Grid was refined on the cylinder and both sides of walls. The size of the grid was $120 \times 303 \times 15$ in isolator section, $438 \times 303 \times 15$ in the test section (without cylinder region), and $465 \times 105 \times 15$ in cylinder region, which had total cells of 3.3×10^6.

3.3.1.3 DES Method for Separation Transition Process Simulation

Separation transition process simulation was implemented by a detached-eddy simulation (DES) code [24] with the Spalart–Allmaras (S–A) RANS model in the near-wall region. The detailed description of this numerical approach can be found in Wang et al. [25]. An isolator of 157 mm in length is settled again to generate the 5-mm-thick turbulent boundary layer from the incoming flow. The flowfield is quasi-two-dimensional; therefore, the spanwise width (Z direction) of the computational domain is set as 10 mm. The grid resolution is set to $\Delta y^+ = 1$ for the near wall regions. Meanwhile, the grid points are coarsened in the X direction for the purpose of saving the computational cost, especially in the pure RANS region (isolator). The grid resolution reaches the highest in the pure LES region (cylinder region) so that the elaborate vortex structures around the backpressure cylinder are able to be well captured. The size of the grid is 100 × 335 × 38 in the isolator section, 552 × 335 × 38 in the test section (without cylinder region), and 516 × 112 × 38 in the cylinder region, which have total cells of 1.05×10^7.

3.3.1.4 DNS Method for Turbulent Boundary Layer Simulation

The detailed transformation features of the boundary layer passing through an expansion corner are simulated by solving the three-dimensional unsteady compressible Navier–Stokes equations directly without any modeling. The in-house DNS code has been previously used for studies of instability, transition, and turbulence in high-speed flows [26]. The governing equations are solved using an entropy-splitting approach for the Euler terms and fourth-order accurate finite differences for spatial discretization. A third order Runge–Kutta scheme is used for marching in time. Details of the governing equations and algorithm can be found in Touber and Sandham [27].

The digital filter inflow approach of Xie and Castro [28] is used. In the present study, the streamwise characteristic length scales for the three velocity components u, v and w are set to $0.65\delta_i$, $0.35\delta_i$, and $0.35\delta_i$ (where δ_i denotes the 99% boundary-layer thickness at the inflow), respectively. Mean and root-mean-square (RMS) profiles are created beforehand from the DNS database of Schlatter and Orlu [29] for incompressible flat-plate TBLs.

The inflow parameters (shown in Table 3.2) are set in accordance with the Ma = 2.7 experiments of Sun et al. [30]. The bottom wall 99% boundary-layer thickness at the inflow δ_i, which is used for all simulations as the reference length, is estimated to be $\delta_i = 5.7$ mm. A full list of cases conducted in the boundary layer simulation is given in Table 3.3 and a sketch of the computational domain is shown in Fig. 3.25.

Table 3.2 Flow conditions for the simulations, including the dimensional boundary-layer (BL) thicknesses and Reynolds number at the inflow [31]

Mach number	Stagnation temperature	Stagnation pressure	BL 99% thickness	Momentum thickness	Reynolds number
M_∞	T_0	P_0	δ_i	θ	Re_θ
2.7	300 K	100 kPa	5.7 mm	0.41 mm	3,600

Table 3.3 Grid numbers and stretching control parameters for different cases [31]

Case	Domain size $L_x \times L_y \times L_z$ (mm)	Grid number $N_x \times N_y \times N_z$	Grid control y_b (mm), N_b β	Grid spacing Δx^+ Δy^+ Δz^+
Flat/Expan2/Expan4	185 × 26 × 15	2305 × 241 × 289	1.85, 88, 2.29	6.5 0.8-9.8 4.1

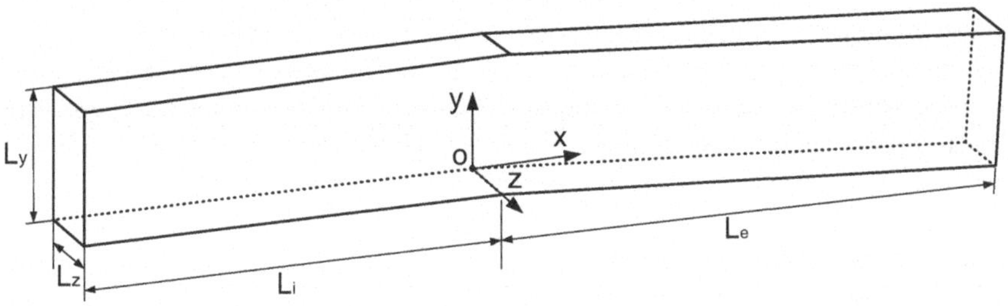

Fig. 3.25 Sketch of the domain used for the simulations [31]

Viscosity is assumed to vary with temperature according to Sutherland's law with a reference temperature set as 122.1 K.

Non-slip boundary conditions were enforced on the bottom wall (shown in Fig. 3.25). The wall temperature is fixed to the stagnation temperature of the inflow. Outflow condition with an integrated characteristic scheme is applied to the top boundary and exit boundary. Periodic boundary conditions are applied in the spanwise direction.

3.3.2 Backpressure Induced Symmetric and Asymmetric Separated Flowfield

Instantaneous flowfields of different separation modes visualized by NPLS system are shown below. All the characteristic structures are illustrated in Fig. 3.26. The intersection of two separation shocks is selected as a reference location for statistics on unsteady behaviors, which will be further discussed in Sect. 3.4.2.2. The dark zone in flowfield is the shadow of the cylinder, which represents both the location and diameter of the back pressure cylinder. Transition zones are depicted in NPLS images to demonstrate the impact of boundary layer transformation on separation

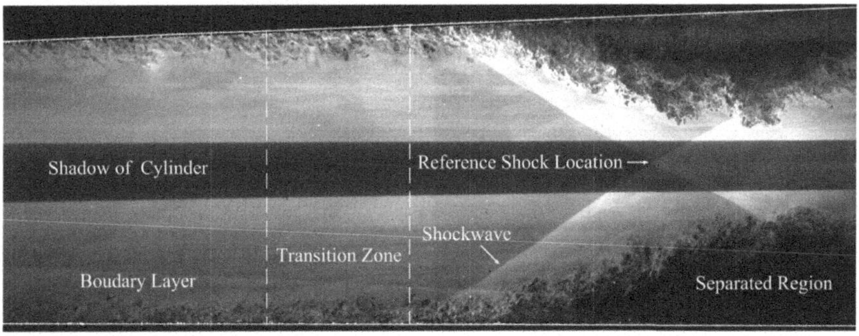

Fig. 3.26 NPLS image of symmetric separation mode, d = 10 mm, symmetric back pressure [16]

patterns, which will be elaborated in Sect. 3.4.2.2. The location of the pressure sensor for dynamic analysis is near the separation shock foot in each case. Numerical results are demonstrated in each case serving as validations. Cold flowfields fully represent typical separation modes of combustion cases [11], which are illustrated as follows.

3.3.2.1 Symmetric Separation

A symmetric separation mode forms when back pressure is relatively low ($P/P_{in} = 4.18$). A cylinder with a diameter of 10 mm is located at the center of flowfield. Back pressure generated by the cylinder forms a bow shock, which leads the boundary layer to separate from both sides of walls. When the separated region appears, the local static pressure rises which results in the forward motion of shockwave. Separated shock becomes gradually weaker as it moves upstream, and finally forms a symmetric separation. The flowfield is quasi-steady in this separation mode.

Computed velocity magnitude contour for symmetric separation mode is shown in Fig. 3.27, and the averaged experimental velocity magnitude contour obtained by PIV is depicted as well. The view section in experiments is demonstrated by the dash-dot-dot line in the computational flow field. Due to the dark zone (a shadow of the cylinder) as shown in the NPLS image, an unreal low-speed region exists in the center of the flow field. Artificial low-speed regions also appear on the edge of boundary layers of both walls, which are caused by optical imaging and the reflected light from side windows. Figure 3.28 illustrates the numerical average streamwise velocity at two characteristic axial locations (nondimensionalized by the height of the test section inlet), which are compared to experimental data. The data of the artificial center low-speed region is omitted. It is observed that both the oblique shocks and separated regions maintain symmetric under relatively low back pressure. However,

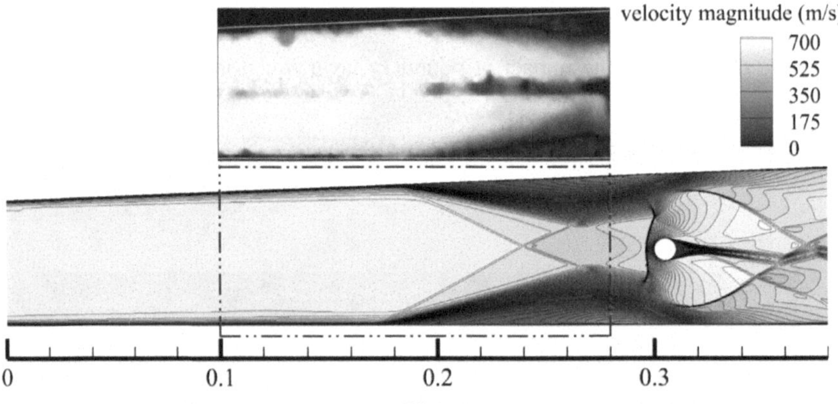

Fig. 3.27 Average velocity magnitude contour of symmetric separation mode [16] (top: experimental data, bottom: calculated result)

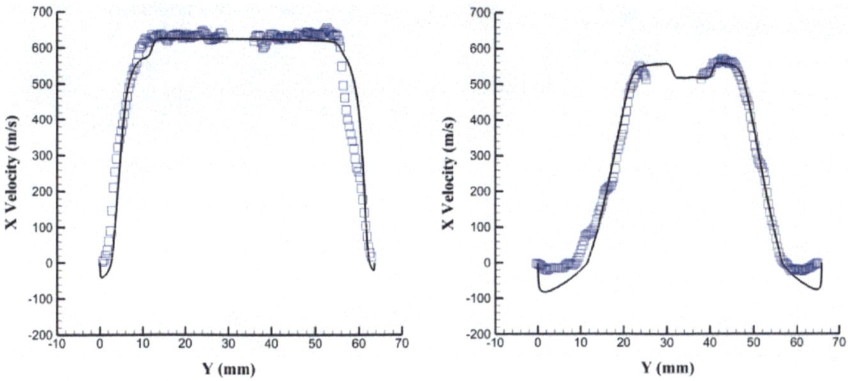

a) X=3.5H$_1$ b) X=4.5H$_1$

Fig. 3.28 Average streamwise velocity of symmetric separation mode [16] (line: calculated result, symbol: experimental data)

velocity profiles show a slight asymmetry downstream (X = 4.5 H_1), which means the separated region on the straight side is a bit larger.

3.3.2.2 Asymmetric Separation

An asymmetric separation mode appears when back pressure is high ($P/P_{in} = 4.69$). The cylinder still locates at the center of flowfield as shown in Fig. 3.29. Symmetric back pressure itself forms an asymmetric flowfield, and a large separated region dominates at the expansion side. The large separated region forces the shock foot at the expansion side to move far more upstream than the straight side. Thus supersonic core flow is suppressed to the straight side, which brings about flow reattachment. Meanwhile, such a separation mode contains prominent unsteady behaviors, and the size of the large separated region varies with the movement of shock foot. Typical separation patterns of two transient flowfields during the unsteady process are illustrated in Fig. 3.29. Figure 3.30 represents the dynamic characteristics of flowfield, and the variation of static pressure near shock foot can be as high as 100%. A broadband low-frequency oscillation is captured by a fast Fourier transform (FFT) analysis of pressure signal.

The asymmetric flowfield is depicted clearly by average velocity magnitude contour, which is shown in Fig. 3.31. The artificial low-speed region (in the downstream center flow) in experimental contour is caused by optical imaging from the cylinder and its upholder. Figure 3.32 illustrates the average streamwise velocity at two characteristic axial locations. Numerical flowfield also shows a large separated region which dominates at the expansion wall, and the oblique shock represents obviously asymmetric structure as well. It is observed from Fig. 3.32a that the flow reattaches at

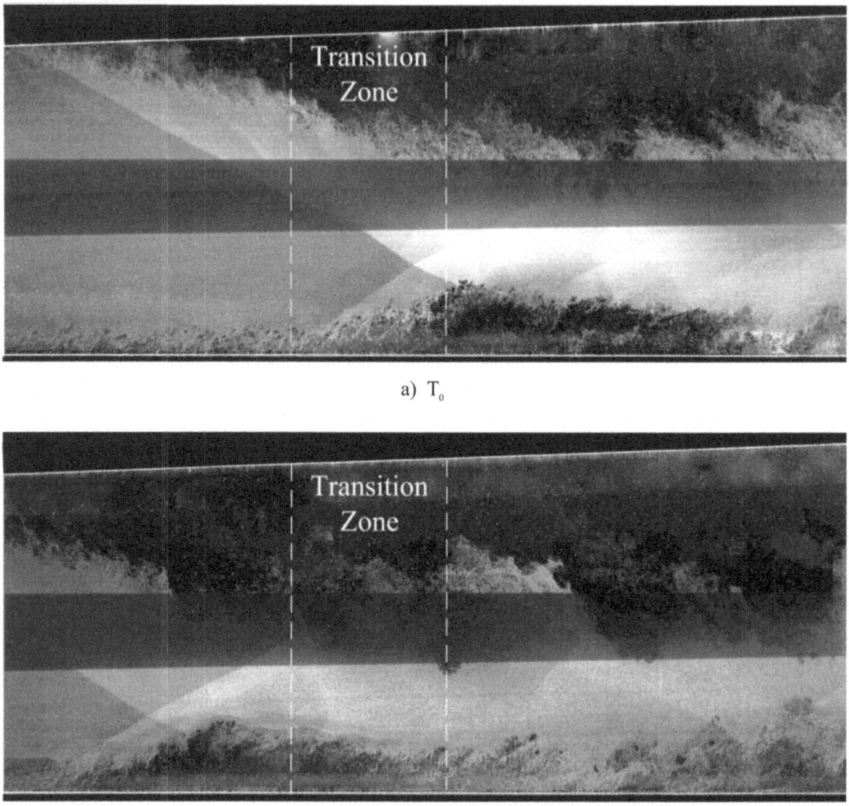

a) T_0

b) T_1

Fig. 3.29 NPLS images of asymmetric separation mode, d = 12 mm, symmetric back pressure [16]

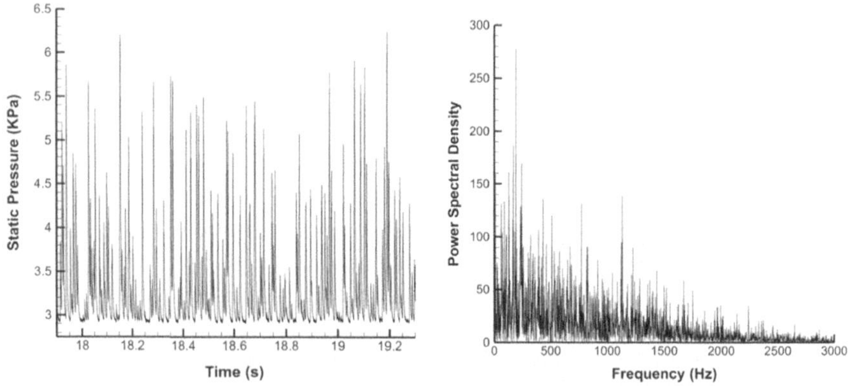

a) History of the pressure signal b) Power spectra of the pressure signal

Fig. 3.30 History and FFT analysis of pressure signal for asymmetric separation mode [16]

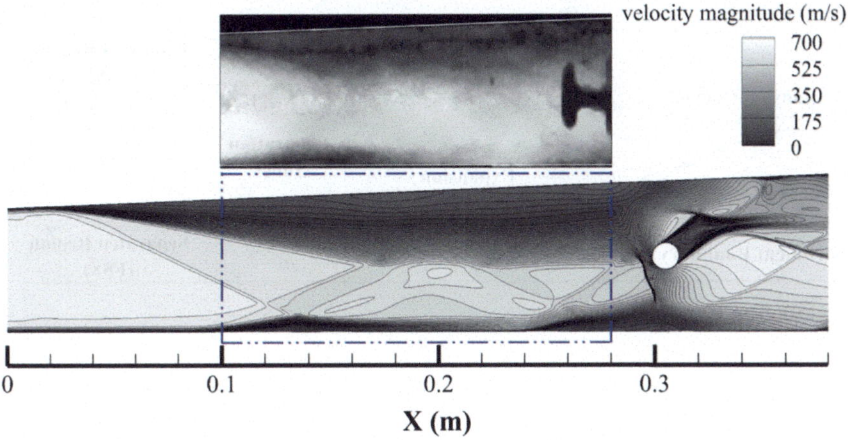

Fig. 3.31 Average velocity magnitude contour of asymmetric separation mode [16] (top: experimental data, bottom: calculated result)

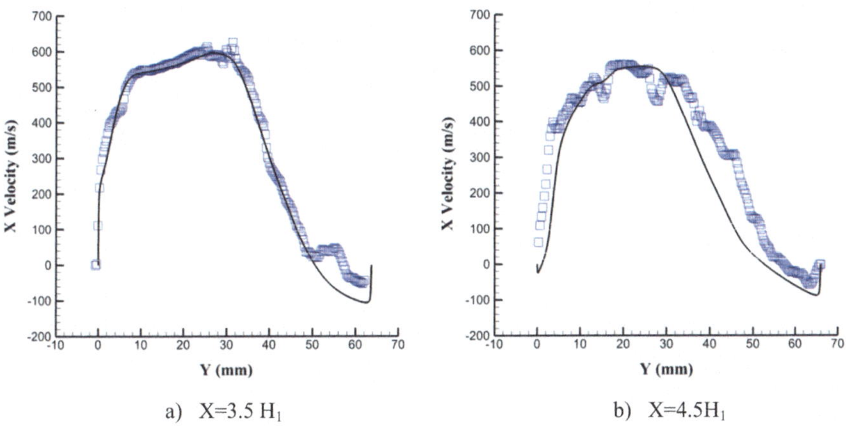

Fig. 3.32 Average streamwise velocity of asymmetric separation mode [16] (line: calculated result, symbol: experimental data)

the straight side. Two small separated regions, which are shown in Fig. 3.31, appear upstream and downstream the reattached region at the straight wall.

3.3.2.3 Flow Structures of Typical Separation Modes

Figure 3.33 depicts the main flow features of two typical separation modes. Outlines of the symmetric separation mode are given in Fig. 3.33a. This mode occurs when separated region locates downstream the expansion wave at a straight wall.

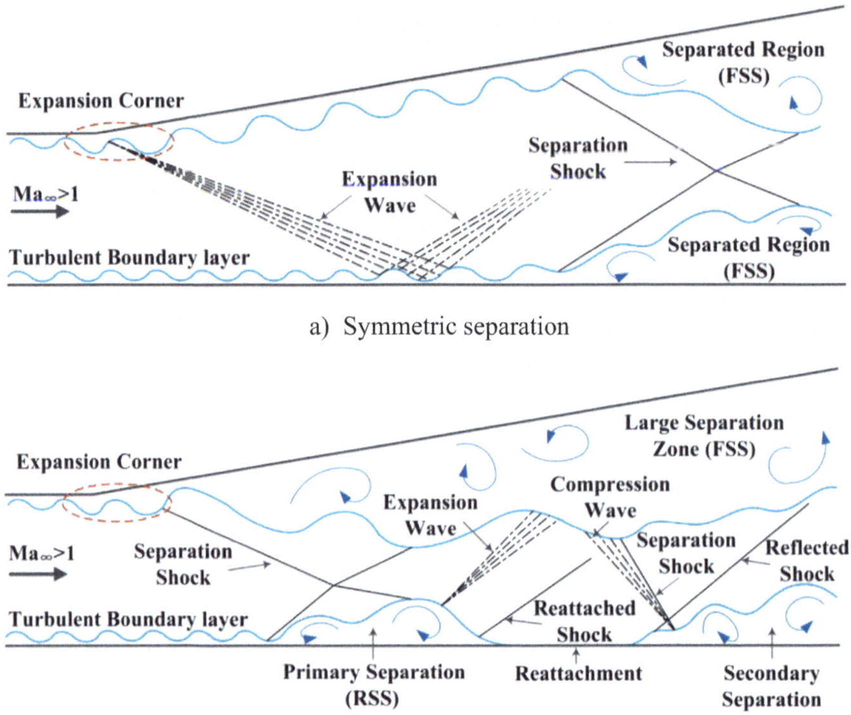

a) Symmetric separation

b) Asymmetric separation

Fig. 3.33 Schematics of two separation modes [16]

The boundary layer with low turbulent intensity after expansion wave at the flat side is illustrated by increased size and decreased number of patterns in the schematic picture. Two balanced free shock separations (FSS) are induced by symmetric separation shock. Boundary layer transformation at the expansion corner is highlighted in both images, which shows the growth in thickness and reduction in turbulence intensity. The boundary layer transformation will be discussed in detail in Sects. 3.3.3 and 3.4.2. To sum up, the flow structure of symmetric separation mode is quite simple. However, complicated wave structures exist in asymmetric separation mode. The large separation zone (FSS) dominates at the expansion wall, which suppresses the supersonic core flow to the straight side. Thus the high-speed flow forces the separation at the flat wall to reattach downstream, which induces a restricted shock separation (RSS). Due to its axial location, the RSS placed on the straight wall is called "primary separation". The large separation at expansion side is able to block sizable part of the main flow (the blockage can be as high as 50% of local height), which persistently compresses the supersonic flow. As soon as the compression wave accumulates to be strong enough and forms a shock wave, a secondary separation appears.

3.3.3 *Mechanism of Asymmetric Separation Based on Boundary Layer Study*

The formation of asymmetric separation is induced by different boundary layer behaviors on expansion and straight walls. The pivotal role which the expansion corner plays in boundary layer transformation is carefully studied in the current section with DNS [31]. As a supersonic flow passes over an expansion corner, it is accelerated through an expansion fan and the boundary layer becomes thicker due to the decrease in density. After the expansion, the boundary layer experiences a relaxation process, during which it exhibits different turbulence features from the incoming equilibrium state. The suppression of turbulence during the expansion process is a significant characteristic of supersonic flow passing an isolated expansion corner, as reviewed by Knight et al. [32].

3.3.3.1 Transformation of Turbulent Boundary Layer Structures Induced by Expansion Corner

The iso-surface of $u_t = 0.55$ given in Fig. 3.34 shows the large scale structures in the outer boundary layer. It is found that the streaks in the inner layer are not affected significantly by the expansion, while the vortex structures in the outer layer respond severely to the expansion. The near-wall quasi-streamwise vortices at the flat side and the expansion side are both preserved, which indicates that the inner turbulence exists in a state of local equilibrium. Meanwhile, in the outer layer of the boundary layer, the coherent structures on the ramp are affected by the upstream flow history, as well as the inner turbulence generation and its exchange with the outer layer. The turbulence on the downstream ramp is especially affected by the history effect of the turbulence decay in the vicinity downstream of the expansion corner. This determines the formation of the two-layer structure in the boundary layer.

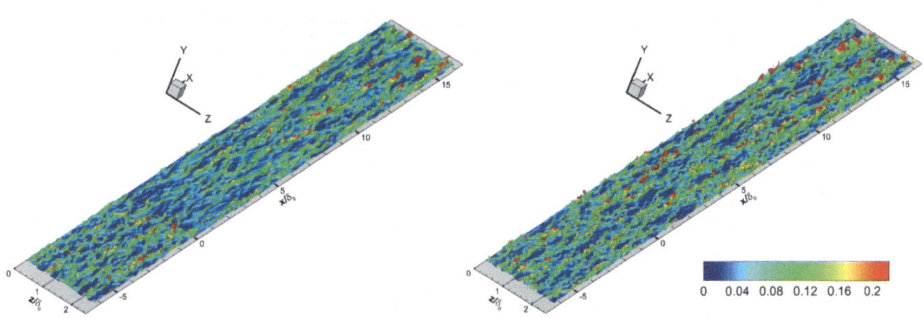

Fig. 3.34 Turbulence coherent structures visualized with the iso-surface of $u_t = 0.55$ and coloured by the wall-normal distance y/δ_0 for the Expan4 case (left) and the flat plate (right) [31]

3.3.3.2 Expansion Effects on Turbulent Perturbations

Figure 3.35 plots the turbulent kinetic energy (TKE) at different wall-normal locations along the streamwise direction. The TKE in the inner layer ($y/\delta_0 = 0.0266$) demonstrates a quick recovery than the outer layer and TKE recovery gets slower when the wall-normal location is larger. Certainly, this is a coarse estimation since the TKE profile is related to y^+ instead of y/δ_0 and y^+ changes along expansion ramp when same y/δ_0 is set with flat plate case.

TKE profiles at different streamwise locations of all cases versus y^+ are shown in Fig. 3.36. It is deduced that due to weakened turbulence in the outer layer, the turbulence exchange of inner layer with the mainstream is greatly suppressed, which results in a turbulent energy accumulation in the inner layer since turbulence in the inner layer exists in a local equilibrium and its production is still ongoing. The turbulence energy in the inner layer cannot transfer through the outer layer since the suppressed exchange in the outer layer blocks the transfer. In summary, the two-layer structure is due to the decay of the turbulence shear stress in the log-law region of the boundary layer downstream of expansion corner, which blocks the exchange between the turbulent structures in the inner layer and the mainstream and finally leads to a quick recovery in inner layer and a slow recovery in outer layer.

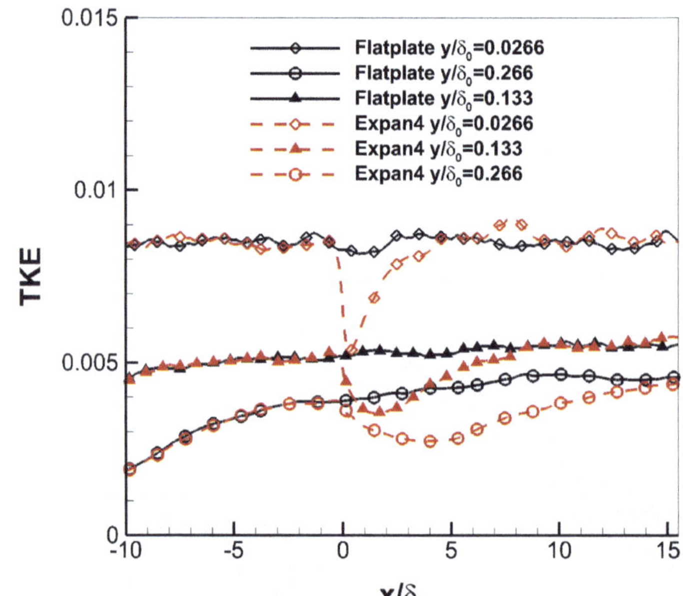

Fig. 3.35 TKE versus x/δ_0 at different y/δ_0 locations of flat plate and Expan4 case [31]

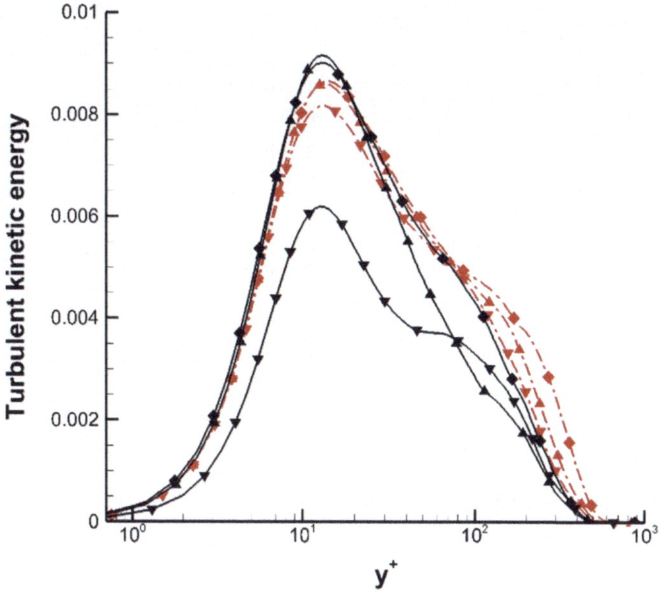

Fig. 3.36 TKE profiles of the flat plate (dash-dotted line) and Expan4 (solid line) at different streamwise locations, including $x/\delta_0 = 0.81$ (symbol ▼), $x/\delta_0 = 5.65$ (symbol ▲), $x/\delta_0 = 15.35$ (symbol ◆) [31]

3.3.3.3 Turbulent Boundary Layer Separation Tendency Affected by Expansion Corner

Previous studies [32, 33] concluded that a fuller streamwise velocity profile increases the ability of a boundary layer to resist flow separation. For a supersonic expansion flow, although the increase in the velocity leads to a fuller velocity profile in the outer part of the boundary layer, the velocity in the inner part of the boundary layer for the expansion cases decreases, which results in a thicker boundary layer compared with the flat plate. It is difficult to quantify the fullness of the boundary layer after equilibrium regeneration on the expansion ramp. The resistance of a boundary layer to an adverse pressure gradient is often evaluated by the shape factor. Usually, an increase in the shape factor of a supersonic boundary layer indicates that flow separation is more likely to occur on the wall.

The shape factor is defined as $H = \delta^*/\theta$, where δ^* represents the displacement thickness and θ the momentum thickness. They are defined as:

$$\delta^* = \int_0^\infty \left(1 - \frac{\rho}{\rho_e}\frac{u}{U_e}\right)dy \tag{3.1}$$

$$\theta = \int_0^\infty \frac{\rho}{\rho_e}\frac{u}{U_e}\left(1 - \frac{u}{U_e}\right)dy \tag{3.2}$$

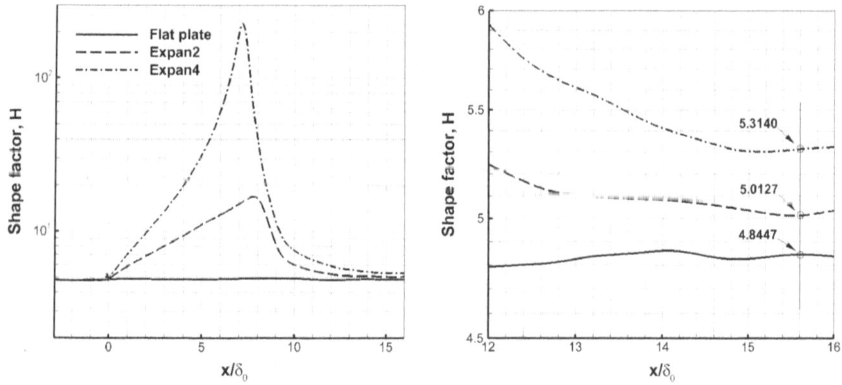

a) shape factor

b) closer view of the shape factor near the outlet

Fig. 3.37 Shape factor as a function of the streamwise location of different expansion cases [31]

Figure 3.37 shows the shape factor as a function of the streamwise location for all three cases. It is clearly seen that the shape factor increases rapidly after the expansion corner, followed by a sudden decrease in the recovery region and a slow increase in the equilibrium section. Figure 3.37b gives a closer view of the shape factor in the equilibrium region. The values of shape factor indicate that the fullness level of velocity profile in the flat plate case is still higher than the 2° and 4° expansion cases. The shape factor increases as the deflection angle are increased. A higher value of shape factor means that the expansion supersonic flow has a lower resistance ability to adverse pressure gradient or flow separations. This explains why large separation always occurred on the expanded wall when the back pressure is large enough.

3.4 Cold Flow Analysis: Separation Transition and Low-Frequency Unsteadiness

Dynamic features of the back pressure induced separation are further studied [23] in the current section. Detached-eddy simulation represents the whole process of the separated region development. Typical simulated transient flow phases are validated by nano-based planar laser scattering images. High-frequency pressure signals are acquired to study unsteady behaviors of the separation shock. The low-frequency unsteadiness in separation is preliminarily discussed, and some controlling measures are put forward.

3.4.1 Symmetric/Asymmetric Separation Transition Under Threshold Backpressure

The elaborate description of the symmetric/asymmetric separation transition process is illustrated by the numerical density gradient magnitude contours below. Due to the limitation of NPLS system, the interval between two images is 500 ms. Thus, the entire transitional flow development cannot be captured in a sequence of experimental images. Only several NPLS images for typical phases are selected to obtain a qualitative validation to numerical results.

The formation of separated regions is named the 'initial separation phase', which is only captured by numerical simulation (due to the limited temporal resolution of experimental apparatus). Moreover, the other four typical phases are described both computationally and experimentally. These five phases together constitute the entire process of separation development. The 'standard symmetric separation phase' appears when the separation zone stays in the 'symmetric separation region' (shown in Fig. 3.45). The separation pattern always sustains symmetric in this phase. However, when the separated regions locate within the 'separation transition zone' (shown in Figs. 3.38, 3.39, 3.40, 3.41, 3.42, 3.43 and 3.45), two completely different separation patterns coexist. As a consequence, two separation phases ('symmetric/asymmetric phase in separation transition') are put forward for distinction. At last, the 'standard asymmetric separation phase' occurs when the separation devel-

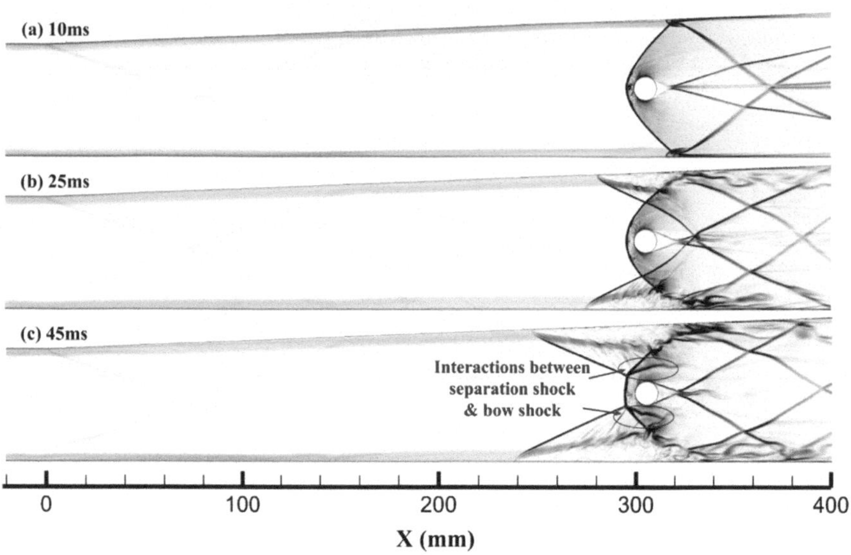

Fig. 3.38 Numerical schlieren of the initial separation phase: the bow shock-induced separation [23]

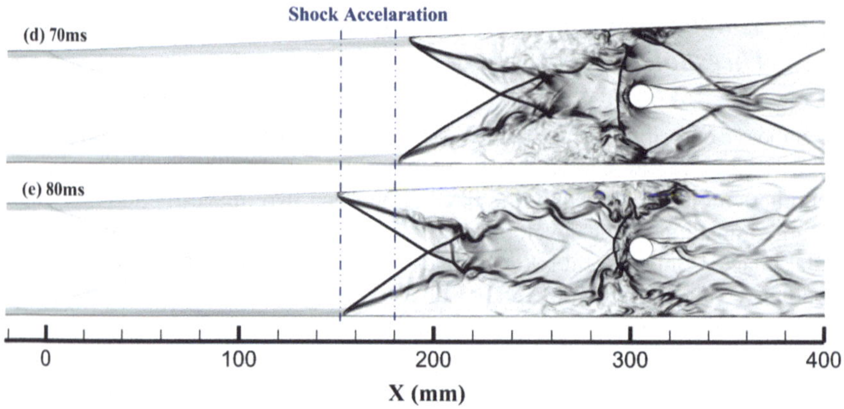

Fig. 3.39 Numerical schlieren of the initial separation phase: the relative acceleration of separation shock on the expansion side [23]

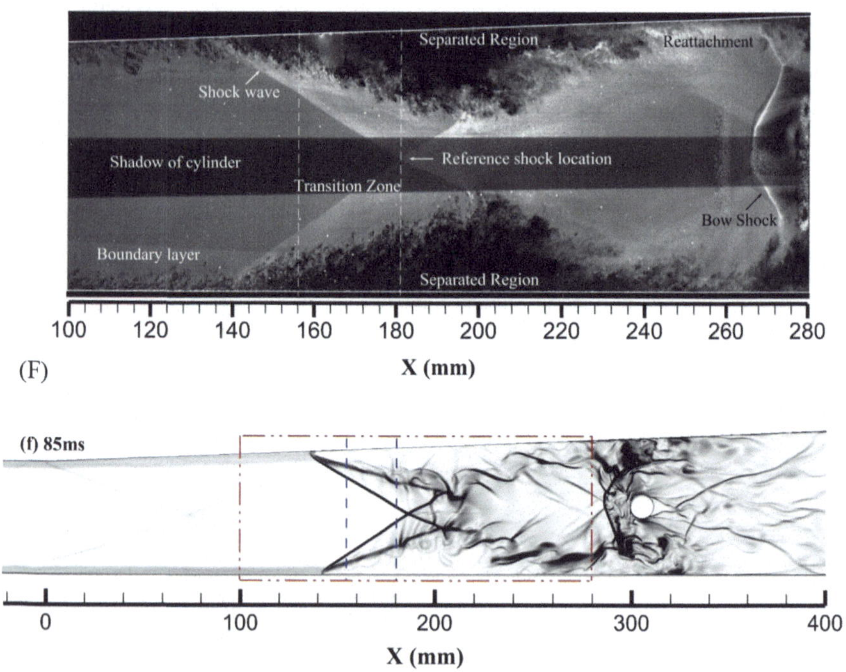

Fig. 3.40 Typical flow structures for standard symmetric separation phase: (F) the experimental NPLS image; (f) the numerical schlieren [23]

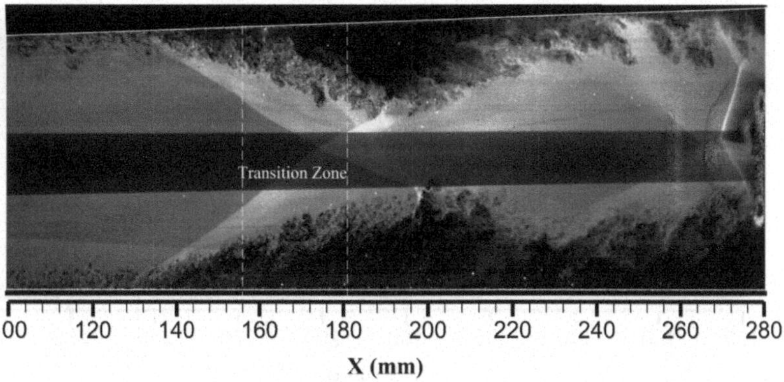

(G$_1$)

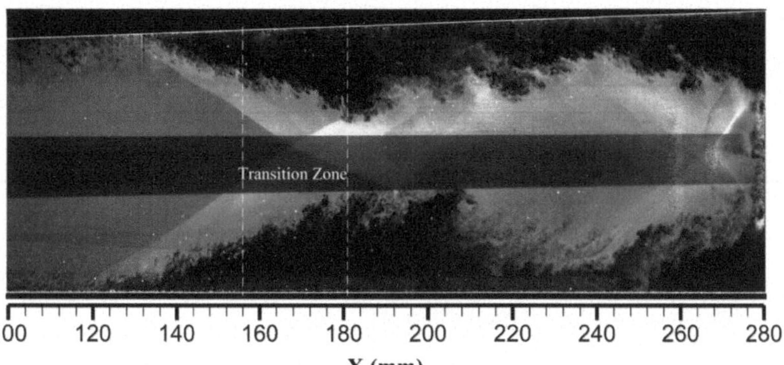

(G$_2$)

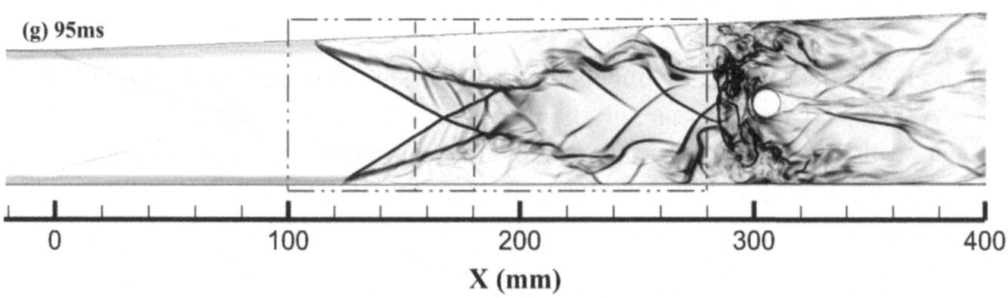

Fig. 3.41 Typical flow structures for a symmetric phase in separation transition: (G$_1$) and (G$_2$) the two experimental NPLS images to show the fluctuation of shock train in the vertical direction; (g) the numerical schlieren [23]

opment reaches the 'asymmetric separation region' (shown in Fig. 3.45). It is a phase where a large asymmetric separated region appears on the expansion wall all the time. The division of 'symmetric separation region', 'separation transition zone' and 'asymmetric separation region' will be introduced in Sect. 3.4.2.2 (Fig. 3.44).

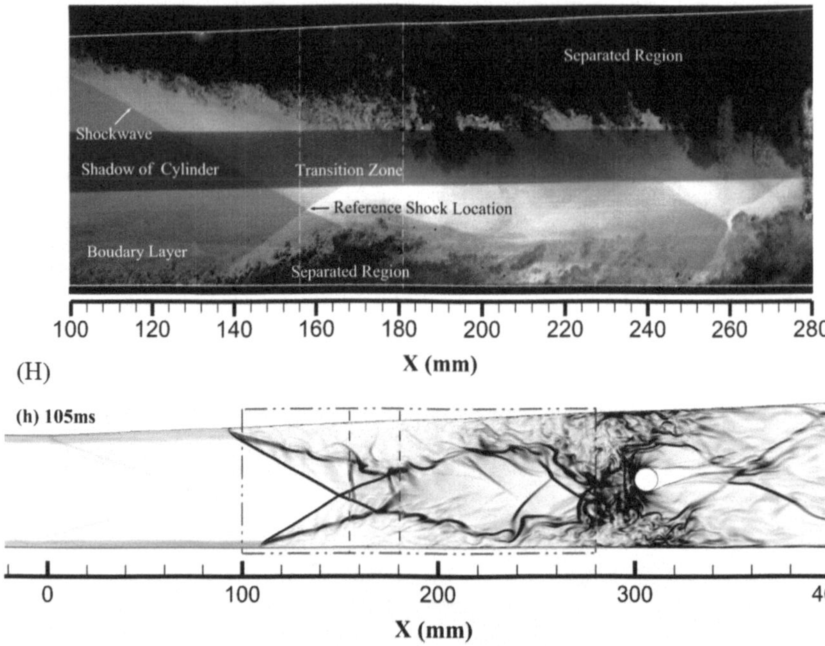

Fig. 3.42 Typical flow structures for asymmetric phase in separation transition: (H) the experimental NPLS image; (h) the numerical schlieren [23]

3.4.1.1 Initial Separation Phase

First of all, the oncoming flow meets the backpressure cylinder, which forms a bow shock. A strong adverse pressure gradient is induced by the bow shock. Consequently, the near wall flows (boundary layer) separate from both sides of walls due to the adverse pressure gradient. It is seen in the contour of 10 ms (Fig. 3.38a) that the turbulent boundary layer has not been well developed yet. For the flowfields later than 25 ms, the turbulent boundary layers upstream of the separation shock sustain nearly the same. This suggests that the transient behaviors of boundary layers disappear. The primary separated regions are the traditional triangular shape. When the separated regions appear, the local static pressure rises which results in the forward motion of shockwave. During the forward motion, the separation shocks from both walls intersect with the bow shock at the center of flowfield. At that time, shear layers appear at the intersections of two separation shocks and the bow shock. These interactions are shown in the contour of 45 ms (Fig. 3.38c). When the shock intersections move upstream, the interactions between two separation shocks and the bow shock are weakened, and the shear layer structures disappear gradually.

An important variation of shock speed happens when the separation shock feet are near the boundary layer transformation region. The boundary layer transformation and its effect on separation development will be carefully discussed in Sect. 3.4.2.

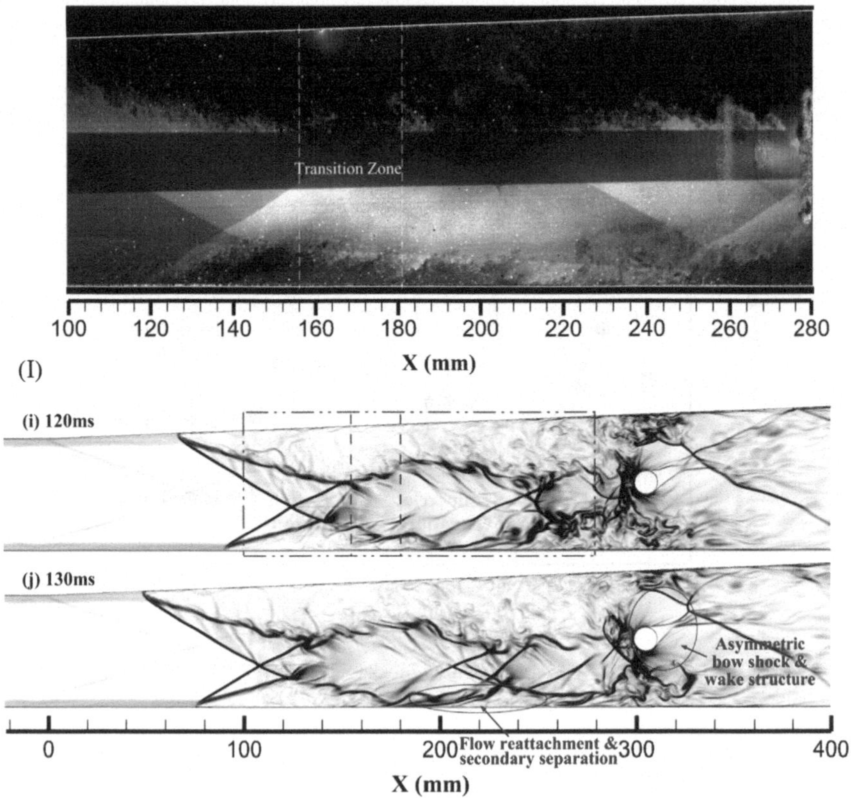

Fig. 3.43 Typical flow structures for standard asymmetric separation phase: (I) the experimental NPLS image; (i) and (j) the numerical schlierens for two well-developed asymmetric separation fields [23]

Upstream of the transformation region, the separation shock of the expansion wall is a bit more than the straight side (shown in the contour of 70 ms, Fig. 3.39d). However, the forward movement of shock foot on the expansion wall accelerates as it passes through the boundary layer transformation region, and surpasses the shock foot on the straight wall when it travels upstream of the transformation region (shown in the contour of 80 ms, Fig. 3.39e). The acceleration of the shock movement on the expansion wall will lead to the final formation of asymmetric separation.

Due to the limitation of the experimental system, the initial separation phase is only depicted by numerical simulation. However, this phase poses a great impact on separation development and configuration transformation. Specifically speaking, the relative shock acceleration is a precursor of the separation transition phenomenon. A detailed quantitative study of the relative axial shock acceleration will be carried out in Sect. 3.4.2.1.

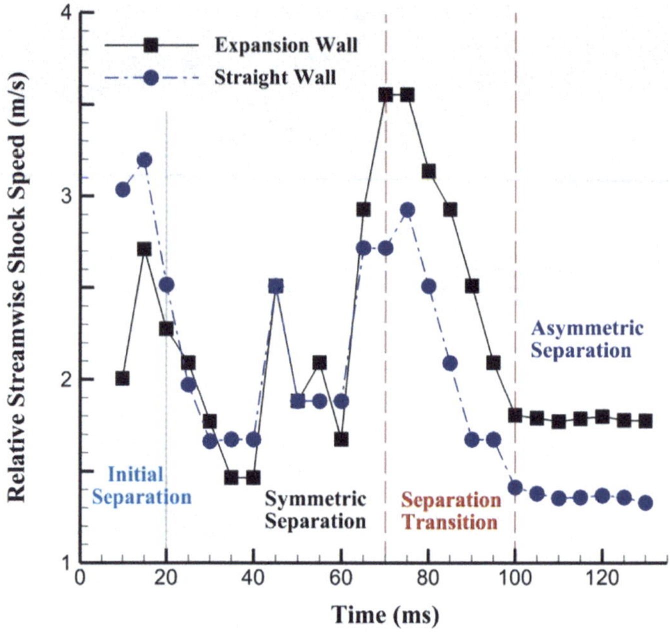

Fig. 3.44 Relative streamwise speed of separation shock foots from two walls during the numerical transition process [23]

3.4.1.2 Standard Symmetric Separation Phase

After the initial separation phase, four characteristic phases in the separation transition process are illustrated with both experimental NPLS images and numerical schlieren images. In NPLS images, the dark zone in the flowfield is the shadow of the cylinder, which represents both the location and diameter of the backpressure cylinder. The artificial vertical lines near the cylinder are caused by optical imaging from the cylinder and its upholder. The experimental view sections are depicted in all the computational flow fields. The separation 'transition zone' shown both in numerical and experimental flow fields is a special region where symmetric and asymmetric separation modes can coexist.

A typical symmetric separation mode always sustains when the intersection of separation shocks locates downstream of the 'transition zone'. The separation pattern behaves completely symmetric, and a bow shock forms near the backpressure cylinder. The two triangular separated regions with reattachment are the typical flow structures of restricted shock separation (RSS), which is opposite to free shock separation (FSS) of the large separated region in asymmetric separation (shown in Figs. 3.42 and 3.43). Downstream of the two symmetric separated regions, the flow is re-accelerated due to the decrease in blockage. Thus, the flow reattachment happens on both sides of the walls. Due to the limitation of our DES approach, the simulated separation development is slightly faster compared with the experimental results.

Consequently, the reattachment phenomenon is not well captured by simulation, although the location of the separation shock intersection is very close to the NPLS image.

3.4.1.3 Symmetric Phase in Separation Transition

During the experiment, the symmetric/asymmetric separation transition occurs when the backpressure is near the threshold (backpressure ratio $P/P_{in} = 4.57$). The diameter of the cylinder (11.8 mm) is slightly smaller than the typical asymmetric separation mode (12 mm).

When the separation shock intersection locates within the 'transition zone', the transformation of separated regions' configuration occurs, and two completely different phases appear. In the symmetric phase, the bow shock near the backpressure cylinder turns sharp, and it becomes weaker than the typical symmetric mode. The shock train is able to fluctuate in the vertical direction in this phase (shown in two NPLS images G_1 and G_2), which means the flow unsteadiness increases.

3.4.1.4 Asymmetric Phase in Separation Transition

In the experimental asymmetric phase of separation transition (Fig. 3.42H), the flow structure behaves exactly like the typical asymmetric separation mode. On the other hand, even though the tendency of asymmetric transformation is clear, the asymmetry of separation shocks in the numerical result is not as obvious as in the experimental image. A probable reason for the discrepancy between computational and experimental results is that the NPLS image may depict the flow structure in reverse transition (from asymmetric to symmetric) process. In the experiment, the symmetric/asymmetric separation transition process is bidirectional. However, the reverse transition phenomenon is not captured by numerical simulation, due to the limitation of present DES code.

3.4.1.5 Standard Asymmetric Separation Phase

When the separation shock intersection travels further upstream, the flow separation turns into the typical asymmetric mode, and a large separated region always dominates on the expansion wall. The large separated region forces the shock foot on the expansion side to move far more upstream than the straight side. Thus, the supersonic core flow is suppressed to the straight side, which leads to the reattachment of flow. Meanwhile, the flow asymmetry can also be illustrated by the configuration of the bow shock. The symmetric bow shock (in Fig. 3.40) turns into asymmetric (in Fig. 3.43) as the asymmetric separation forms.

3.4.2 *Mechanism of Separation Pattern Transition*

3.4.2.1 Variation of Separation Shock Speed

The relative streamwise speed of separation shock feet from both walls is calculated to show the forward movement of separation shocks in a quantitative way. The 'relative streamwise shock speed' is the speed of separation shock relative to the free stream velocity in the streamwise direction. Specifically speaking, it is a velocity of the upstream development of the separation shock. The relative streamwise shock speed can be treated as a precursor of the separation configuration transformation.

As is illustrated in Fig. 3.44, the shock on the straight side moves a little faster when the initial separation appears. From then on, the forward motions of two separation shocks are nearly synchronized between 20 and 60 ms. Undoubtedly, the separation patterns will remain symmetric in this period of time. After that, sharp rises occur in the relative streamwise velocity of separation shocks from both walls, and the forward movement of separation shock foot on the expansion wall is always faster than on the other side. The symmetric/asymmetric separation transition happens exactly in this period of time (70–100 ms). When the asymmetric separation forms, forward movements of two separation shocks are both slowed down, and the relative streamwise shock speed of the expansion side is still faster. As a consequence, the separation pattern stays asymmetric after the transition process. Actually, such a shock speed variation is induced by boundary layer profile transformation, which will be further discussed in Sect. 3.4.2.2.

3.4.2.2 Separation Transition Induced by Boundary Layer Transformation

Figure 3.45 is drawn based on the quantitative analysis of boundary layer transformation, and some experimental results are gained from Gao et al. [16]. Moreover, the data of the separation transition experiment (circle symbols) have been added. The streamwise boundary layer thickness, shape factor and turbulent kinetic energy are all demonstrated in Fig. 3.45, which is able to represent the boundary layer transformation in an all-around way. Figure 3.45 also shows the experimental relationship between the shape and streamwise location of separated regions, which serves as a validation to the analysis of boundary layer separation tendency. The reference shock location (intersection of two separation shocks) mentioned in Sect. 3.3.1.1 is obtained to make statistics of different separation modes in each case of experiments. The origin locates at the intersection of the flowfield centerline and the downstream boundary of the experimental view section, which is depicted in Fig. 3.24a. The average height of test section H is chosen to non-dimensionalize the results. The X-axis illustrates axial locations of separated regions, and the absolute value along the

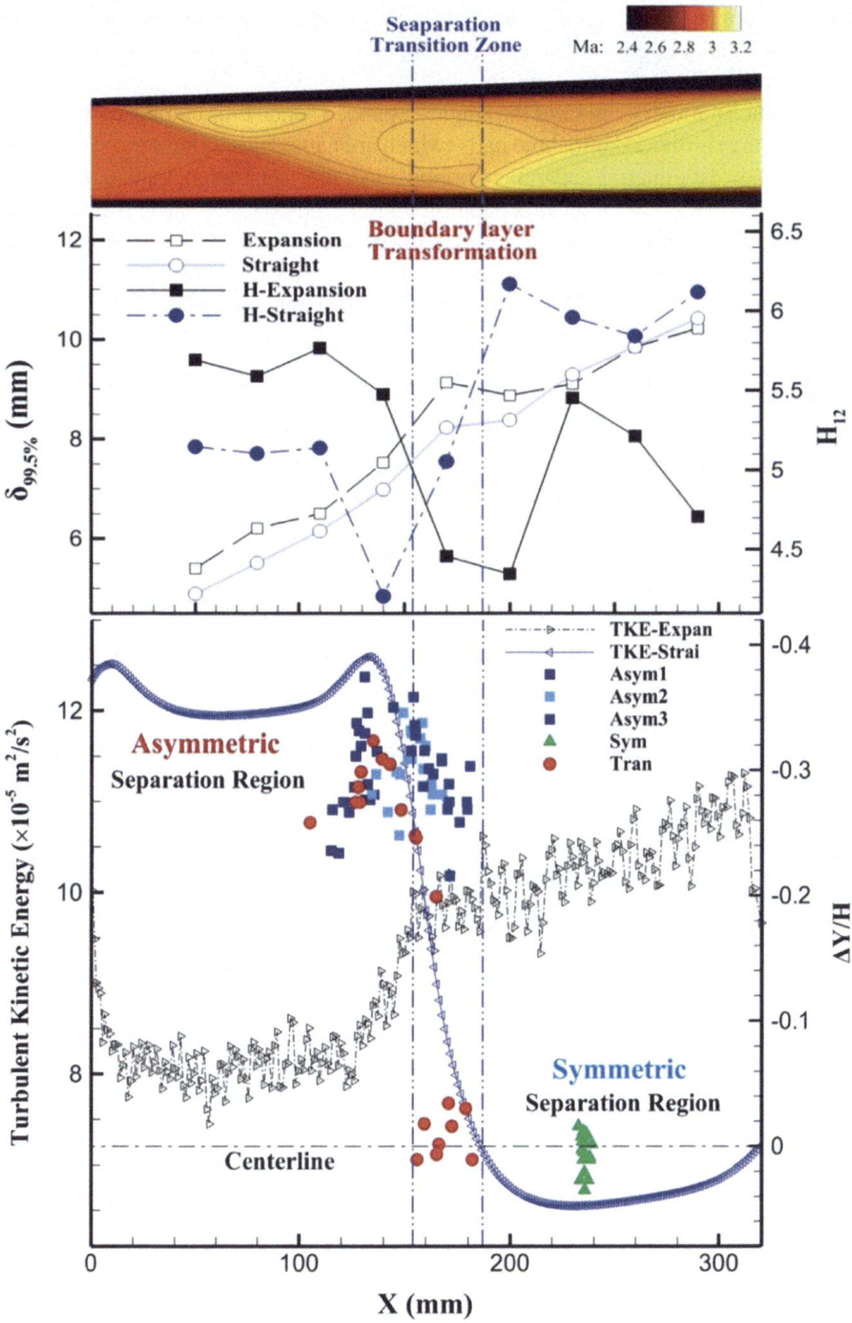

Fig. 3.45 The impact of boundary layer transformation on separation transition [23]

ΔY-axis indicates the degree of flowfield asymmetry. A negative value of the ΔY-axis means that a larger separated region appears on the expansion wall, whereas a positive value stands for the opposite.

The Mach number contour of the single-expanded supersonic combustor (without backpressure) is divided into three parts, which are named the 'symmetric separation region', the 'asymmetric separation region' and the 'separation transition zone', respectively. In the 'asymmetric separation region', the flow near the expansion wall is of higher Mach number. On the contrary, the flow near the straight wall has a higher Mach number in the 'symmetric separation region'. An expansion wave (generated by the expansion corner) reaches and reflects on the expansion wall, which causes a boundary layer transformation. Such a transformation region is called the 'separation transition zone'.

The turbulent kinetic energy of the expansion wall undergoes a sharp decrease around the expansion corner, so it is lower than the straight side in the front part of the combustor. When the expansion wave generated from upstream reaches the straight wall, a sudden fall of turbulent kinetic energy happens on the flat side. The reflected expansion wave on the straight wall enables the flow to be further accelerated. Specifically speaking, the expansion of flow on the straight side is twice over the flow on the expansion side. Meanwhile, the turbulent kinetic energy of the expansion side recovers slightly because of the weak compression effects. This compression is caused by the over expansion on the straight side (the local flow expansion of the straight side exceeds the total expansion of geometry). The compression induced turbulence production is a typical phenomenon in turbulent boundary layers, which has also been observed in the supersonic case [34]. Thus, an interlaced turbulent kinetic energy distribution happens, which is the very location of 'separation transition zone'.

It is clearly observed from Fig. 3.45 that the expansion wave poses a great impact on the boundary layer transformation, which further results in the transition of two different separation modes. The shape factor of the boundary layer on the expansion wall is always larger than the straight side in the 'asymmetric separation region'. Meanwhile, the boundary layer thickness is also larger on the expansion side. Thus, the boundary layer is easier to separate on the expansion wall in this 'asymmetric separation region'. Consequently, when the backpressure is high enough, an asymmetric large separated region always appears on the expansion wall.

The incident and reflection expansion waves on the straight wall induce a sharp rise of the shape factor on this side, which means the fullness of velocity/momentum profile is greatly weakened. At the same time, the shape factor of the expansion wall falls gradually, and an interlaced shape factor distribution appears in the 'separation transition zone'. This interlaced distribution means that the separation tendency of turbulent boundary layers from both walls exchanges in this region. Thus, it is the interlaced shape factor distribution that leads to the bidirectional transition of separation patterns. The boundary layer on the expansion wall is still thicker than the straight side in the 'separation transition zone'. Specifically speaking, the difference of boundary layer thickness between the two walls reaches a maximum in the 'separation transition zone', which causes the relative streamwise acceleration of separation shock on the expansion wall.

When it comes to the 'symmetric separation region', the boundary layer thicknesses from both walls are almost the same. These nearly equal boundary layer thicknesses enable the separated region to stay symmetric. Meanwhile, the larger shape factor on the straight wall makes the separation shock on this side located slightly upstream than the expansion side. The shape factor of the straight wall slightly changed in the 'symmetric separation region', whereas an apparent decrease of shape factor occurs on the expansion side.

The quantitative analysis of boundary layer separation tendency uncovers the key impact factor of the symmetric/asymmetric separation transition phenomenon: the fullness of the velocity/momentum profile in the boundary layer.

3.4.2.3 Schematic of Separation Transition

Figure 3.46 depicts a schematic of the separation transition mechanism. When the supersonic flow passes through an expansion corner, the boundary layer of the expansion wall becomes thicker [35, 36] due to the decrease of density. Meanwhile, the favorable pressure gradient brings about the reduction of turbulent kinetic energy (TKE) and Reynolds shear stresses [37, 38] in the boundary layer, which blocks the momentum transportation from the mainstream to the near-wall region. A recent DNS study [31] finds that the streamwise velocity actually reduces in the near-wall region, even though the velocity of the main flow increases compared to the flat plate cases. As a result, the fullness of the velocity and momentum profile in the boundary layer is weakened, especially in the inner layer, which leads to a higher value of boundary layer shape factor distribution along the expansion wall. The thicker boundary layer and the weakened velocity/momentum profile fullness enable the effects of backpressure to accumulate faster (forms a relative separation shock acceleration) on the expansion wall, so it always forms an asymmetric separation under high backpressure.

In the 'symmetric separation region', the boundary layer thicknesses of both walls are nearly the same, and the boundary layer along the straight wall has a larger

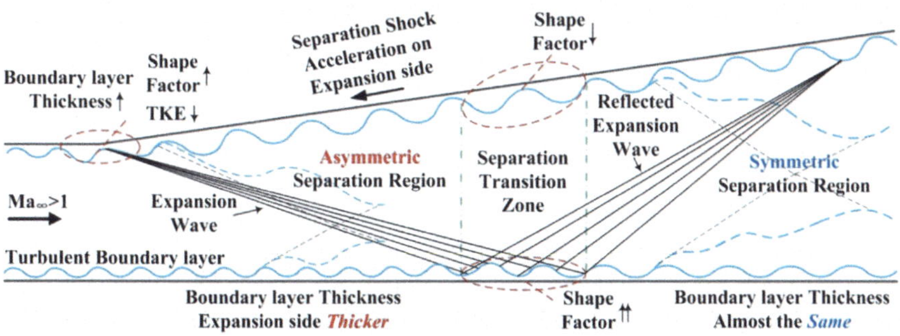

Fig. 3.46 Schematic of separation transition mechanism [23]

shape factor with lower turbulent kinetic energy. Consequently, a balanced symmetric separation appears, and the separation shock foot on the straight wall locates slightly upstream than the expansion side. This accounts for the symmetric separation mode under low backpressure.

The expansion wave arrives at the straight wall and reflects, which results in a sharp rise of boundary layer shape factor on the straight side. Thus, an interlaced boundary layer shape factor distribution (separation transition zone) between the expansion and straight walls is formed. The ‘separation transition zone’ divides the quasi-steady symmetric separation mode and the unsteady asymmetric separation mode. It is a region where the separation tendency of turbulent boundary layers from two walls exchanges. As a result, a bidirectional separation transition phenomenon happens when the separation locates within this region (under the threshold backpressure).

3.4.3 Low-Frequency Unsteadiness in the Separated Flowfield

Low-frequency unsteadiness appears in several operating conditions of the separated flowfield. Frequency and amplitude features vary under different separation modes, which are introduced in the current section. A preliminary study of low-frequency unsteadiness mechanism is carried out in the last Sect. 3.4.3.3.

3.4.3.1 Low-Frequency Unsteadiness of Different Separation Modes

The flow unsteadiness is only analyzed in unsteady and quasi-steady separation modes. Both the asymmetric separation mode and the transition separation mode are induced by symmetric backpressure, which has been introduced in Sects. 3.3.2.2 and 3.4.1. The quasi-steady separation mode is induced by asymmetric backpressure, which will be further discussed in Sect. 3.4.4.1.

Locations of pressure sensors for fast Fourier transform (FFT) analysis are selected within the intermittent region in unsteady cases (signal 5 in transition mode and signal 2 in typical asymmetric mode). Because the case with asymmetric backpressure is quasi-steady, the oscillation of separation shock is so small that none of the pressure sensors locates within the intermittent region. Thus, the pressure signal in the asymmetric backpressure case (signal 6) records the static pressure history in primary separated flow.

Figure 3.47 represents dynamic characteristics of three separation modes. The variation of static pressure near the shock foot can be as high as 80–100% in unsteady cases (transition mode and asymmetric mode). It is seen in Fig. 3.47a that the amplitude of shock oscillation is fluctuant, and intermittent behaviors exist in the unsteady process. The high-amplitude oscillation can be twice as low amplitude oscillation

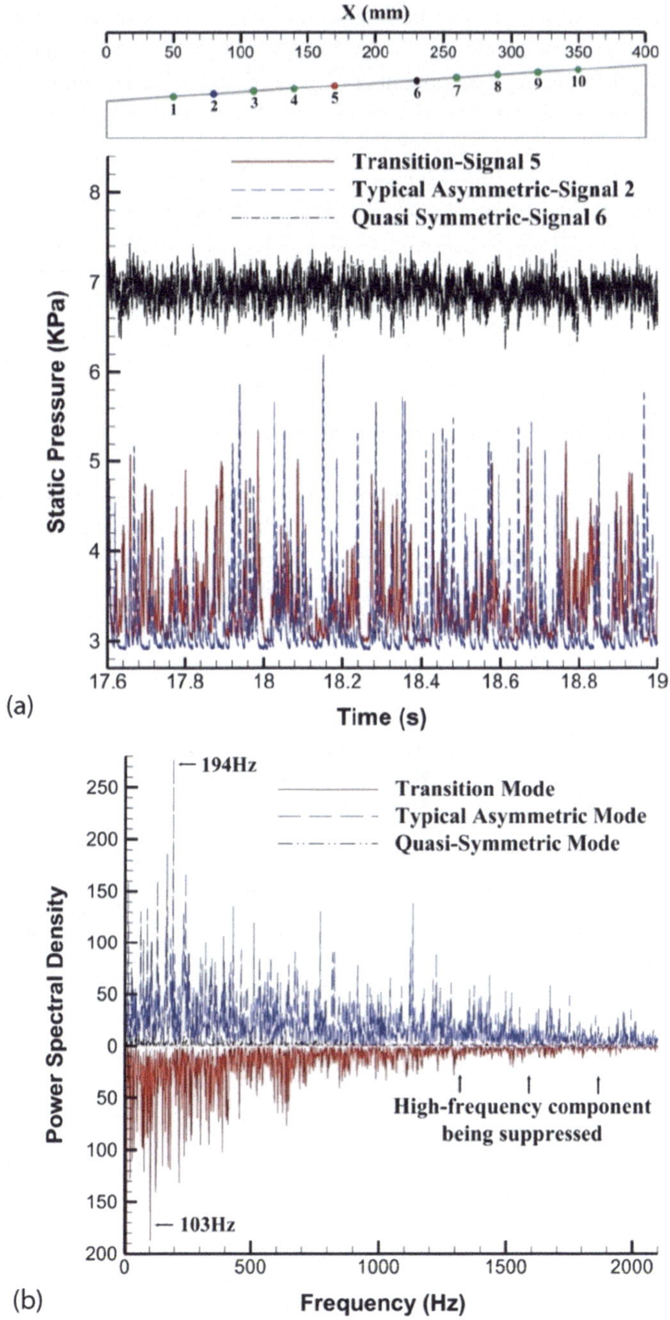

Fig. 3.47 Pressure signals for different separation modes [23]: **a** oscillation history; **b** power spectra

in static pressure variation. The FFT analysis in Fig. 3.47b depicts that a broad-band low-frequency oscillation occurs in unsteady cases. In transitional separation mode, the main frequency components range from 50 to 650 Hz, and the peak frequency of oscillation is 103 Hz. Compared with the typical asymmetric separation mode (194 Hz), this peak frequency is obviously lower. Moreover, high-frequency components (higher than 1000 Hz) are suppressed. This phenomenon is consistent with the theory in low-frequency unsteadiness of shockwave boundary-layer interaction, which states that a larger-scale motion of the separation shock foot leads to a lower oscillation frequency [39]. There are no apparent unsteady behaviors in quasi-symmetric separation mode (with asymmetric backpressure), which is also shown in Fig. 3.47b.

3.4.3.2 Amplitude of Separation Shock Oscillation

There are many methods of shock train leading-edge detection, such as the pressure ratio method [40], the pressure increase method [40], and the recently reported differential pressure method [41]. Here we choose the standard deviation method [41] to capture amplitude features of different separation modes. Normalized standard deviations of the static pressure for all pressure sensors are calculated, which is illustrated in Fig. 3.48.

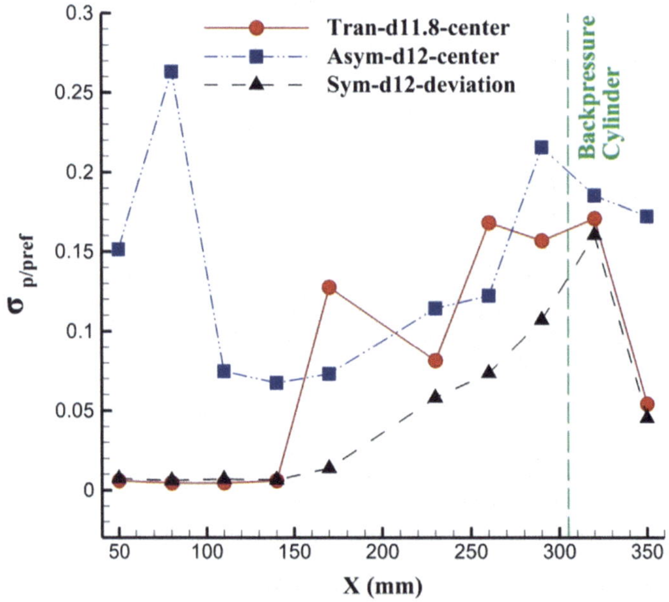

Fig. 3.48 Normalized standard deviation of the pressure signals for different separation modes [23]

The standard deviation of the pressure signal implies the amplitude of the static pressure oscillation. For the purpose of normalizing the value of standard deviation, the static pressure of inflow is selected as the reference pressure. As seen in Fig. 3.48, the normalized standard deviation curves from all three separation modes reach a high value near the backpressure cylinder (axial location X = 305 mm). This is induced by the bow shock and its unsteadiness. The peak in asymmetric separation mode indicates that an intense unsteadiness lies in the separation shock foot on the expansion wall. The sharp rise and fall of the standard deviation curve in asymmetric mode suggests that the intermittent region of separation shock foot locates within the axial location from 50 to 110 mm. When it comes to the quasi-symmetric separation mode with asymmetric backpressure, the unsteadiness of flow can be neglected upstream from the streamwise location of 230 mm. It is not until the region near the backpressure cylinder that the unsteadiness of the flowfield becomes apparent. From this standard deviation curve, it can also be concluded that the separation shock of quasi-symmetric mode behaves quasi-steady. However, in the separation transition mode, the standard deviation curve is widened and no manifest peak value appears. The first rise at X = 170 mm indicates that the separation shock reaches this axial location intermittently. Moreover, the obviously lower value compared with the peak from asymmetric mode implies that the energy of shock oscillation is dispersive. That is to say, the intermittent region is broadened in the streamwise direction.

3.4.3.3 Preliminary Statistical Study on Low-Frequency Unsteadiness

Since the separated related unsteady flow is evolved from shockwave boundary layer interaction (SWBLI), two unsteady separation cases presented in the current chapter (asymmetric and transition) have been compared with several typical SWBLI results. The length of the intermittent region is chosen as the characteristic length to show the amplitude of shock oscillation. Meanwhile, the peak frequency of power spectra (pressure signal) is selected to depict temporal behaviors of separation shock. As is illustrated in Fig. 3.49a, the profile of intermittent region length L_i versus peak frequency F_{max} is quasi-linear in log coordinates. Although separation cases in the current chapter have evidently larger intermittent regions and lower peak frequencies, the quasi-linear relationship is still observed. The quasi-linear profile (in log coordinates) in Fig. 3.49a suggests that an inversely proportional relationship exists in these two parameters.

The dimensionless frequency (Strouhal number) is a canonical parameter in low-frequency unsteadiness analysis of SWBLI problems. In traditional SWBLI cases, the shock foot undergoes larger-scale motion and lower oscillation frequencies as the scale of separation increases [39]. This trend has also been reported in forced oscillations of shock train by Xiong et al [44]. However, the Strouhal number varies little for different inflow conditions and separation patterns. Following the analysis in SWBLI problems, Strouhal numbers of asymmetric and transition separation cases have been calculated as well. It is seen in Fig. 3.49b that values of Strouhal numbers from cases in the current chapter are within the range of typical SWBLI results. These

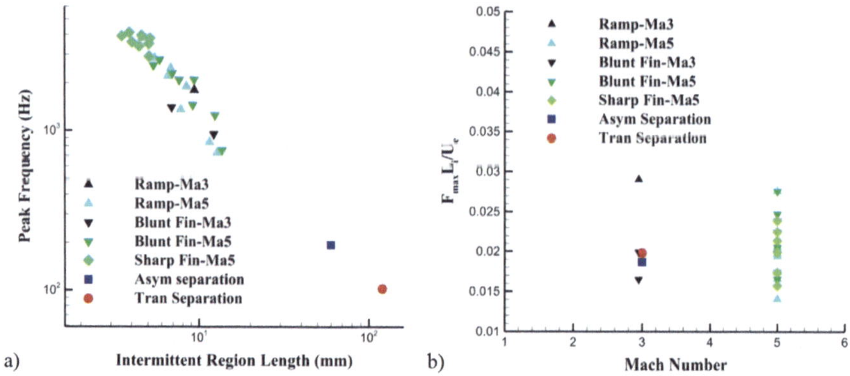

Fig. 3.49 Low-frequency shock motion of several different turbulent separated flows [23]: **a** intermittent region length L_i versus peak frequency F_{max} in log coordinates; **b** Strouhal number scaling based on L_i, F_{max} and free stream velocity U_e. (The typical SWBLI results are obtained from studies of Dolling et al. [42] and Gonsalez et al. [43])

results in Fig. 3.49b imply that a similar mechanism may lie in these unsteady flow issues. Though the preliminary statistical study has gained interesting results, there is still a long way to go in uncovering the specific mechanism of this broadband, low-frequency shock unsteadiness.

3.4.4 Control of Unsteadiness

Unsteady phenomena are usually unwelcome in scramjets. As a result, two measures are put forward in the current section to make a heuristic study of unsteadiness controlling.

3.4.4.1 Asymmetric Back Pressure

When the cylinder deviates from the center of flowfield, separation remains quasi-symmetric even under high back pressure. The cylinder in Fig. 3.50 is exactly the same size as in Sect. 3.3.2.2 (back pressure ratio $P/P_{in} = 4.69$), but it deviates 8 mm from the center location. Asymmetric back pressure restrains the forward movement of shock train and development of the separated region. Meanwhile, the dynamic features of the flowfield are greatly weakened. The power spectral density of FFT analysis in Fig. 3.51 is one order of magnitude smaller than the asymmetric separation mode in Sect. 3.3.2.2, which indicates that flowfield is quasi-steady.

Figure 3.52 illustrates the experimental and computational average velocity magnitude contour for asymmetric back pressure. Just as in cases from Sects. 3.3.2.1 and 3.3.2.2, artificially low-speed region still exists in the downstream center flow due to

Fig. 3.50 NPLS image of symmetric separation mode, d = 12 mm, asymmetric back pressure

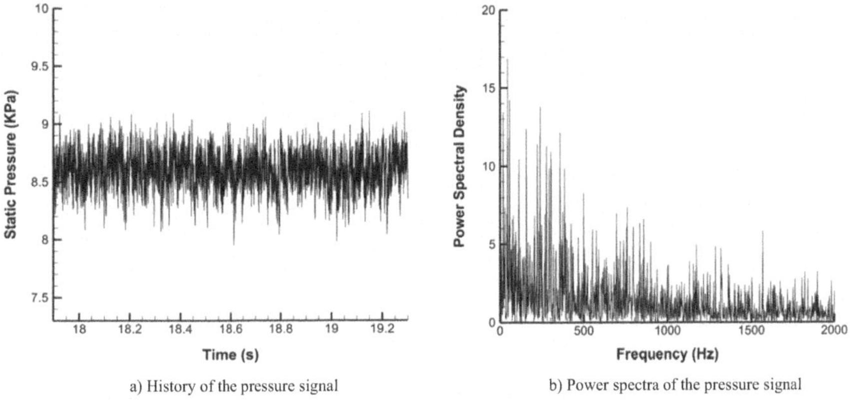

Fig. 3.51 History and FFT analysis of pressure signal for symmetric separation mode [16]

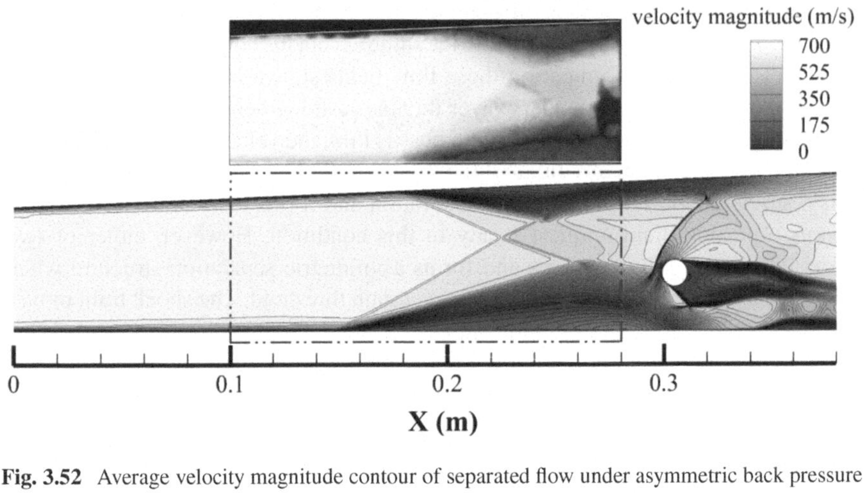

Fig. 3.52 Average velocity magnitude contour of separated flow under asymmetric back pressure [16] (top: experimental data, bottom: calculated result)

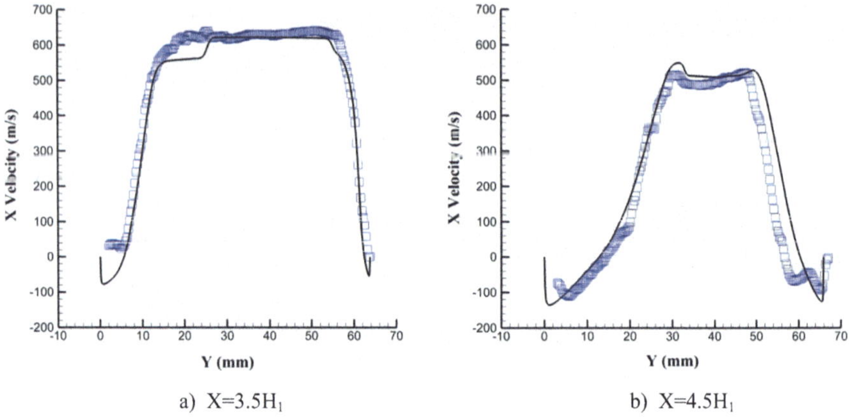

Fig. 3.53 Average streamwise velocity of separation under asymmetric back pressure [16] (line: calculated result, symbol: experimental data)

the optical imaging from the cylinder. Numerical and experimental average streamwise velocity profiles at two characteristic axial locations are shown in Fig. 3.53. It is observed that all these velocity profiles show slight asymmetry between two separated regions. Due to the asymmetric back pressure, the separated region at the straight side is a bit larger than that at the expansion side. However, no reattachment is found on either side of walls, which means this separation mode is not a typical asymmetric separation.

3.4.4.2 Decrement of Boundary Layer Thickness

Processes of transient flow with different boundary layer thickness [45] are shown in Fig. 3.54. The simulation condition is all the same except the boundary layer thickness of inlet (1 and 3 mm). Comparing these flow fields shown in Fig. 3.54, it appears that the increment of the boundary layer thickness strengthens the instability of the flowfield. When the boundary layer thickness is thin, the velocity profile of the fluid close to the wall is relatively full and the flow is more resistant to the backpressure. This restricts the development of the separation and forms of symmetric separated regions. The flow field is quasi-steady in this condition. However, either of two separated regions fully develops and forms asymmetric separation structure when the boundary layer thickness increases to a certain threshold. The shock train moves upstream and then oscillates in large amplitude along the flow direction.

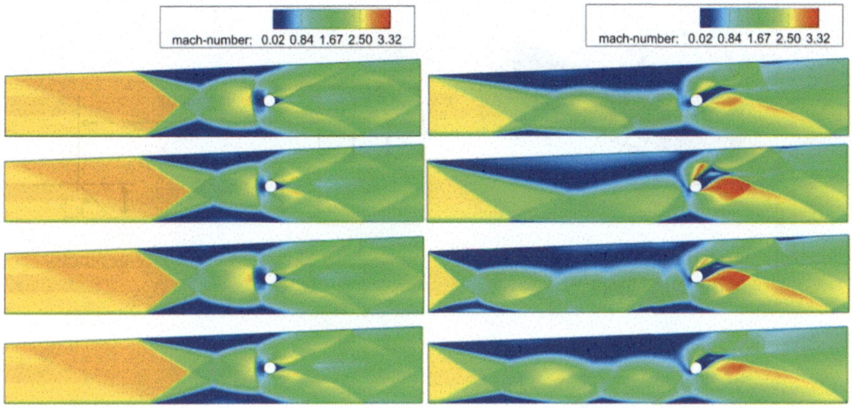

a) boundary layer thickness of inlet: 1mm b) boundary layer thickness of inlet: 3mm

Fig. 3.54 The unsteady flowfield with a different boundary layer thickness of inlet (d = 12 mm) [45]

3.5 Validation on Reactive Flows with Different Geometry

Based on four sections above, it is found that a typical kind of asymmetric and unsteady combustion in scramjets is dominated by back pressure induced separation. However, whether this flow dominating instability is universal in supersonic combustion remains suspectable. For the purpose of proofing the universality of flow dominating instability, experiments [46] are implemented in a combustor with different geometry and inflow conditions.

3.5.1 Experimental Facility

Experiments are performed on the direct-connected supersonic combustion facility, as shown schematically in Fig. 3.55. The air is heated by an air/O_2/alcohol vitiator which could heat the air up to approximately 1429 K and the mass fraction of O_2 in

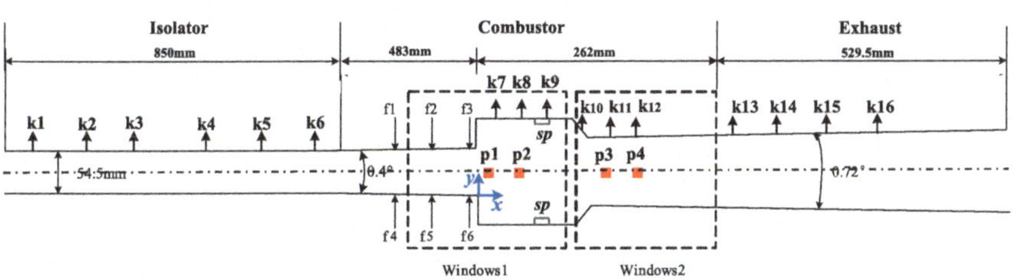

Fig. 3.55 Schematic of combustor and distribution of transducers [46]

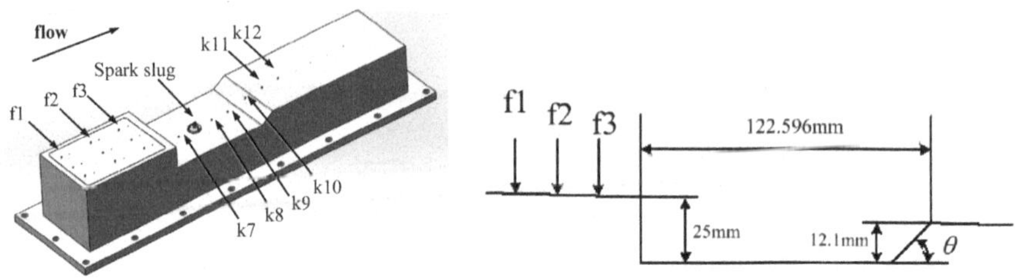

a) Integrated fuel injector and cavity flame holder.

b) Sketch of the cavity geometry.

Fig. 3.56 Fuel injection schemes and cavity geometry [46]

the blend air is added up to 23%. The total pressure is 1980 kPa and the mass flow rate is 2 kg/s. A two-dimensional nozzle (Ma 2.92 in the combustor entrance) was used to accelerate the blend air to supersonic flow, simulating flight Mach number of 6.0 with a static pressure of about 54.5 kPa. The scramjet flow path consists of a heat-sink rectangular isolator and a rectangular combustor with dual parallel recessed-cavities and flush-mounted injectors normal to the wall. The isolator has a rectangular cross-sectional area with a height of 54.5 mm, a width of 75 mm, and a length of 850 mm. The combustor consists of two diverging sections and the divergence angles are 0.4° and 0.72°, respectively. The combustion products are exhausted into the atmosphere through the exhaust section with slightly expansion angle of 0.72°.

Symmetric dual parallel recessed-cavities flameholder are located on the divergent top and bottom walls separately. Figure 3.56 shows the integrated injection unit and cavity flameholder. The cavity has a forward-facing ramp that could effectively interact with the shear layer originating from the cavity leading edge. In each run of the experiment, the fuel is injected simultaneously from both sides at the same position (e.g. injected from f3 and f6). The depth of the cavity is 25 mm, length to depth ratio (L/D) is 5 and the aft angle is 45°, as shown in Fig. 3.56b. The leading edge of the cavity is located 483 mm downstream of the combustor entrance. And the height of the flow path at the cavity leading edge is 67 mm. The spark plug located at the base of the cavity was used as the baseline ignition source. Gaseous ethylene was used as the main fuel and could be injected form location f1/f2/f3, which are 95, 60 and 25 mm upstream the cavity leading edge. There are 3 orifices with a diameter of 1.5 mm on each injection location. Meanwhile, there are also injection ports with a diameter of 2 mm located 85, 50 and 15 mm upstream the cavity leading edge. Different injection sites and the orifices are selected to evaluate the effects of injection location and fuel/air equivalence ratios on the combustion.

Wall static pressure along the facility is captured by an electronic pressure scanning system at 1000 Hz at sixteen locations (k1–k16) in the isolator and combustor, as shown in Fig. 3.55. There are 4 positions (p1–p4) on side wall available for 2 high-frequency pressure transducers (PCB model 064B02) depending on the experimental requirements. The acquisition frequency of the high-frequency sensor is 500 kHz. High-speed flame luminosity images are captured 4000 frames per second with an exposure time of 0.2 ms to investigate the flame oscillations. And the resolution of the image is 1024 × 512 pixel which could cover window1 and window2 as shown in Fig. 3.55. Schlieren system is used to visualize the shock waves in the combustor with the image resolution of 896 × 848 pixel which covered only window1. The exposure time was 1 μs and the frame rate was 10000 fps.

The test sequence for each run started with the steady vitiator operating condition. Hydrogen was used as pilot fuel and was injected into the main flow at the entrance of the combustor on the top wall. The spark plug started working after the injection of hydrogen, then followed the injection of ethylene. Pilot fuel was closed with spark plug after ethylene fuel was ignited. The camera and PCB were triggered simultaneously before the injection of ethylene.

3.5.2 Variation of Combustion Modes Under Different Equivalence Ratios

Similarly to Sect. 3.1.2, combustion mode transfers from scramjet mode to ramjet mode as the increment of equivalence ratio. Due to the lower inflow Mach number and weaker geometric expansion compared with previous experiments, the symmetric combustion in scramjet mode combustion is not observed.

3.5.2.1 Scramjet Mode Combustion with Asymmetric Flame: Strong Unsteadiness

The scramjet mode with asymmetric flame in current section qualitatively reproduces the dynamic combustion in Sect. 3.1.3.

It is clearly seen in Figs. 3.57 and 3.58 that typical structures of combustion flowfield are very close to Sect. 3.1.3. An asymmetric jet-wake stabilized flame is able to appear at either side of walls. When this jet-wake flame appears, the large separated region suppresses the supersonic core flow to the other side. As a result, a shear layer stabilized flame occurs. During the dynamic process, the shock train oscillates with large amplitude in the axial direction. Meanwhile, the switch of the

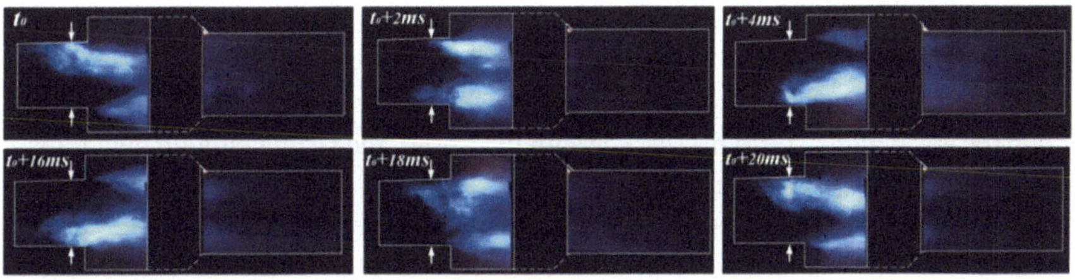

Fig. 3.57 Flame luminosity images of unsteady scramjet mode combustion, $\Phi = 0.27$

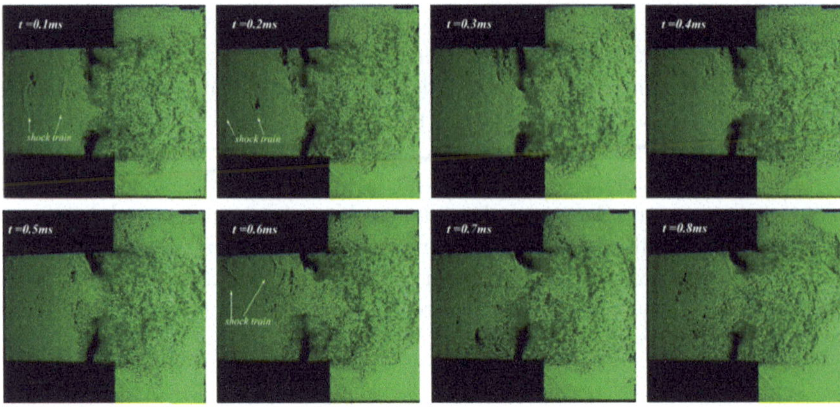

Fig. 3.58 Schlieren images of unsteady scramjet mode combustion, $\Phi = 0.27$ [46]

asymmetric jet-wake flame is accompanied by the vertical movement of the shock train. Still similar to Sect. 3.1.3, the switch of the asymmetric flame is intermittent, not periodically. This intermittent flame switch phenomenon will be further discussed in Sect. 3.5.3.

3.5.2.2 Transition Mode Combustion with Asymmetric Flame: Weak Unsteadiness

As the increment of equivalence ratio, the combustion mode in combustor gradually transfers from scramjet mode to ramjet mode. During this ram-scram transition process, a special mode is defined, which is named as "transition mode" (also called dual mode). In this operation condition, the shock train appears in the flame stabilized region occasionally. The averaged leading-edge of shock train locates further upstream due to the stronger heat release compared with scramjet mode. In this transition mode, the asymmetric flame structure still exists, however, the switch of two flames from both walls are rather rare. The flow unsteadiness is still obvious, but it is much weaker than the scramjet mode (Fig. 3.59).

3.5.2.3 Ramjet Mode Combustion with Symmetric Flame: Quasi-steady

When the equivalence ratio passes a certain threshold, the strong enough heat release will choke the flame stabilized region to pure subsonic. This operation condition is called the ramjet mode. It is demonstrated in the schlieren image of Fig. 3.60, the shock structure completely disappear in the current combustion mode. Both fuel injections and flames from two walls are generally symmetric, and the combustion flowfield is quasi-steady.

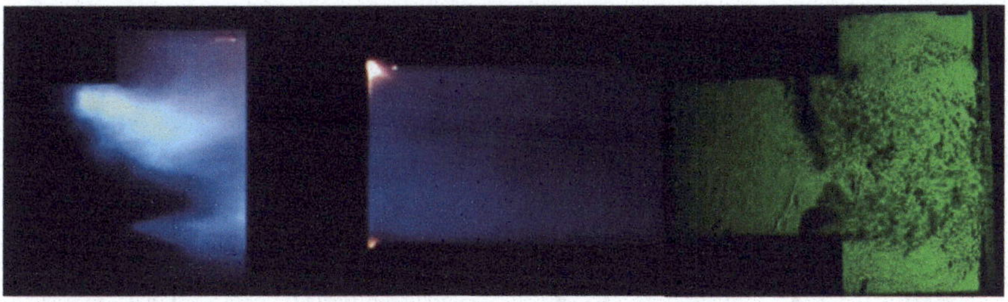

a) Jet-wake flame on the upwall

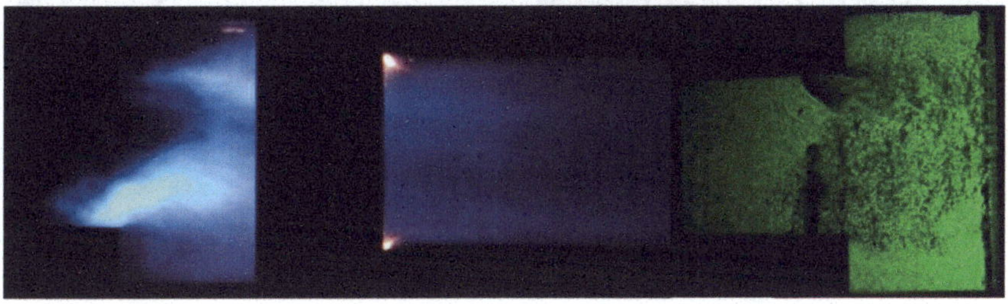

b) Jet-wake flame on the downwall

Fig. 3.59 Flame luminosity and schlieren images of unsteady transition mode combustion, $\Phi =$ 0.47 [46]

Fig. 3.60 Flame luminosity and schlieren images of quasi-steady ramjet mode combustion, $\Phi =$ 0.70 [46]

3.5.3 Quantitative Descriptions of Unsteady Combustion

Flame front oscillation histories for different combustion modes are illustrated in Fig. 3.61. These profiles are calculated based on the same procedure as Fig. 3.7. It is depicted that the unsteadiness of flowfield gradually decreases as the increment of the equivalence ratio. The switch of asymmetric flame in the case of $\Phi = 0.47$ is relatively infrequent compared with the case $\Phi = 0.27$. The symmetric flame structure of ramjet mode can also be illustrated in the profile of flame front oscillation history (Fig. 3.62).

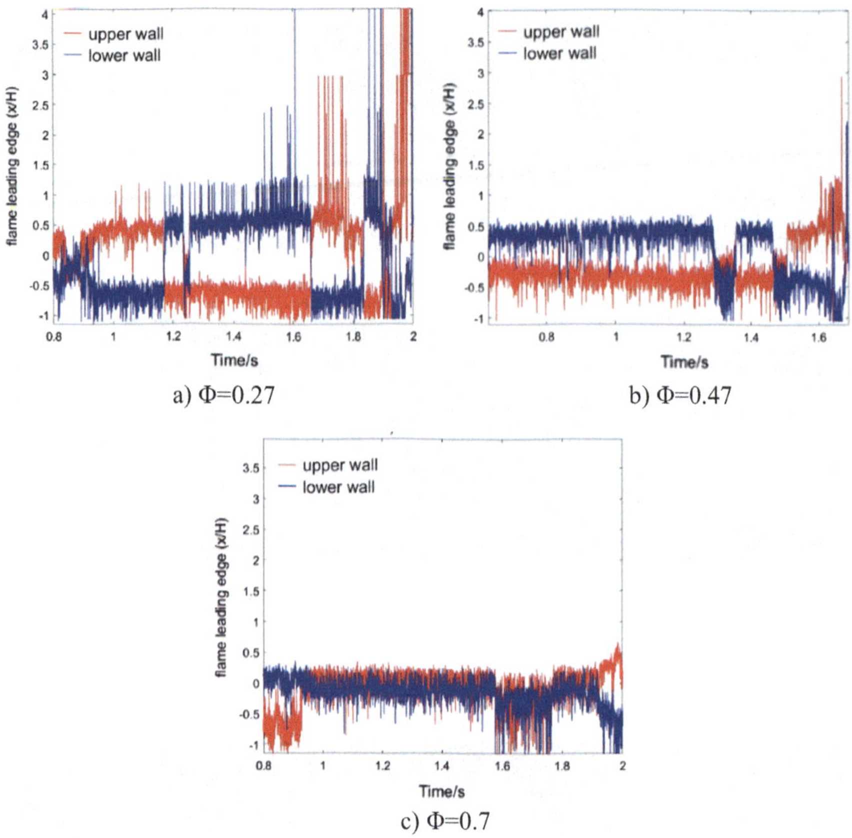

Fig. 3.61 Flame front oscillation histories for different combustion modes [46]

The time-averaged static pressure distribution and the related standard deviation along the upper wall centreline are demonstrated in Fig. 3.61. Both of these pressure profiles are normalized by the total pressure of inflow P_{in}^{*}. The axial location is normalized by the local height H of combustor at the leading edge of dual parallel cavities. The origin of X-axis locates rightly at these two leading edges. From these standard deviation profiles, it is also clearly seen that the unsteadiness of combustion flowfield decreases as the increment of equivalence ratio. In the ramjet mode, the symmetric flame in subsonic region behaves quasi-steady in general. Fluctuations mostly concentrate in the upstream location of dual cavities (x/H about −6.5), which is the leading-edge of shock train.

In general, the dynamic combustion mode in Sect. 3.1.3 is qualitatively reproduced in a different combustor under a different inflow condition. This suggests that the flow dominating instability is a universal source in unsteady supersonic combustion.

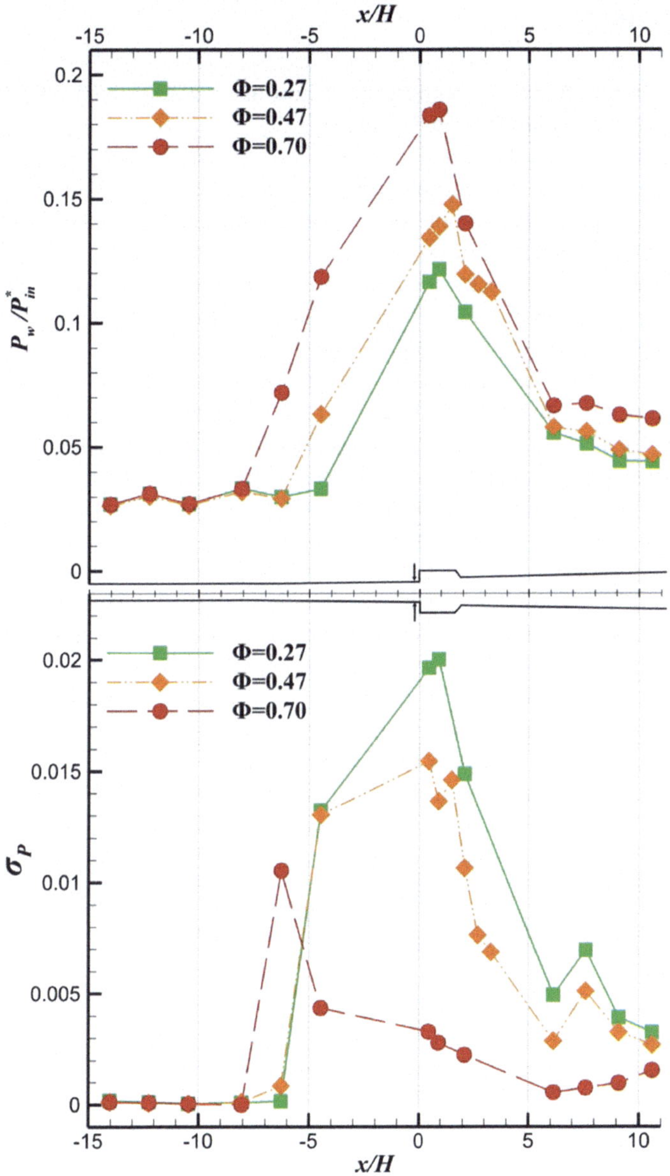

Fig. 3.62 Normalized averaged static pressure and standard deviation of pressure signals for different combustion modes [46]

3.6 Summary

In this chapter, a study of combustion unsteadiness dominated by backpressure induced separation is implemented systematically.

For a single-expanded rectangular supersonic combustor with dual parallel cavities and near-cavity fuel injections, flowfield structures generally transform from symmetry into asymmetry as the increment of equivalence ratio [11]. Under an intermediate equivalence ratio [3], intermittent dynamic combustion occurs with a high-amplitude pseudo-shock oscillation in the streamwise direction. A decoupling analysis [15] is carried out to discover the key impact factor of this unsteady combustion. It is found that flame, fuel jet as well as cavity flameholder do not play a key role in this issue, and only the backpressure is the pivotal factor that really matters.

Thus, cold flow analysis [16] is applied, and typical flame structures are fully reproduced by the backpressure induced separated flowfield. Specifically speaking, a symmetric separation under low backpressure generates symmetric combustion, while an asymmetric separation under high backpressure results in asymmetric combustion. It is found that the thicker boundary layer, the weaker velocity/momentum profile fullness (larger shape factor), and the lower turbulent kinetic energy on the expansion wall altogether leads to the asymmetric separation. Based on the detached-eddy simulation, the whole process of symmetric/asymmetric separation transition [23] (occurs under threshold backpressure) is captured. Boundary layer separation tendency analysis shows that an interlaced shape factor distribution of boundary layers from both walls accounts for the switch of separation modes.

Typical dynamic combustion behaviors mentioned above are reproduced in a combustor with a different geometry [46], which operates under a different inflow condition. It is concluded that the flow dominating instability is a universal source in unsteady supersonic combustion.

References

1. Mohieldin, T. O., Tiwari, S. N., & Olynciw, M. J. (2001). Asymmetric flow-structures in dual mode scramjet combustor with significant upstream interaction. In *37th AIAA/ASME/SAE/ASEE Jet Propulsion Conference & Exhibit* (p. 3296). AIAA.
2. Rodriquez, C. (2001). Asymmetry effects in numerical simulation of supersonic flows with upstream separated regions. In *39th Aerospace Sciences Meeting and Exhibit* (p. 84). AIAA.
3. Gao, T., Liang, J., Sun, M., & Zhong, Z. (2017). Dynamic combustion characteristics in a rectangular supersonic combustor with single-side expansion. *Proceedings of the Institution of Mechanical Engineers, Part G: Journal of Aerospace Engineering, 231*(10), 1862–1872.
4. Sun, M.-B., Zhong, Z., Liang, J.-H., & Wang, Z.-G. (2014). Experimental investigation of supersonic model combustor with distributed injection of supercritical kerosene. *Journal of Propulsion and Power, 30*(6), 1537–1542.
5. Ouyang, H., Liu, W., & Sun, M. (2016). Parametric study of combustion oscillation in a single-side expansion scramjet combustor. *Acta Astronautica, 127*, 603–613.
6. Zhong, Z., Wang, Z., & Sun, M. (2015). Effects of fuel cracking on combustion characteristics of a supersonic model combustor. *Acta Astronautica, 110*(1), 1–8.

7. Sun, M.-B., Geng, H., Liang, J.-H., & Wang, Z.-G. (2008). Flame characteristics in supersonic combustor with hydrogen injection upstream of cavity flameholder. *Journal of Propulsion and Power, 24*(4), 688–696.
8. Ming-bo, S., Jing, L., Hai-yan, W., Jian-han, L., Wei-dong, L., & Zhen-guo, W. (2011). Flow patterns and mixing characteristics of gaseous fuel multiple injections in a non-reacting supersonic combustor. *Heat Mass Transfer, 47*(11), 1499–1516.
9. Westbrook, C. K., & Dryer, F. L. (1981). Simplified reaction mechanisms for the oxidation of hydrocarbon fuels in flames. *Combustion Science and Technology, 27*, 31–43.
10. Zhong, Z., Wang, Z., Sun, M., & Wang, H. (2016). Combustion characteristics in a supersonic combustor with ethylene injection upstream of dual parallel cavities. *Proceedings of the Institution of Mechanical Engineers, Part G: Journal of Aerospace Engineering, 230*(13), 2515–2522.
11. Sun, M., Zhong, Z., Gao, T., Wang, H., & Liang, J. (2017). Asymmetric combustion characteristics of transverse ethylene injection in a rectangular supersonic combustor with single-side expansion. In *21st AIAA International Space Planes and Hypersonic Technologies Conference* (pp. 2016–4759). AIAA.
12. Wang, H., Wang, Z., & Sun, M. (2013). Experimental study of oscillations in a scramjet combustor with cavity flameholders. *Experimental Thermal and Fluid Science, 45*, 259–263.
13. Ben-Yakar, A., & Hanson, R. K. (2001). Cavity flame-holders for ignition and flame stabilization in scramjets: An overview. *Journal of Propulsion and Power, 17*(4), 869–877.
14. Doerffer, P., Charles, H., & Dussauge, J. P. (2010). *Unsteady effects of shock wave induced separation*. Springer.
15. Gao, T. Y., Liang, J. H., & Sun, M. B. (2018). Decoupling analysis on oscillation of separated region in a supersonic combustor with single-side expansion. *Journal of Propulsion Technology, 9*(10), 2381–2393.
16. Gao, T., Liang, J., Sun, M., & Zhao, Y. (2017). Analysis of separation modes variation in a scramjet combustor with single-side expansion. *AIAA Journal, 55*(4), 1307–1317.
17. Zhao, Y.-H., Liang, J.-H., Zhao, Y.-X., & Zhang, Y.-J. (2016). Research on passive control of jet in supersonic crossflow base on micro-vortex generator. *Journal of Propulsion Technology, 37*(5), 801–806.
18. Zhenguo, W., Yilong, Z., Yuxin, Z., & Xiaoqiang, F. (2015). Prediction of massive separation of unstarted inlet via free-interaction theory. *AIAA Journal, 53*(4), 1108–1112.
19. Zhao, Y., Yi, S., Tian, L., & Cheng, Z. (2009). Supersonic flow imaging via nanoparticles. *Science in China Series E: Technological Sciences, 52*(12), 3640–3648.
20. Zhao, Y., Liang, J., & Zhao, Y. (2016). Non-reacting flow visualization of supersonic combustor based on cavity and cavity–strut flameholder. *Acta Astronautica, 121*, 282–291.
21. Zhao, Y., Liang, J., & Zhao, Y. (2016). Vortex structure and breakup mechanism of gaseous jet in supersonic crossflow with laminar boundary layer. *Acta Astronautica, 128*, 140–146.
22. Quan, P., Yi, S., Wu, Y., Zhu, Y., & He, L. (2015). Experimental investigation on the effects of swept angles on blunt fin-induced flow. *AIAA Journal, 53*(9), 2805–2810.
23. Gao, T., Liang, J., & Sun, M. (2017). Symmetric/asymmetric separation transition in a supersonic combustor with single-side expansion. *Physics of Fluids, 29*(12), 126102.
24. Wang, H., Wang, Z., Sun, M., & Qin, N. (2013). Combustion characteristics in a supersonic combustor with hydrogen injection upstream of cavity flameholder. *Proceedings of the Combustion Institute, 34*(2), 2073–2082.
25. Wang, H., Wang, Z., Sun, M., & Qin, N. (2013). Numerical study on supersonic mixing and combustion with hydrogen injection upstream of a cavity flameholder. *Heat Mass Transfer, 50*(2), 211–223.
26. Sandham, N. D., Schülein, E., Wagner, A., Willems, S., & Steelant, J. (2014). Transitional shock-wave/boundary-layer interactions in hypersonic flow. *Journal of Fluid Mechanics, 752*, 349–382.
27. Touber, E., & Sandham, N. D. (2009). Large-eddy simulation of low-frequency unsteadiness in a turbulent shock-induced separation bubble. *Theoretical and Computational Fluid Dynamics, 23*, 79–107.

28. Xie, Z. T., & Castro, I. P. (2008). Efficient generation of inflow conditions for large-eddy simulation of street-scale flows. *Flow, Turbulence and Combustion, 81*, 449–470.
29. Schlatter, P., & Orlu, R. (2010). Assessment of direct numerical simulation data of turbulent boundary layers. *Journal of Fluid Mechanics, 659*, 116–126.
30. Sun, M. B., Zhang, S. P., Zhao, Y. H., Zhao, Y. X., & Liang, J. H. (2013). Experimental investigation on transverse jet penetration into a supersonic turbulent crossflow. *Science China Technological Sciences, 56*(8), 1989–1998.
31. Sun, M., Hu, Z., & Sandham, N. D. (2017). Recovery of a supersonic turbulent boundary layer after an expansion corner. *Physics of Fluids, 29*(7), 076103.
32. Knight, D. D., Yan, H., Panaras, A. G., & Zheltovodov, A. A. (2003). Advances in CFD prediction of shockwave turbulent boundary layer interactions. *Progress in Aerospace Sciences, 39*, 121–184.
33. Delery, J. (1985). Shock wave/turbulent boundary layer interaction and its control. *Progress in Aerospace Sciences, 22*, 209–280.
34. Wang, Q.-C., Wang, Z.-G., & Zhao, Y.-X. (2016). On the impact of adverse pressure gradient on the supersonic turbulent boundary layer. *Physics of Fluids, 28*(11), 116101.
35. Arnette, S. A., & Samimy, M. (1995). Structure of supersonic turbulent boundary layer after expansion regions. *AIAA Journal, 33*(3), 430–438.
36. Arnette, S. A., Samimy, M., & Elliott, G. S. (1998). The effects of expansion on the turbulence structure of compressible boundary layers. *Journal of Fluid Mechanics, 367*, 67–105.
37. Dussauge, J. P., & Gaviglio, J. (1987). The rapid expansion of a supersonic turbulent flow: role of bulk dilatation. *Journal of Fluid Mechanics, 174*, 81–112.
38. Smith, D. R., & Smits, A. J. (1991). The rapid expansion of a turbulent boundary layer in a supersonic flow. *Theoretical and Computational Fluid Dynamics, 2*(5), 319–328.
39. Clemens, N. T., & Narayanaswamy, V. (2014). Low-frequency unsteadiness of shock wave/turbulent boundary layer interactions. *Annual Review of Fluid Mechanics, 46*(1), 469–492.
40. Le, D. B., Goyne, C. P., & Krauss, R. H. (2008). Shock train leading-edge detection in a dual-mode scramjet. *Journal of Propulsion and Power, 24*(5), 1035–1041.
41. Xiong, B., Wang, Z., Fan, X., & Wang, Y. (2016). A method for shock train leading edge detection based on differential pressure. *Proceedings of the Institution of Mechanical Engineers, Part G: Journal of Aerospace Engineering, 231*(1), 61–71.
42. Dolling, D. S., & Murphy, M. T. (1983). Unsteadiness of the separation shock wave structure in a supersonic compression ramp flowfield. *AIAA Journal, 21*(12), 1628–1634.
43. Gonsalez, J. C., & Dolling, D. (1993). Correlation of interaction sweepback effects on the dynamics of shock-induced turbulent separation. In *31st Aerospace Sciences Meeting & Exhibit* (p. 776). AIAA.
44. Xiong, B., Wang, Z., Fan, X., & Wang, Y. (2017). Response of shock train to high-frequency fluctuating backpressure in an isolator. *Journal of Propulsion and Power, 33*(1), 1–9.
45. Gao, T. Y. (2015). *Investigations on unsteady behaviors of separated region in a rectangular supersonic duct with expansion*. Graduate School of National University of Defense Technology, National University of Defense Technology, Changsha.
46. Gao, G. Y. (2018). *Mechanism research on flame instability in supersonic combustor with dual cavities*. Graduate School of National University of Defense Technology, National University of Defense Technology, Changsha.

Chapter 4
Cavity Ignition in Supersonic Flows

Scramjets are promising propulsion systems for hypersonic flight vehicles. Development of scramjets relies on sufficient understanding of the complex mixing, ignition and combustion processes inside a scramjet combustor. Current scramjets usually require flame holding devices which provide subsonic environments to facilitate flame ignition and stabilization, because the ignition delay time of hydrocarbon fuels is long relative to their residence time in the combustor. The most common flame holding device is a cavity with a slanted rear wall. This flame holder scheme is generally combined with transverse injection in supersonic combustors.

The ignition process in supersonic flows is complicated, and depends on many factors such as the local fuel equivalence ratio, the flame holder configuration, the fuel auto-ignition ability, the ignition energy and the flow conditions near the ignition location. For cavity-based supersonic combustors, a major concern regarding the ignition is how to establish stable subsonic combustion regions around and inside cavities, which motivates researches on chemical and physical phenomena on cavity ignition. This chapter introduces our studies through four aspects, involving diverse ignition methods, flame behaviours during ignition, ignition mechanisms and auto-ignition effects.

4.1 Ignition Processes Under Different Ignition Methods

Interactions between turbulence and chemical reactions are vital for ignition process. In cavity-based scramjet combustors, the core flow of supersonic inflows usually has a velocity of 1000 m/s, and the velocity of the recirculating flow inside cavities often varies from 0 to 200 m/s. The core flow and the shear layer over cavities trigger strong turbulent dissipation which challenges the initial flame propagation, although cavities provide an ignition-friendly environment. Therefore, the cavity ignition process presents many unique characteristics. These characteristics may even change as different ignition methods are used, which is not fully understood

M. Sun et al., *Unsteady Supersonic Combustion*,
https://doi.org/10.1007/978-981-15-3595-6_4

yet. This section sketches typical ignition processes with respect to diverse ignition methods.

4.1.1 Spark Ignition

Among many methods for cavity ignition, spark ignition is still the most widely used. The main shortage for spark ignition is the low ignition energy. Spark ignition is often used to ignite cavity-based scramjets that utilize hydrogen or gaseous hydrocarbons as fuels. For liquid kerosene that requires larger ignition energy, however, employing spark ignition without aid of other methods is less reported.

Figures 4.1 and 4.2 demonstrate a typical spark ignition process via high-speed imaging and schlieren in a direct-connected test facility fueled by hydrogen [1]. The inflow conditions were $Ma = 1.92$ with stagnation state $T_0 = 846$ K and $P_0 = 0.7$ MPa. As shown in Fig. 4.1, the initial flame kernel induced by spark appears around the igniter which is near to the front wall of the cavity. The injection valve opens at t = 0.0 ms, and simultaneously the spark plug in the upper cavity is triggered. The spark brightness is identified in Fig. 4.1b and c with white circles. Before t = 0.5 ms the initial flame kernel grows slow, but the flame propagation suddenly speeds up, as shown in images of t = 0.5 ms and t = 0.75 ms. In this short period, the flame speed approximately reaches 400 m/s. A reasonable explanation for this high flame speed is that the flame kernel penetrates the cavity shear layer and ignites the fuel in the shear layer, and the high-speed shear layer transfers flame

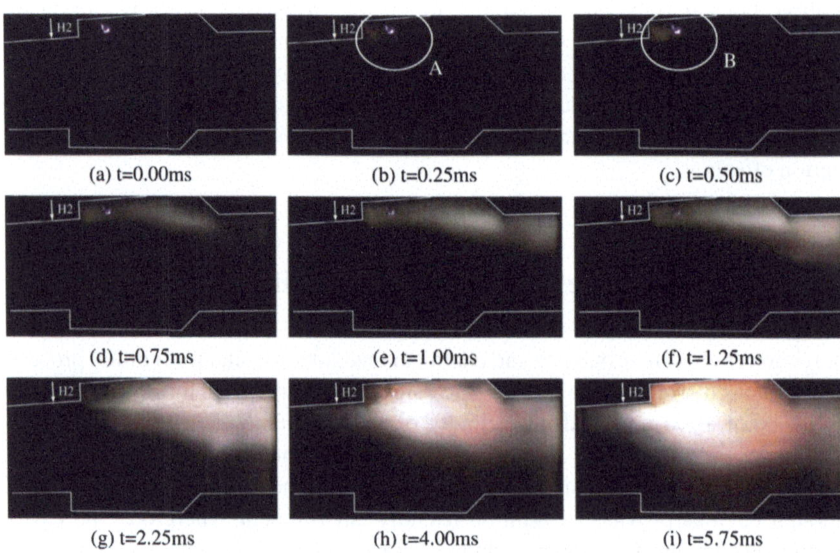

Fig. 4.1 High-speed images of spark ignition process of hydrogen [1]

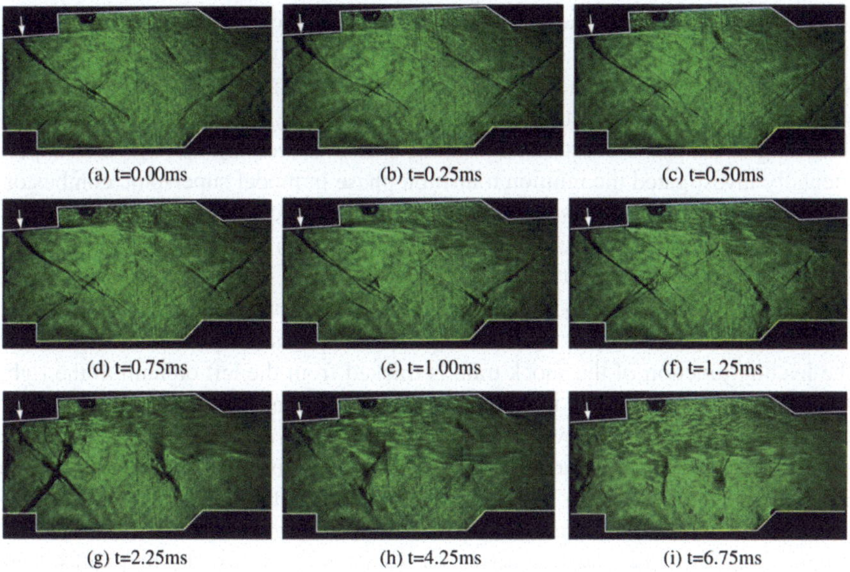

Fig. 4.2 High-speed schlieren images of spark ignition process of hydrogen around the T2 cavity with T2 ignition [1]

downstream very quickly. Images of t = 0.75 ms to t = 2.25 ms show that after the fuel downstream of the upper cavity being ignited, the flame propagates upstream due to the pressure increase caused by heat release. Meanwhile, the flame spreads transversely, igniting the fuel in the core flow. During this period, the flame appears distortion which might result from shockwave incidence from the bottom cavity. At t = 4.00 ms, the flame begins stabilizing and the flame front extends to the injection location.

Schlieren images during the ignition process are shown in Fig. 4.2. Figures 4.1 and 4.2 are from two individual ignition experiments whose operating conditions are identical. In Fig. 4.2 the cavity shear layer and heat release region can be recognized clearly. The shock trains and flame bounds change hugely owing to unsteadiness of the ignition process.

4.1.2 Piloted Ignition

Piloted ignition is widely applied in practical scramjets in which utilizing spark ignition only is unable to achieve reliable ignition, for example when the fuel is kerosene. Piloted ignition refers to establishing initial combustion with the help of pilot flame which is formed by igniting chemically active fuels, such as hydrogen and ethylene. In piloted ignition, the spray usually starts as immediately as the pilot flame is established. Thus the spray and the pilot injection coexist for a short period. The pilot injection terminates once the spray flame is stabilized.

Compared with direct spark ignition in the cavity, piloted ignition could broaden ignitable conditions. This is because the pilot flame not only supplies vast heat and active radicals, but also induces pre-combustion shock trains upstream of the cavity, which enhances mixing of the spray and the core flow. The mixing enhancement pertaining to the precombustion shock trains was observed in Ref. [2] which experimentally investigated the ignition transition phase in model supersonic combustor. In the experiment three types of ignition processes were compared: spark ignition of ethylene, spark ignition of kerosene, and piloted ignition with kerosene as the spray fuel and ethylene as the pilot fuel.

The left column of Fig. 4.3 records the density field variation around the cavity, and the right column corresponds to a region in the isolator upstream of the cavity. The leading junction of the shock train is tracked from the left column to the right column, as indicated by the time steps. Ethylene is transversely injected from the location denoted by the arrow, under a injection pressure of 2 MPa. The fuel trace and the bow shock are clear in the image of t = 0 ms. There exist irremovable reflected shockwaves in the isolator, which are attributed to small gaps between different sections. But those shockwaves are too weak to affect the bulk velocity of supersonic primary flow. Following the spark at t = 0 ms, the first shockwave intersection appears at t = 3.5 ms, and then it is pushed upstream to the isolator, as

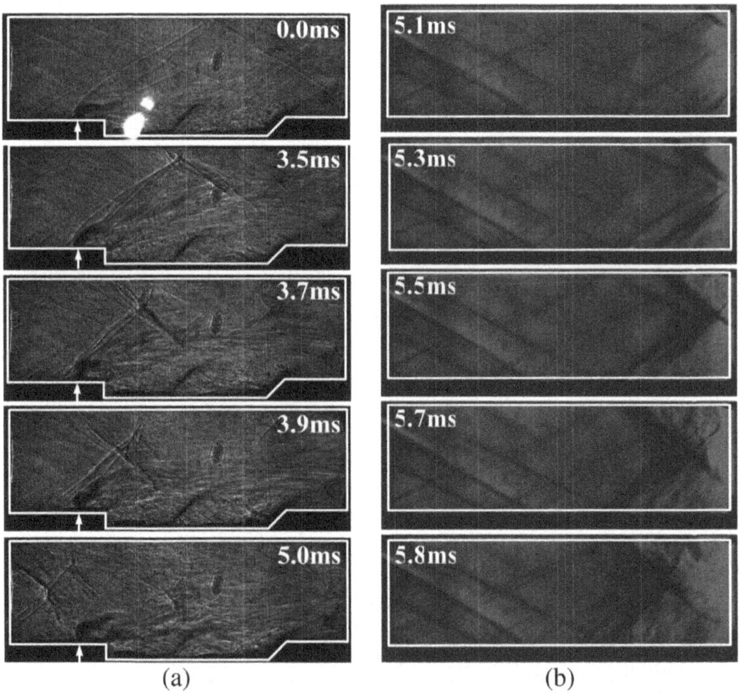

Fig. 4.3 Schlieren images showing evolution of precombustion shock train during spark ignition of ethylene [2]

shown in the right column. The speed at which the shockwave intersection moves upstream is estimated to be 50–100 m/s, according to the camera's frame rate. This speed is close to the flame propagation speed in conditions of high temperature and turbulence, but much less than the detonation speed (1 km/s or above). The moving of shockwave intersection is caused by separation of boundary layers on walls, and this separation results from local heat release. It is then conclude that the initial flame development is a subsonic flame propagation dominated by expansion of the cavity flame.

Figure 4.4 reports the density field change in the isolator during piloted ignition. The camera setup is similar to that used in Fig. 4.3a. At t = 0 ms, the leading shockwave induced by the cavity flame has reached the right side. Nevertheless, the leading shockwave is pushed down when the fuel touched the downstream ethylene flame owing to heat-absorbing effect of cold fuel droplets. At t = 1.8 ms, the kerosene flame first appears. From images of t = 1.8 ms to 3.2 ms, the effect of the precombustion shock train on the spray can be clearly observed. The spray lifts and the fuel evaporates immediately when swept by the leading shockwave, which enhances the mixing and combustion processes. This effect is also verified by the experimental

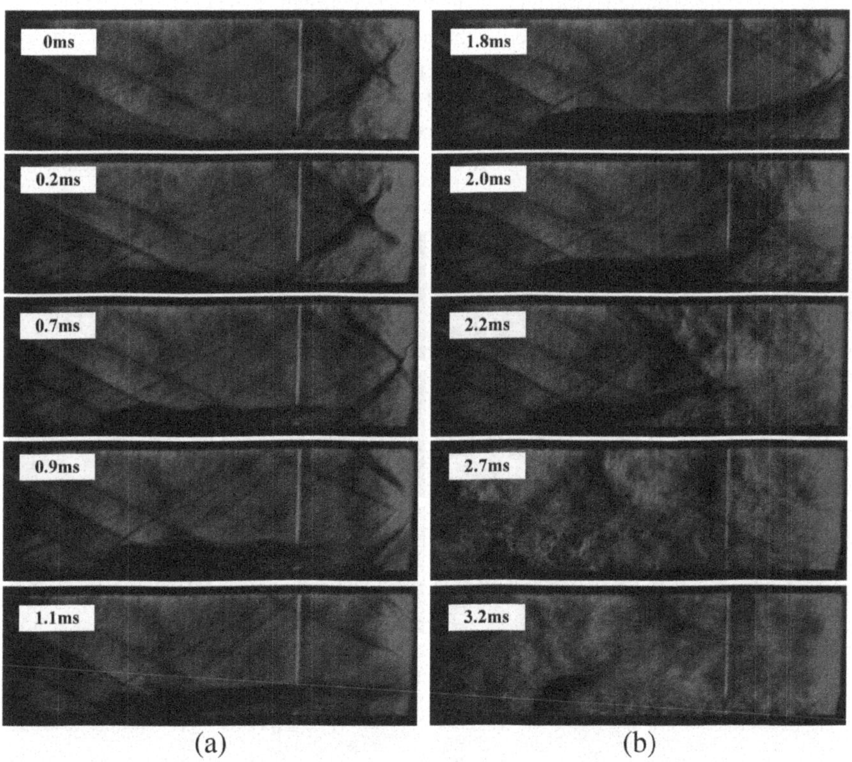

Fig. 4.4 Schlieren images showing evolution of piloted ignition with kerosene as the spray fuel and ethylene as the pilot fuel [2]

cases that a weak flame without a generated shock train cannot lead to a successful piloted ignition event [2].

4.1.3 Gliding-Arc-Discharge (GAD) Ignition

For the spark ignition discussed in Sect. 4.1.1, the discharge produces thermal equilibrium plasmas so that the ignition is caused by heating of fuel-oxidizer mixture in a small-volume around the arc channel. In recent decades, nevertheless, utilizing non-equilibrium plasmas for ignition has gained increasing interest. Non-equilibrium plasmas could effectively generate active radicals substantially reducing the ignition delay time, which greatly facilitates combustion organization in scramjet engines.

GAD refers to plasmas generated between separated electrodes in a fast gas flow. The gliding arc is stretched in the flow field, changing from the equilibrium state to the non-equilibrium state. In the non-equilibrium state, up to 70–80% of the gliding arc power is dissipated, which means that the GAD has the potential to efficiently stimulate nonequilibrium chemical reactions significant for the ignition process.

Figure 4.5 illustrates a typical ignition process using the GAD [3]. The experiment was implemented in a cavity-based scramjet combustor. The inflow had a stagnation temperature of 1600 K and a stagnation pressure of 1.65 MPa. The GAD power supply had an average power of 2 kW and a maximum peak voltage of 20 kV.

In Fig. 4.5 multiple flame kernels first appear but are then blown off, from t = −8075 μs to t = −8025 μs. At t = 0 μs, a flame kernel appears again and

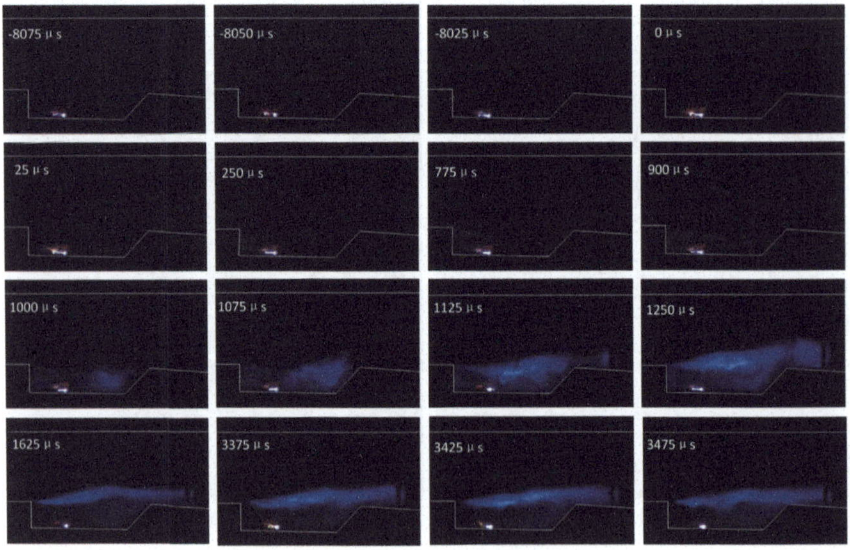

Fig. 4.5 High-speed images for gliding-arc-discharge ignition [3]

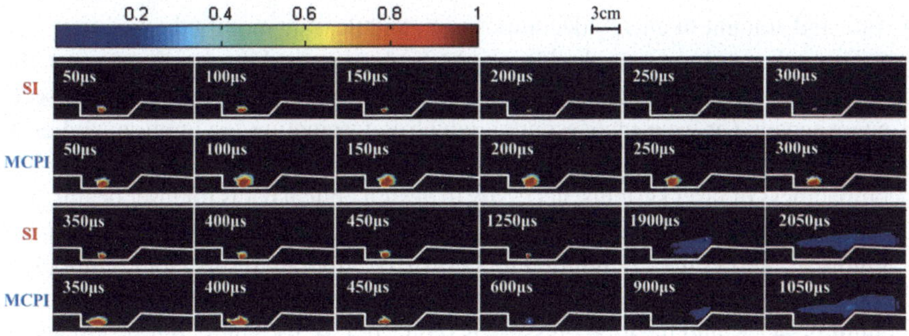

Fig. 4.6 Comparison of CH* chemiluminescence images related to spark ignition and multichannel plasma ignition [4]

then successfully spreads. The flame kernel is wrinkled anti-clockwise which agrees with the flow pattern inside the recirculation zone. The flame further propagates, filling the cavity at t = 1075 μs. Then the fuel in the cavity shear layer is ignited at t = 1125 μs, following that flame in the cavity begins to dissipate at t = 1250 μs. Finally, the flame stabilizes at t = 1625 μs. During the whole ignition process, the GAD continuously induces flame kernels inside the recirculation zone, which speeds up the ignition process. It costs 131 ms to form a stabilized flame from the discharge beginning.

It is experimentally observed that the GAD triggers initial flame kernels when it reaches the maximum length. This phenomenon is consistent with the state transition and energy dissipation of GAD. As above mentioned, the GAD in non-equilibrium state could generate active radicals helpful for formation of initial flame kernels. Also, the instantaneous power of GAD increases as the arc is continuously stretched, according to the voltage and current records during the ignition process.

A relatively large kernel can be realized by utilizing multi-channel plasma igniter (MCPI) powered by a conventional ignition power supply [4]. Fig. 4.6 compares the flame kernel developments of spark ignition (SI) and MCPI in a scramjet combustor with inflow conditions of $Ma = 2.52$, stagnation pressure $P_0 = 1.6$ MPa and stagnation pressure $T_0 = 1600$ K. Figure 4.6 shows that the flame kernel of MCPI is much larger than that of SI. For MCPI the flame takes 900 μs to fill the cavity and reach the shear layer, but for SI the time is 1000-μs longer.

4.1.4 Laser-Induced Plasma (LIP) Ignition

Ignition through LIP can be classified as laser thermal ignition, laser-induced photochemical ignition, laser-induced resonant breakdown ignition, and laser-induced spark ignition [5]. Non-resonant breakdown is the most common, because it does not require a wavelength to photo-dissociate particular target species. Compared with ignition methods based on electrical discharge, LIP ignition can precisely control the rate and amount of energy decomposition, and flexibly modify the ignition position. Through synchronizing the laser and advanced optical diagnostic systems, the instantaneous ignition process could be investigated intensively in the time scale of nanoseconds (or even picoseconds), which is fascinating for ignition studies in a supersonic flow. Disadvantages of LIP ignition chiefly include complication and incompactness of laser systems, necessity to design optical paths for laser beams, and impossibility to provide uniform energy decomposition in a relatively large volume.

Figure 4.7 shows the LIP ignition process in a cavity-based supersonic combustor with ethylene as the fuel [6]. The inflow's Mach number is 2.1, the stagnation temperature is 947 K and the total pressure is 0.65 MPa. A 1064-nm laser beam, which came from a Q-switched Nd:YAG laser source running at 10 Hz and had 940 mJ

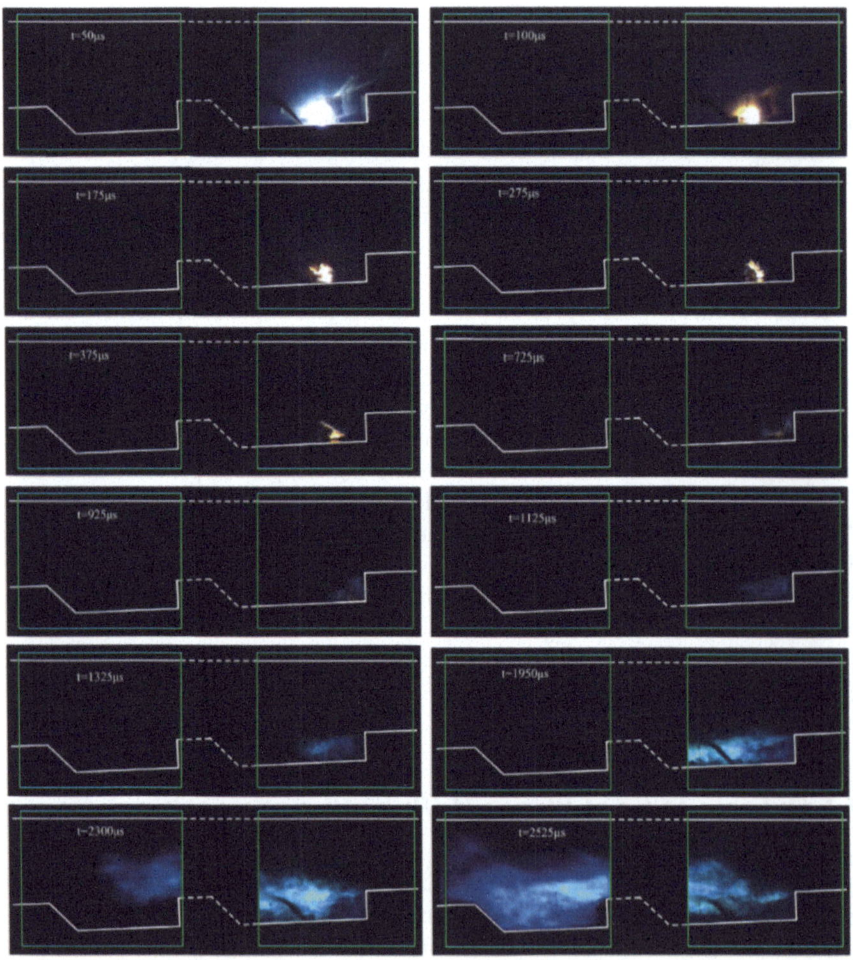

Fig. 4.7 High-speed images for evolution of the flame kernel induced by LIP ignition [6]

per pulse, was focused into the cavity for ignition. The laser was directed into the upstream cavity at t = 0 μs.

The image of t = 50 μs in Fig. 4.7 shows a bright and white light emitted in the bremsstrahlung process of LIP. The ionized plasma is almost round at the start, but then wrinkles due to the turbulence, as shown in the image of t = 100 μs. The yellow light arises, because the combustion chain reactions are initiated and the flame kernel is formed. From t = 275 μs to 375 μs, the top of the flame kernel is entrained into the cavity shear layer, and then quenches out due to the high strain rate. The left flame kernel rotates anti-clockwise owing to the entrainment of circulation zone inside cavity. The flame kernel reaches to the front wall at t = 725 μs and is stabilized at the corner for about 200 μs. At t = 925 μs, the flame begins to spread downstream in the cavity via the bottom of the shear layer. Then the flame fully fills the cavity at t = 1950 μs. When the shear layer impinges on the cavity ramp, the burning combustible mixture in cavity shear layer is transported into the core flow, following that the cavity shear layer stabilized flame is formed. The images of t = 2300 μs to t = 2950 μs further show the establishment of stabilized flame in the cavity downstream.

Ignition process by two-pulse laser-induced plasma is shown in Fig. 4.8 [7]. In the experiment, the second laser pulse was triggered 200 μs after the first one. This 200-μs delay approximately equals to the time in which the flame kernel produced by the first laser pulse was transported to the cavity corner. Note that the flame structures

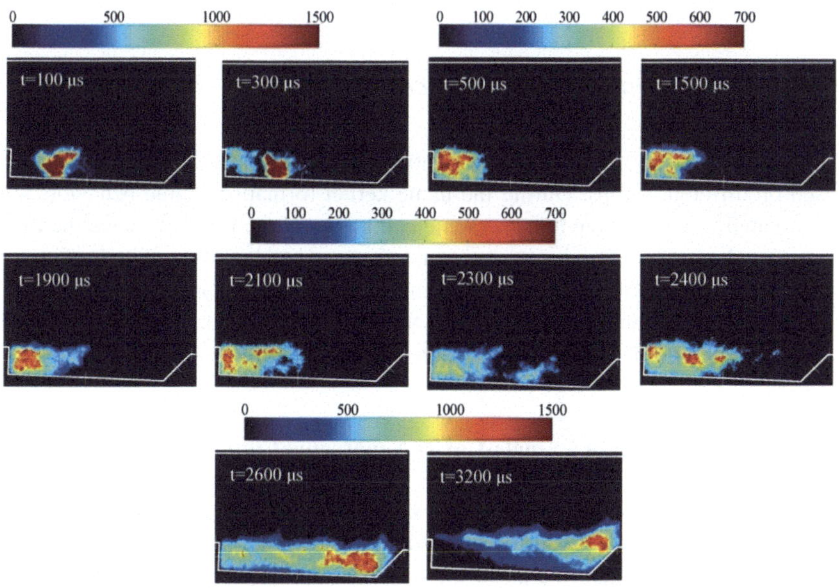

Fig. 4.8 Instantaneous distribution of OH* chemiluminescence during ignition by dual-laser laser-induced plasma [7]

represented by OH* chemiluminescence were obtained from different test runs, due to low frame rate of the used ICCD camera.

The ignition process resembles that using a single laser pulse. The main difference is that the two laser pulses are combined to form the initial flame kernel. As reported in Fig. 4.8, the two kernels merge together at $t = 500$ μs. The flame kernel triggered by the second laser pulse is weaker than that by the first one, because the first pulse improves the temperature and reduces the flow density in the laser focus, decreasing the absorbed energy of the second pulse. From $t = 500$ μs to 1900 μs, the flame in the recirculation zone then enters into a quasi-stable state. Further, the flame spreads downstream to fill the whole cavity from $t = 1900$ μs to 2600 μs. Finally, the flame is stabilized along the shear layer over the cavity, as shown in the image of $t = 3200$.

4.2 Flame Behaviors During Ignition

Since the ignition process is affected by many factors and partially stochastic, the flame usually behaves differently from one ignition event to another, even with identical configurations. The flame behaviors during ignition directly determine ignition results and closely correlate to the ignition mechanism. In this section, studies of flame behaviors during processes of the flame kernel formation and the initial flame propagation are introduced accordingly.

4.2.1 Experimental and Numerical Setups

The studies about the flame kernel formation and the initial flame propagation are based on different setups. During the flame kernel formation, flame behaviors are investigated in a single-cavity supersonic combustor ignited by LIP, since the rate, amount and position of ignition energy injection can be altered flexibly. During the initial flame propagation, nonetheless, flame behaviors are studied in a single-cavity combustor and a multi-cavity one, both with the spark ignition. Configuration used in these studies are outlined in this section for reference.

4.2.1.1 LIP Ignition in a Single-Cavity Supersonic Combustor

A direct-connected test facility composed of an air heater, a supersonic nozzle and a scramjet model combustor was used for experiment. Through the air heater burning pure ethanol and oxygen, the air at a mass flow rate of 1 kg/s was continuously heated to a stagnation temperature of $T_0 = 1650$ K and a stagnation pressure of $P_0 =$ 2.6 MPa. The mole fraction of oxygen in the vitiated air was 21%. The outlet of air heater was equipped with a two-dimensional Larval-nozzle to accelerate the heated air to $Ma = 2.92$.

Shown in Fig. 4.9 is the model scramjet engine. A constant-area 315 mm long isolator with height of 40 mm and width of 50 mm was directly connected to the exit of nozzle, followed by a 512-mm-long combustor whose lower wall diverges at an angle of 2.25°. A 220 mm long expansion section with a single-side expansion angle of 3° was at the end of the combustor. One cavity was installed on the lower wall.

As illustrated in Fig. 4.10, the cavity depth $D = 15$ mm, cavity length to depth ratio $L/D = 7$, and the aft ramp angle was 45°. Ethylene at room temperature was injected through 2-mm-diameter orifice at 10-mm upstream of the leading edge along the central plane. The injection pressure of ethylene before was 2.5 MPa. The overall equivalent ratio was kept at 0.152 for all the tests. A Q-switched Nd:YAG laser system (Vlite-500) was used for ignition. The laser system provided 532 nm laser pulses with a 10-ns width, a 12-mm diameter and 350-mJ maximum pulse energy. Nonetheless, only 85% of the laser pulse energy passed through the lens and the quartz window of the combustor. The laser pulses were focused by a convex lens (f = 150 mm) to produce the plasma for ignition. The breakdown region denoted by red spots had a cone shape. The shape size given in the figure was estimated when the laser pulse energy approximately equalled 300 mJ. There existed three different ignition positions that all of them located in the central plane, with a 5-mm distance above the cavity bottom. Along the streamwise direction, their distances to the front wall of the cavity were 22.5 mm, 45.0 mm and 67.5 mm, respectively. Hereafter, the three positions are separately referred to as "P1", "P2" and "P3".

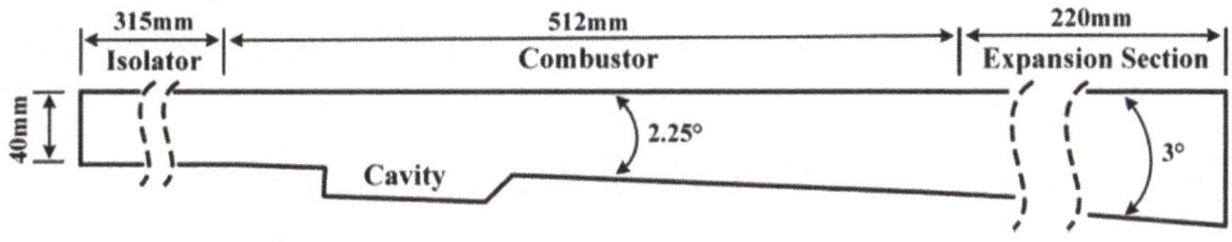

Fig. 4.9 Schematic of the model scramjet engine and the cavity [8]

Fig. 4.10 Schematic of the cavity and the optical configuraiton [8]

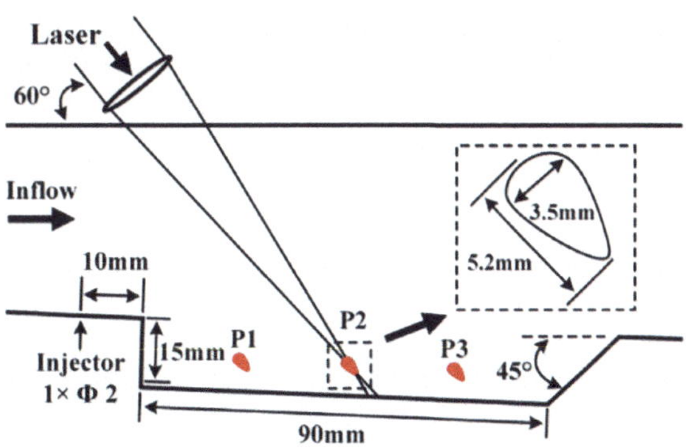

For the purpose of marking flame locations, CH* and OH* chemiluminescence were captured simultaneously. The CH* chemiluminescence was recorded by a high-speed camera (Photron SA-Z) with a Nikon 85 mm f/1.8 lens and a bandpass filter (centered at 430 nm, with a 10 nm FWHM). The frame rate of the camera was set at 25,000 frames per second (fps) with an exposure time of 40 μs, and the spatial resolution was 240-μm per pixel. An intensified charge-coupled device (ICCD) camera equipped with a UV lens (95 mm focal length and f/4.1) and a bandpass filter (centered at 311 nm, with a 10-nm FWHM) was applied to acquire OH* chemiluminescence. The ICCD camera operated at 3 fps with a shutter time of 2 μs, with a spatial resolution of 250-μm per pixel. As the cameras were mounted on the same side, neither of them oriented normal to test section precisely. The consequent image distortion was then corrected by an image processing program.

4.2.1.2 Spark Ignition in a Single-Cavity Supersonic Combustor

Main components of the direct-connected test facility and their functions have been given in the last section, and in this section only the differences are mentioned. As shown in Fig. 4.11, The cavity flame-holder along was mounted on the bottom wall which had an expansion angle of 1°. The plug was located in the middle of the cavity. The impulse excitation energy of the plug was 5.0 J and the excitation frequency was 50 Hz [9]. A quartz window with a 162-mm length and a 38-mm width was embedded into the top wall of the combustor. Meanwhile another four quartz windows with a 148-mm length and a 90-mm width were embedded into the side walls of the combustor, as plotted in Fig. 4.12. The cavity bottom length L and aft ramp angle A were 80 mm and 45° respectively. The model scramjet combustor width was 50 mm. The cavity bottom length L and aft ramp angle A were respectively 80 mm and 45°. The model scramjet combustor had a width of 50 mm. As the leading edge height D1 of 20 mm is taller than the rear wall height D2 of 16 mm, this kind of cavity is often called the real-wall-expansion cavity [10–13].

The experiment involved three cavity geometries A, B and C with rear wall heights of 16 mm, 12 mm and 8 mm, respectively. Two groups of porthole injectors, represented by red spots in Fig. 4.12, were located 10-mm and 30-mm upstream of the front wall of the cavity, separately. Each group consisted of three injectors evenly spaced along the spanwise direction.

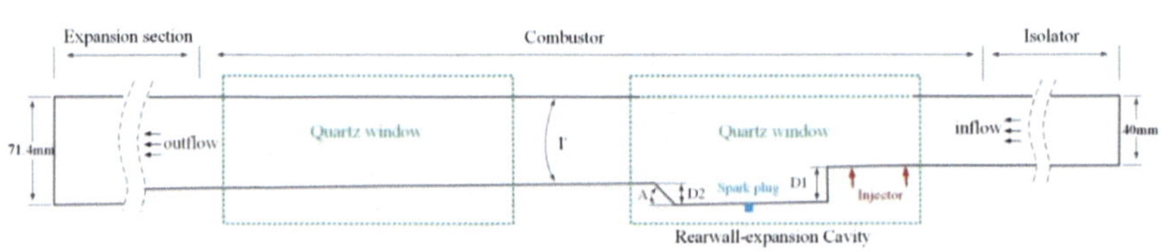

Fig. 4.11 Schematic of the scramjet engine and the cavity [12]

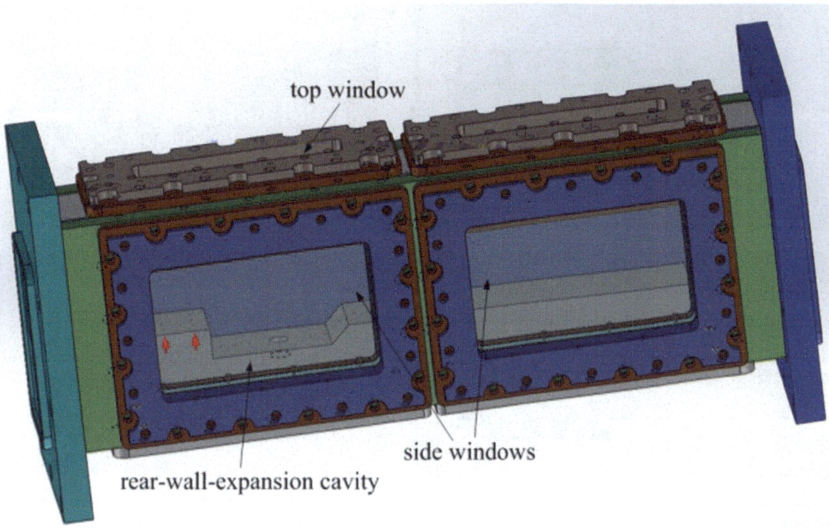

Fig. 4.12 Rear-wall-expansion cavity combustor [14]

The ignition and flame propagation were visualized by two high-speed cameras, one camera capturing flame chemiluminescence from the side view while the other from the top view. The cameras worked at 20,000 fps with a shutter time of 1/20683 s and an aperture number of 1.4. The pressure along the combustor's bottom wall was recorded by 32 static pressure taps in a sample frequency of 100 Hz with ±0.5% uncertainties.

4.2.1.3 Spark Ignition in a Multi-cavity Supersonic Combustor

Figure 4.13 shows the two cavity configurations adopted in the direct-connected test facility. Flame was observed through the glass windows embedded in the side wall, by a high-speed camera with an imaging frequency of 5,000 fps and an exposure time of 1/6 ms. In the parallel dual-cavity configuration, two cavities were oppositely installed on the top and bottom walls. Nevertheless, the tandem dual-cavity configuration adopted two cavities which adjoined along the streamwise direction. Sharing the same size, all the cavities had a cavity depth of 8 mm, a length-to-depth ratio of 7 and an rear wall angle of 45°. Hydrogen injectors were located in the central plane with a 10-mm distance upstream of the leading edges. Each injector exit had a dimeter of 2 mm. It is noticed that there were two injectors for the parallel dual-cavity configuration but only one for the tandem dual-cavity configuration. The injection pressure in the former configuration was twice that in the latter, in order to keep the same global equivalence ratio.

The combustion processes corresponding to the two configurations were also numerical investigated. Figure 4.14 plots the xy-plane projection of the combustor

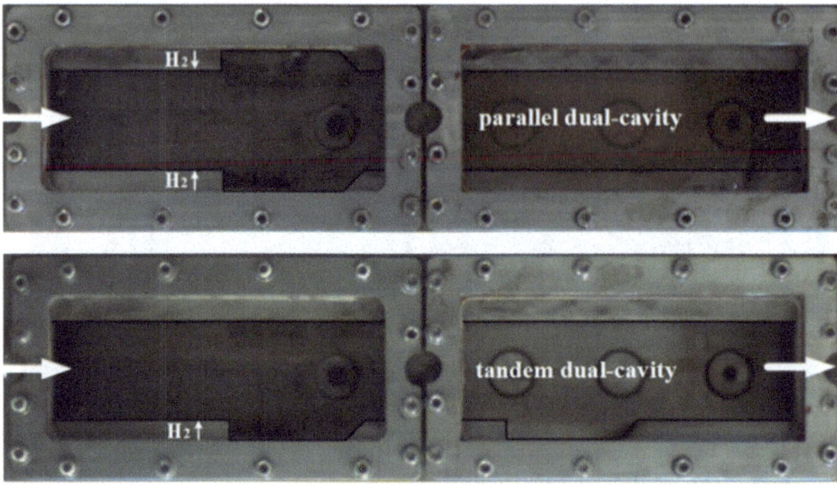

Fig. 4.13 Supersonic combustors with tandem and parallel dual-cavity [15]

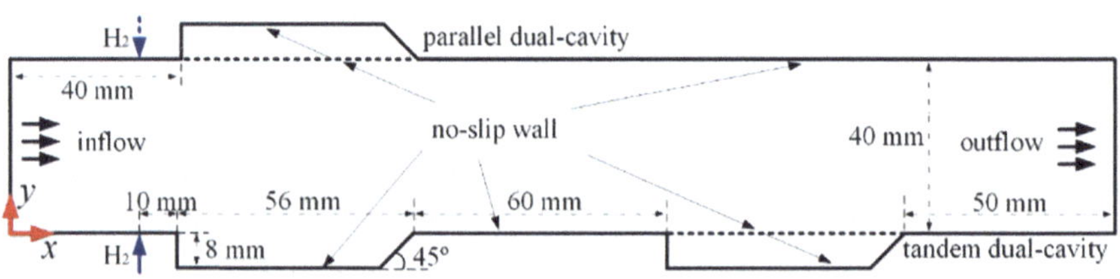

Fig. 4.14 Schematic of computational domain of dual-cavity combustor [15]

in the simulation with key sizes. Along the z-axis the size of the combustor was 25 mm. The total grid points were about 10.3 million and 11.5 million for the parallel dual-cavity and the tandem dual-cavity configurations, respectively. In focused computational region, the grid had a resolution of $\Delta x^+ \approx 1$–50, $\Delta y^+ \approx 1$–30 and $\Delta z^+ \approx 20$–50, on basis of the wall stress τ_w at the inlet bottom floor. It may be coarse for a wall-resolved LES, but is suitable for a hybrid approach. The inflow boundary layers on the top and bottom walls had a thickness of 3 mm, which aimed to simulate the flow condition at the isolator exit in experiments.

In the simulation, the fifth-order Weighted Essentially Non-Oscillatory (WENO) scheme [16] was adopted for inviscid fluxes, and viscous fluxes were calculated by the 2nd-order central scheme. Temporal integration was performed by a second-order dual time-step approach, with the inner iteration achieved by a lower-upper symmetric Gauss-Seidel (LU-SGS) method. The Spalart-Allmaras Reynolds-Averaged Navier-Stokes (RANS) model and the Yoshizawa sub-grid scale (SGS) model were blended in the hybrid RANS/LES method. One equation S-A RANS model and one equation Yoshizawa SGS model for LES were used in near-wall regions and the regions away from the solid wall, respectively. The turbulence modelling process

Table 4.1 Experiment and simulation condition

	Ma	T_0/K	P_0/MPa	Y_{O_2}	Y_{H_2O}	Y_{N_2} and Y_{CO_2}	Y_{H_2}
Air inflow	2.52	1486	1.6	0.2338	0.0622	0.704	0.0
Hydrogen jet	1.0	300	0.63, parallel 1.26, tandem	0.0	0.0	0.0	1.0

was detailed in Ref. [17]. This simulation choose the H_2-Air reaction mechanism proposed by Jachimowski [18] and including 9 species and 19 steps. An assumed sub-grid Probability Density Function(PDF) closure model was adopted the LES region. The modelling approach had been detailed in Ref. [19] and tested in previous works [20, 21]. The detailed NS equations could be referred to Ref. [15].

The air stream and fuel jet parameters for experimental and numerical studies are reported in Table 4.1. In the simulation, the CO_2 (mainly produced by the air heater) in the air flow was treated as nitrogen, since the adopted reaction mechanism neglects nitrogen oxidation. Based on Table 4.1, the calculation is normalized by the following reference parameters, $L_{ref} = d = 2$ mm (d denotes the injector exit diameter), $U_{ref} = u_{inf} = 1368.2$ m/s, $T_{ref} = T_{inf} = 753.8$ K, $\rho_{ref} = \rho_{inf} = 0.3832$ kg/m^3, and $p_{ref} = \rho_{ref} U_{ref}^2 = 715.3$ kPa.

4.2.2 *Formation of the Flame Kernel*

The supersonic combustor is ignited at various ignition energies and positions. The CH* and OH* chemiluminescence are then captured to investigate the effects of ignition energy and position on formation of flame kernel. Section 4.2.1.1 gives the experimental setup.

4.2.2.1 Effect of Ignition Energy

The effect of ignition energy on ignition was studied with the ignition position at P2. A representative set of CH* chemiluminescence images are shown in Fig. 4.15 with ignition energies $E_1 = 303.6 \pm 7.4$ mJ and $E_2 = 230.7 \pm 6.0$ mJ. The intensity of each pixel is normalized by the maximum intensity of all the pixels during the ignition process.

As shown in Fig. 4.15, for the ignition energy E_1, the flame kernel at 40 μs is bright and located in the cavity middle. The flame kernel follows the flow direction in the recirculation zone, and by 120 μs it has moved to the leading edge, becoming smaller and less bright. After being anchored there, its strength grows slightly from 120 to 200 μs. Then it begins propagating to the downstream at 200 μs, and has filled the cavity by 560 μs. After 80 μs, the flame kernel spreads into the mainstream from

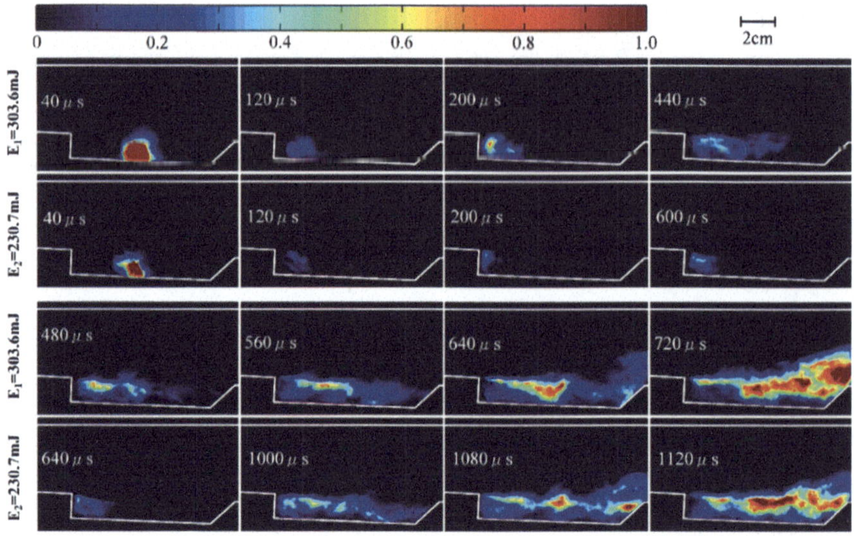

Fig. 4.15 CH* chemiluminescence images of ignition processes with ignition energies E_1 = 303.6 mJ and E_2 = 230.7 mJ [8]

the rear wall. At last, the combustor settles into the quasi-stable burning process by 720 μs.

When the ignition energy is reduced to E_2 the initial flame kernel becomes smaller. It takes 1120 μs for the flame kernel to fully fill the cavity and propagate to the mainstream, which is 400 μs longer than the ignition process of E_1. It is also noteworthy that the flame kernel is suppressed in the cavity during most time of the ignition process, because the high-speed flow and consequent high strain rate in the mainstream. In the quasi-stable burning process, besides, the flame intensity in the shear layer is much stronger than that elsewhere, because the air-ethylene mixture is rich in the shear layer.

4.2.2.2 Effect of Ignition Position

As represented by red spots in Fig. 4.10, three different ignition positions P1, P2 and P3 were used to investigate the effect of ignition position on formation of the flame kernel. The corresponding tests are then abbreviated as "Case P1", "Case P2" and "Case P3", respectively. All the tests shared an ignition energy of E_1 and a laser pulse frequency of 3 Hz.

A series of typical images are given in Fig. 4.16 to depict the flame spread processes with ignition positions at P1, P2 and P3. In Case P1 the initial flame kernel is adjacent to the leading edge, but it does not propagate downstream until 440 μs after the laser pulse, which is 200 μs later than the Case P2. Then it takes another 520 μs for the flame kernel to fully fill the cavity. By 1040 μs, a quasi-stable burning

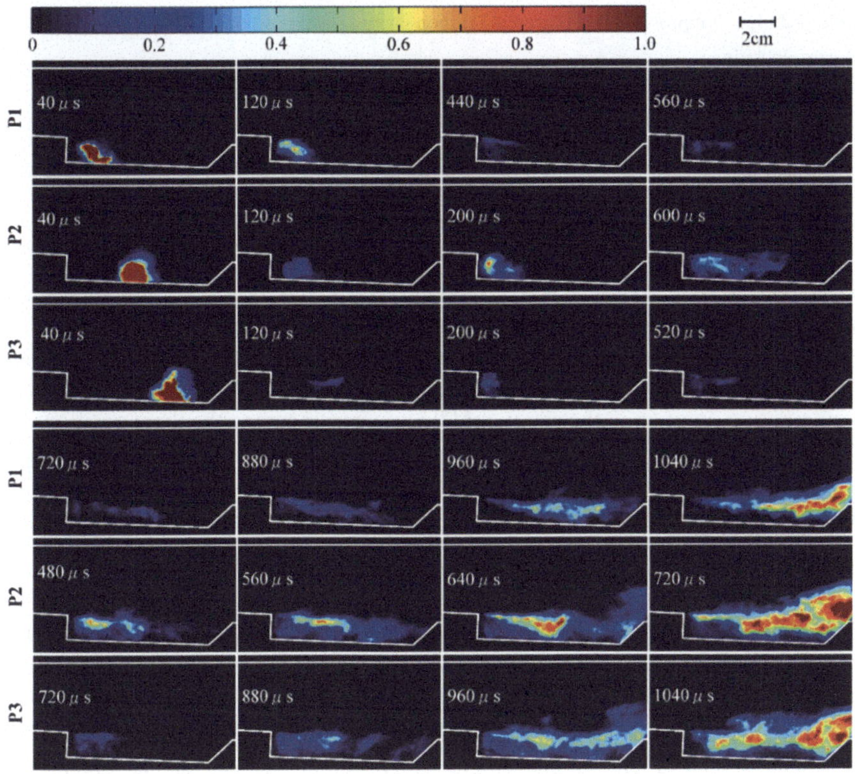

Fig. 4.16 CH* chemiluminescence images of ignition processes with ignition positions at P1, P2 and P3 [8]

process is achieved finally. If the ignition position is altered to P3, the initial flame kernel is the largest among these three cases. However, as the flame kernel spreads upstream, it shrinks in size and intensity strikingly, and arrived at the leading edge of the cavity by 200 μs. During the subsequent 320 μs, the flame kernel is anchored there without significant growth in size. Nevertheless, it begins propagating rapidly at 720 μs. Additionally, the ignition position does not influence the flame structure in the quasi-stable burning process.

During the experiment the laser system operated at 3 Hz, which means that for two successive laser pulses the time interval was about 333 ms. Since this interval is much longer than the typical residence time (~1 ms) in scramjet combustors, the flowfields at different ignition moments were independent of each other, provided that the combustor was not ignited by the former laser pulses. Analyzing successful and unsuccessful ignition events provides more insights. Table 4.2 reports ignition probability for 4 cases and 32 tests (each case was repeated for 8 times), where η_i ($i = 1$, 2, and 3) represents the probability that the combustor is successfully ignited by the i-th laser pulse. Assume η_s as the ignition probability for a single laser

Table 4.2 Ignition probability at different ignition energies and positions

Energy/position	η_1	η_2	η_3	η_s
303.6 ± 7.4 mJ/P2	0.875	0.125	0	0.87
230.7 ± 6.0 mJ/P2	0.375	0.500	0.125	0.42
303.6 ± 7.4 mJ/P1	0.500	0.375	0.125	0.50
303.6 ± 7.4 mJ/P3	0.625	0.375	0	0.62

pulse, then $\eta_s = \eta_1$, $(1 - \eta_s)\,\eta_s = \eta_2$, and $(1 - \eta_s)^2\,\eta_s = \eta_3$. According the values of η_1, η_2 and η_3 given in the table, the ignition probability η_s for each case can be estimated via the least square method. As the ignition energy increases from 230.7 to 303.6 mJ, the estimated value of η_s is doubled. This indicates that with the current configuration, ignition in the cavity middle is superior among the three positions.

4.2.3 Flame Propagation in the Single-Cavity Supersonic Combustor

With the experiment setup described in Sect. 4.2.1.2, fifteen ignition cases are conducted as listed in Table 4.3 and each case is executed twice for ensuring repeatability. In the table, A, B and C represent three cavities with different rear wall heights. Each cavity pertains to five cases, in order to study the features of combustion flowfield and the effects of equivalence ratio.

Table 4.3 Experimental test conditions

Case	$P_i\,(MPa)$	ϕ	Ignition state
A1	0.96	0.14	√
A2	1.30	0.20	√
A3	2.05	0.31	√
A4	2.69	0.40	√
A5	3.56	0.53	√
B1	0.94	0.14	√
B2	1.28	0.19	√
B3	1.94	0.29	√
B4	2.66	0.39	√
B5	3.46	0.53	√
C1	1.03	0.15	×
C2	1.44	0.22	×
C3	2.03	0.30	√
C4	2.74	0.41	√
C5	3.58	0.54	√

Previous numerical studies [10, 22] have revealed that the flowfield structure in the combustion with the rear-wall-expansion cavity would change greatly, and the cavity shear layer moves towards the bottom wall of the cavity, resulting in stronger turbulent dissipations and a smaller recirculation zone in the cavity. Thus, in the rear-wall-expansion cavity, the ignition environment would be exacerbated. As listed in Table 4.3, C1 and C2 failed to form a stable flame in the combustor, which indicates that for cavity C, it would be rather hard to achieve successful ignition when the equivalence ratio is lower than 0.3. This also demonstrates that the cavities A and B own a better ignition performance than C under low equivalence ratios.

4.2.3.1 Flame Propagation Routine

Figure 4.17 taken from the top view illustrates the flame propagation routine. The initial flame first propagates towards the leading edge of the cavity. After a short time, the initial flame grows stronger and propagates back towards the rear wall of the cavity. Finally, a steady flame is formed in the cavity. From t = 0.6 ms to t = 0.7 ms, it is noted that the initial flames near two side walls propagate faster than those in the middle part. This might be attributed to integrated effects of local equivalence ratios and turbulent flowfield in the cavity, which should be further investigated. The flame propagation routines of cases A3 and B3 present similar behaviors with that of case C3. However, the corresponding time spans of establishing a stable flame in the cavity are different, as shown in Fig. 4.18.

Figure 4.18 shows the difference of flame propagation among cases A3, B3 and C3 at t = 0.8 ms. All the three cases achieve successful ignitions at the equivalence ratio of 0.30. Nonetheless, both the flame propagation distance and the flame luminosity decrease from the top down, which indicates that lowering the rear wall height reduces

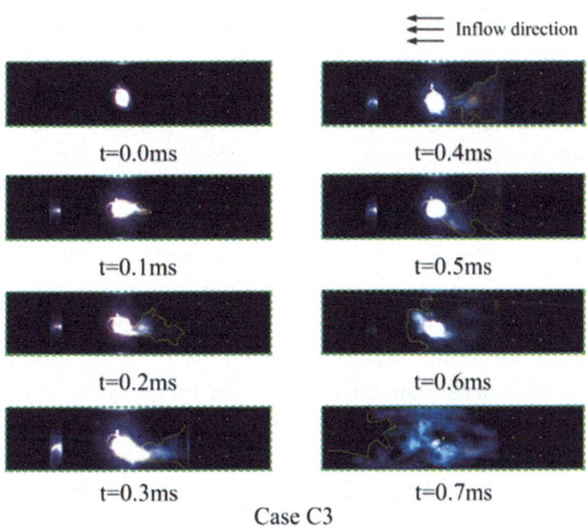

Fig. 4.17 Flame luminosity images (top view) of the flame propagation process of case C3 [14]

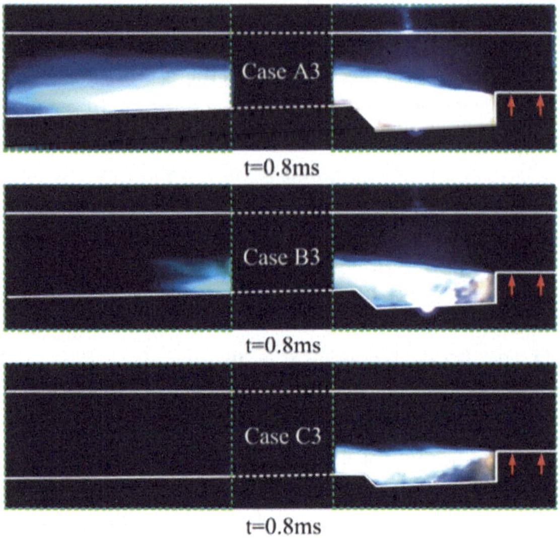

Fig. 4.18 Comparison of the flame propagation of cases A3, B3 and C3 at t = 0.8 ms [14]

the initial flame propagation. One possible reason is that lowering the rear wall height decreases the recirculation zone in the cavity and reduces the fuel entrained into the cavity. Therefore, the cavity with a large rear wall height is more favorable for ignition and flame propagation in a cavity-based supersonic combustor.

4.2.3.2 Reaction Zones Flashback Phenomenon

During the flame propagation process of cases A5, B5 and C5 with equivalence ratios close to 0.53, the reaction zone flashback phenomenon is observed. Figure 4.19 gives an example for the phenomenon captured in case C5. The brightest zones correspond to the most intense reaction zones. From t = 1.2 ms to t = 1.5 ms, the flame far downstream the cavity (the left side of the left window) grows thick, then the most intense reaction zones move upstream gradually. From t = 2.7 ms to t = 3.3 ms, the flame above the cavity becomes thick. As a result, a more evenly distributed flame is established in the combustor.

Widely existing in real scramjets, the reaction zones flashback phenomenon is a complex physical process closely related to the boundary layer separations. The phenomenon is probably induced by the upstream propagation of the combustion region with boundary-layer separation in the downstream weak combustion region [23]. It starts where the shockwave from the cowl lip crossed over the weak combustion region downstream the cavity at the initial phase of the ignition process. The transition is driven by the combustion-induced pressure rise. So the phenomenon usually occurs in the flame propagation process. Also, the combustion-induced pressure rise increases as the equivalence ratio becomes larger, which explains why the flashback phenomenon is clearly captured in cases A5, B5 and C5.

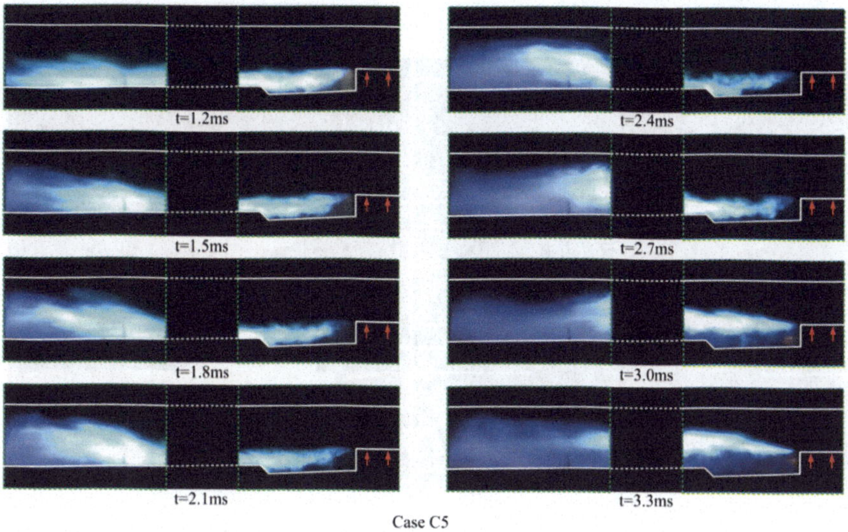

Fig. 4.19 Reaction zone flashback phenomenon during the flame propagation process of case C5 [14]

4.2.4 Flame Propagation in the Multi-cavity Supersonic Combustor

The ignition transients of hydrogen flame in a model scramjet combustor with parallel and tandem dual-cavity are separately simulated, along with the experimental observation on flame structure evolution. The numerical and experimental setups are given in Sect. 4.2.1.3.

4.2.4.1 Transient Process of Flame Stabilization in A parallel Dual-Cavity

Before reaching robust combustion, the cavity-stabilized jet flame would experience a transformation with the flame base moving around the cavities. For the parallel-dual cavity, a movement of flame observed experimentally is reproduced in simulation, as shown in Fig. 4.20.

In the transient process the reaction zone moves upstream, which is evident both from the flame luminosity images captured by a high-speed camera and the instantaneous temperature contours obtained in the simulation. Figure 4.20 shows three stages of flame structure during the stabilization process. Stage 1 relates the state just after the successful ignition, and stage 3 presents a periodic steady state. The time span for the three experimental images is 0.8 ms, and in the simulation changing

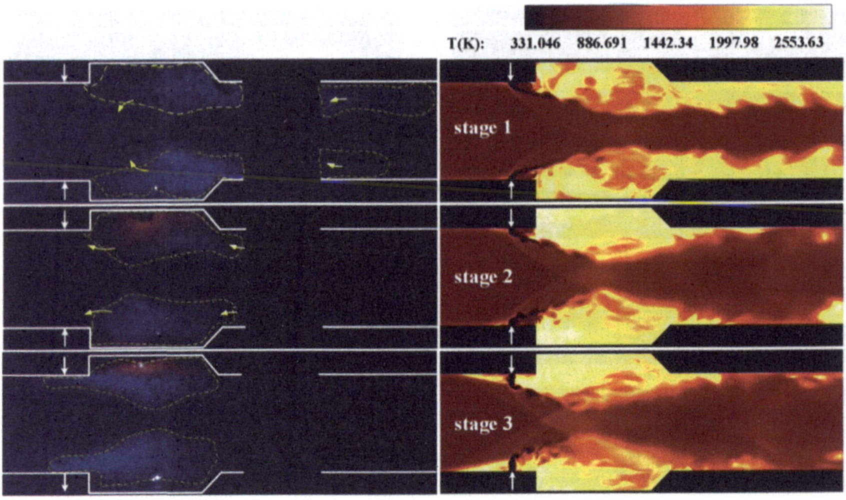

Fig. 4.20 Flame luminosity images captured by experiment (left) and instantaneous temperature contours in the central plane obtained by calculation (right) [15]

from stage 1 to stage 3 costs 487.5 $L_{ref}/U_{ref} \approx 0.713$ ms. Therefore, the time scale for the transient process is of the order of magnitude of 1 s.

In regard of the transient process, the simulation coincides well with the experiment. The yellow arrowheads marked on the flame luminosity images indicate tendency of flame development. The flame stabilized by the cavities eventually extends its leading front to the region around the jet exit, and then a small combustion zone emerges and stabilizes in the recirculation zone upstream of the jet exit. The existence of this combustion zone makes the hydrogen jet penetrate deeper into the core flow. During the flame stabilization transient process, the reaction zone downstream of the cavity is reduced, and at the dual-cavity the flames gradually bulge to compress the core flow passage. It is observed that the flame structure shown in stage 3 is quasi-steady on the given condition for both the experiment and the simulation, and the temperature of the main combustion zones exceeds 2400 K. Hot products within cavities contributes heat and active radicals to fuel-jet mixing regions.

The spatiotemporal evolution of the non-dimensional heat release rate (HRR) flux in Fig. 4.21 explicitly reveals the movement of reaction zone upstream towards the leading edge of the cavity during the transient. HRR flux is denoted as

$$Flux_{HRR}(x) = \frac{1}{\Delta t}\int_{0}^{\Delta t}\oint_{A(x)} HRR(x, y, z, t)\mathrm{d}y\mathrm{d}z\mathrm{d}t$$

where Δt represents a time step and $A(x)$ the combustor sectional area at location x. HRR flux helps to quantify the evolution of combustion intensity and heat release in

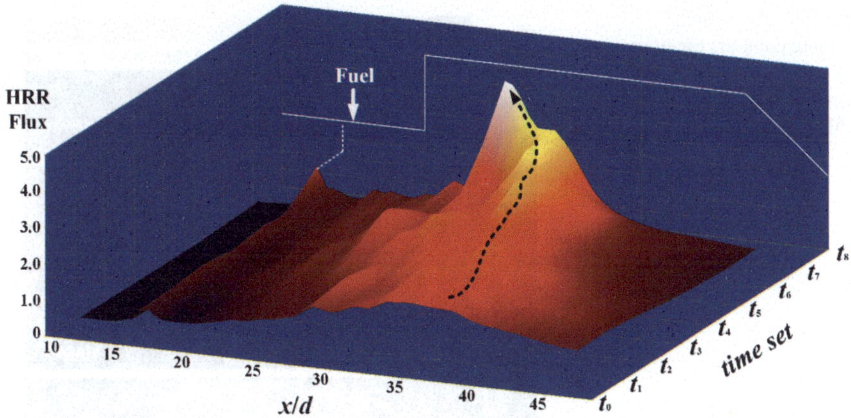

Fig. 4.21 Spatiotemporal distribution of non-dimensional heat release rate flux along the streamwise direction [15]

the combustor. In Fig. 4.21, the time set begins from t_0 to t_s between which the whole transient process is captured, and t_s is in stage 3 after which there exists no remarkable change in the distribution and magnitude of HRR flux, i.e., a quasi-steady combustion state has been reached. The HRR rate is computed with $\Delta t = 55.25\ L_{ref} / U_{ref}$.

As plotted in Fig. 4.21, the magnitude of heat release rate increases rapidly during the upstream extending of the cavity-stabilized flame zone. HRR flux starts from 1.0 at t_0, and then grows to about 4.0 at t_s, arriving at a 'heat peak'. The HRR flux history indicates the combustion becomes stronger during the transient process. After t_4, the combustion mainly concentrates in the middle of the cavity, and then spreads over the front part. In addition, there is an increment of HRR flux around the Hydrogen jet exit, which is attributed to the ignition and flame stabilization upstream of the injection location.

Figure 4.22 illustrates the streamlines and pressure contours of different stages in the transient process. The first subfigure presents the central plane in stage 1 when the hydrogen jet is successfully ignited and the early-stage flame zone is being established. In stage 1, the pressure of the flowfield rises slightly, and the recirculation flow within the cavities is weak. At this moment, the flame anchors at the rear part of the cavity and squeezes the streamlines, thus the core flow fails to reattach the combustor floor downstream of the parallel cavities.

The second subfigure corresponds to stage 2 where the transient process is undergoing. The pressure rises around the cavities and a quite high-pressure region (240–280 kPa) appears in the core flow. The streamlines evidently show that a major vortex grows under/above the cavity shear layer and predominates the flow in the recirculation zones. The vortex cores near the high-pressure region compress the incoming flow. In contrast to stage 1, the core flow in this stage reattaches the rear walls.

The third subfigure presents an approximately steady state of the cavity-stabilized flame. An region with higher pressure can be observed in the core flow, indicating

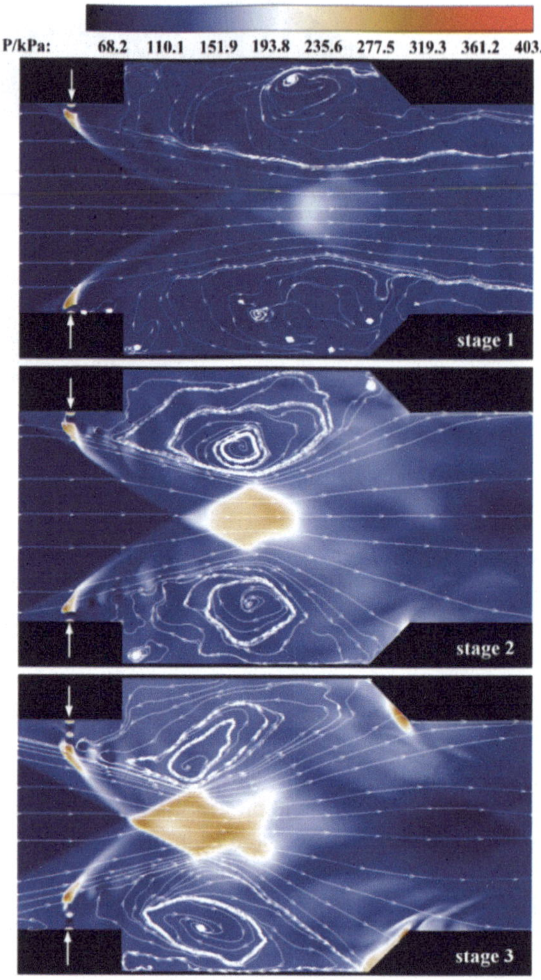

Fig. 4.22 Calculated pressure contours together with streamlines in the central plane of three stages [15]

stronger combustion overall. The high-pressure region, compared with that at stage 2, moves toward the combustor inlet. This movement coincides with the combustion region change shown in Fig. 4.20. Similarly, the predominating vortices in the recirculation zones experience a transformation and then lift the cavity shear layers. It is noteworthy that the flow and combustion in parallel cavities perform asymmetric characteristics provided with the same geometry, mesh and simulation conditions for the two cavities.

The profiles of streamwise velocity and temperature related to the transient process are shown in Fig. 4.23. The figure includes the data at axial locations $x/d = 20$ (the leading edge of the cavity) and $x/d = 44$ (the bottom edge of the rear wall), in order to emphasize the variations of core flow from stage 1 (blue) to stage 3 (red). From Fig. 4.23a and c, the faster and cooler inlet gas is compressed from $y/d = 2 \sim 18$ at

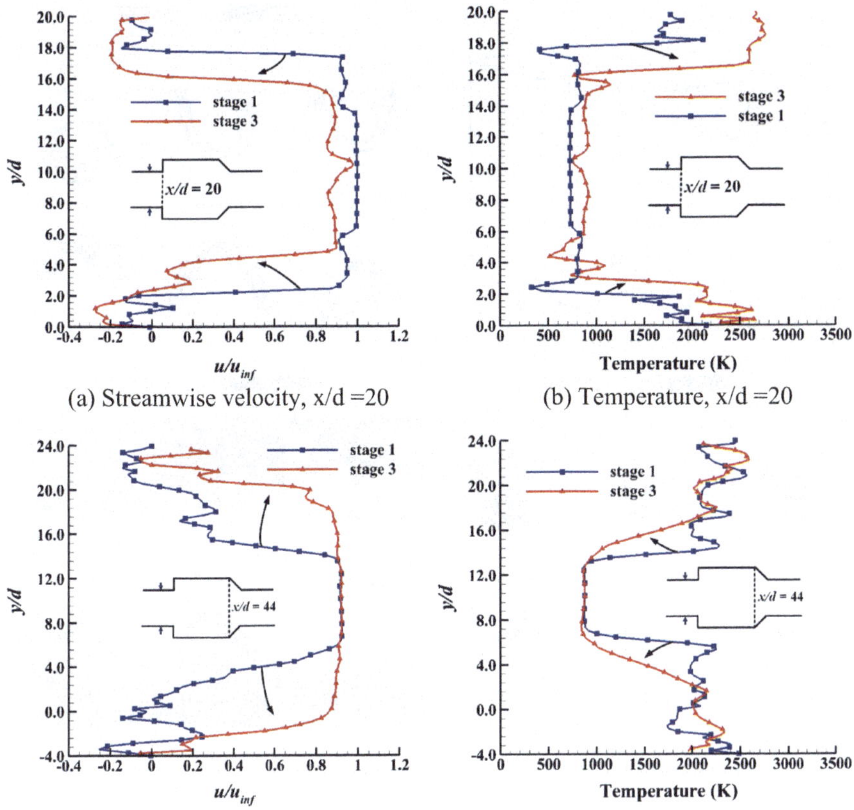

(a) Streamwise velocity, x/d =20 (b) Temperature, x/d =20

(c) Streamwise velocity, x/d =44 (d) Temperature, x/d =44

Fig. 4.23 Calculated profiles of streamwise velocity and temperature at two streamwise locations in the centralplane [15]

stage 1 to $y/d = 4 \sim 16$ at stage 3, and at the same time the core flow expands near the rear wall of the cavity. As plotted in Fig. 4.23c, the streamwise velocity at stage 1 regains 90% of u_{inf} in $y/d = 5$–15, but at stage 2 the range varies to $y/d = 0$–20. Figure 4.23d further denotes that the hot combustion region (around 2500 K) gets closer to the cavity floor. The whole behaviors are consistent with the observation in Fig. 4.20.

Figure 4.24 shows composition concentrations and non-dimensional HRR in the spanwise plane at the middle of the parallel cavities. The region with high HRR mainly concentrates in the jet shear layer. Further, both the region size and the HRR rise as the fuel jet penetrates deeper during the transient process. As plotted in the bottom left subfigure, hydrogen concentration gets weaker and the hydrogen spreads more widely around the cavities, denoting much fuel is consumed upstream of this spanwise plane and more fuel is entrained into the cavities. According to the bottom right subfigure, it is noted that the OH radicals chiefly appear in the recirculation

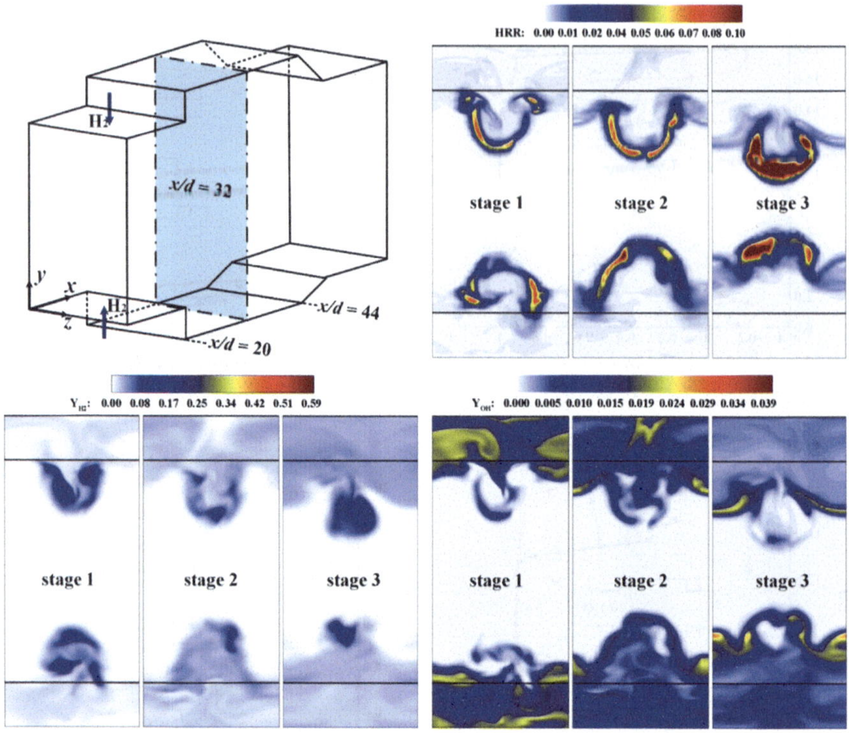

Fig. 4.24 Heat release distribution and OH, H_2 mass fractions in plane $x/d = 32$ [15]

zones at stage 1, then the concentration decays quickly within the two cavities during stage 2, and eventually in the shear layers are found the OH radicals at stage 3. This shift reflects the transient behaviors within recirculation zones.

Based on the above discussion from Figs. 4.20, 4.21, 4.22, 4.23, 4.24, it is illustrated in Fig. 4.25 that the mechanism of the transient process from ignition to flame stabilization, with respect to the dual-cavity. The flame evolution during the transient process may be attributed to positive feedbacks among the three sub-processes, involving (1) the strong heat release and hot products generating from jets and shear layers, (2) the recirculation zones where the major vortices stir whilst transfer active radicals, and (3) highly-pressurized combustion zones extending and compressing the incoming core flow. During the whole transient process, reaction keeps robust in the shear layers and the high-temperature reactants are continuously transferred into the recirculation zones. The initial flame caused by successful ignition is anchored near the rear parts of the parallel cavities, which helps to accumulate active radicals. Meanwhile, the major vortices are being built, and then accelerate the transport of energy and reactive chemicals into the fresh premixing fuel/air gas around the leading edges. Also, high pressure in the recirculation zones further enhances the combustion in them. As time pushes forward, therefore, larger robust combustion zones

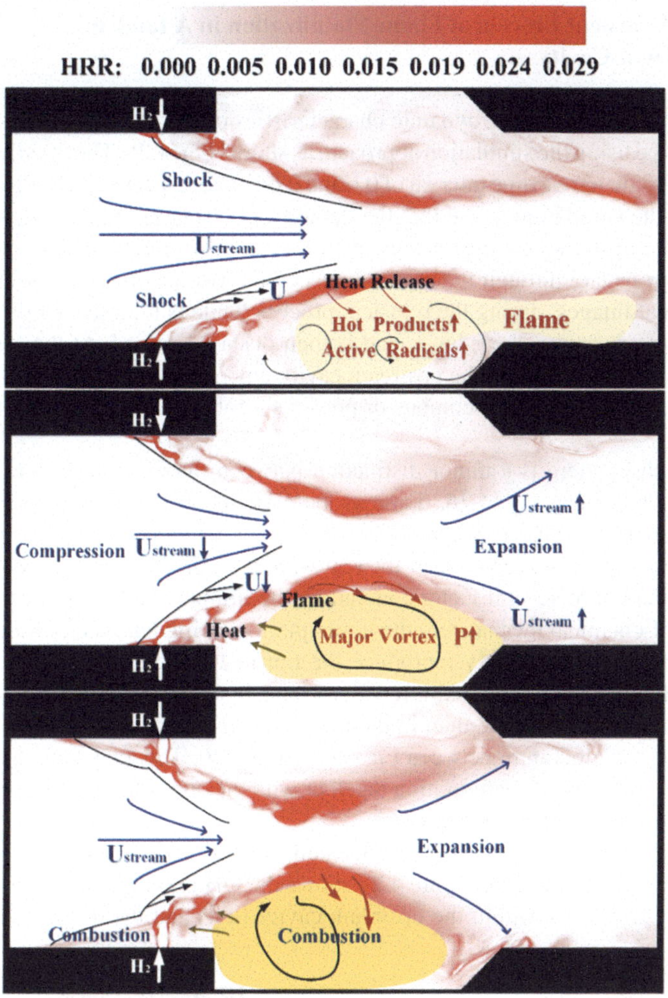

Fig. 4.25 Schematic of flame stabilization transient process in parallel dual-cavity [15]

are established in the parallel cavities. With the interaction of flame and shockwave developing from the two cavities, strong reaction regions quickly move upstream along the shear layers, consuming more fuel within a short distance downstream of the jet exits. On the other hand, the expanding recirculation zones also compress and slow down the core flow, in turn enhancing the premixed or partially-premixed combustion in the jet wakes. The process witnesses a quick lift of jet penetration, and a significant upstream movement of strong combustion zones in the dual-cavity. After that, dual-cavity-stabilized flame climbs over the leading edges, finally getting stabilized in the separation zones of the two jet exits.

4.2.4.2 Transient Process of Flame Stabilization in A tandem Dual-Cavity

For the tandem dual-cavity, dynamic characteristics of the flame transients are also well captured using the simulation approach, as shown Fig. 4.26. The three images on the left and right have a time interval of 0.6 and 0.665 ms, respectively. Both the experiment and the simulation reveal that the flame front displaces from the downstream cavity to the upstream cavity, during which the 'stabilization base' for combustion seems to propagate through the region between the two cavities. According to the experimental images, during the transient process flame is initially stabilized along the downstream cavity shear layer, and is then pushed forward. At the same time, a flame packet anchored by the upstream cavity emerges, as plotted in the left middle image. After that, in the combustor presents a robust flame throughout the two cavities.

Nonetheless, there is a minor distinction between the experiment and the simulation. As shown in Fig. 4.26, flame in the experiment is totally invisible in the upstream cavity at stage 1, but in the simulation the flame is evident that there is a small high-temperature region near the rear wall. This distinction may be attributed to a combination of several factors, including the limited resolution in experiment, the reduced-chemical mechanism, the mesh quality in simulation, etc. Although the simulation fails to accurately reproduce the transient process, it roughly captures the related phenomena. The major reaction zones in the tandem dual-cavity show a relatively lower temperature than in the parallel dual-cavity, via the comparison of instantaneous temperature between Figs. 4.18 and 4.26. This may demonstrate that the parallel dual-cavity configuration together with the opposite fuel injections could strengthen combustion around cavities.

Existence of OH radicals in the hydrogen/air mixture can be viewed as an evidence for presence of combustion. Figure 4.27 shows the OH mass fraction distribution in the plane $y/L_{ref} = -1$ within the upstream cavity (left) and the downstream cavity

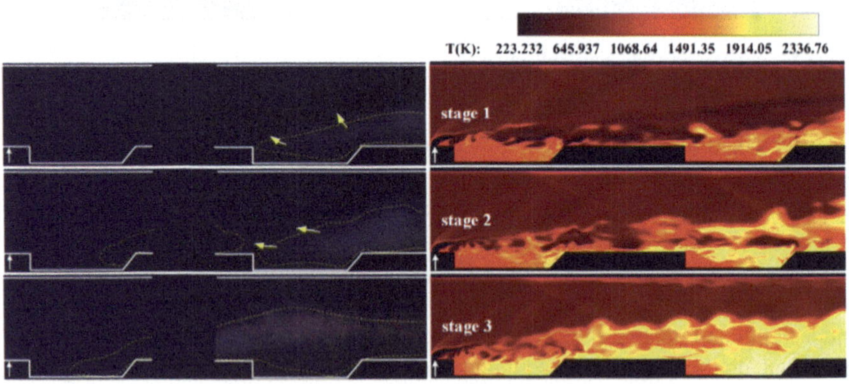

Fig. 4.26 Flame luminosity images captured by experiment (left) and instantaneous temperature contours in the central plane obtained by calculation (right) [15]

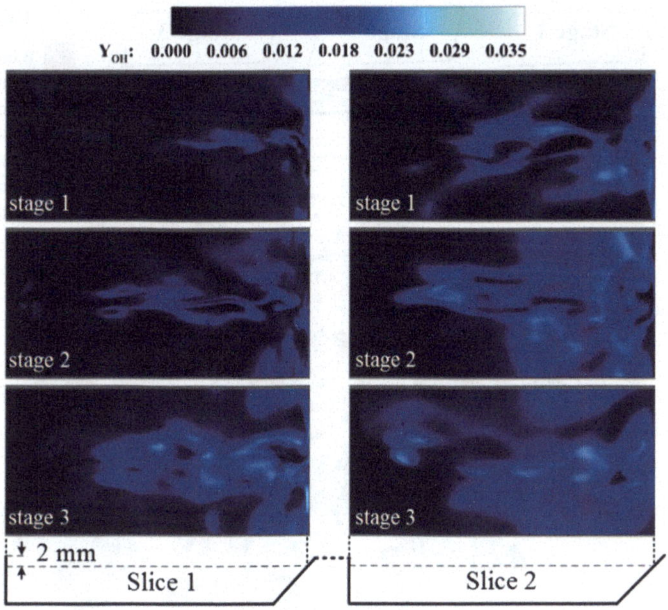

Fig. 4.27 The development of OH mass concentration in two vertical slices with the tandem cavity [15]

(right), respectively. Both the planes have a 2-mm distance below the cavity lips, as labelled in the figure. The OH mass fraction variation clearly depicts how the combustion is organized within the recirculation zones of the two cavities during the transient process. Specifically, there is a plenty of OH radicals in the downstream plane, indicating the initial flame observed both in the experiment and the simulation. From stages 2 to 3 the transient process occurs with the combustion region expanding spanwisely and toward the front parts of the two cavities.

In order to display the translational flame motion in the tandem dual-cavity, Fig. 4.28 shows the 2000 K temperature iso-surfaces where the colors represent the HRR. Variation of the high-temperature iso-surface between different stages delineates the development of main reaction regions from the downstream cavity to the upstream cavity. During the transient process, high-temperature regions caused by jet flame quickly roll up, enlarging themselves streamwisely and spanwisely. It takes less than 1 ms for the flame to conquer the upstream cavity. At stage 3, a strong combustion is stabilized by the two tandem cavities, and the high temperature regions further expand along the streamwise direction from the upstream cavity to the computational combustor exit. The HRR magnitude shown on the iso-surfaces reveals evolution of chief heat release regions. At stage 1, the heat release mainly concentrates near the downstream cavity and comes from initial flame packets. Then the strong heat release regions quickly move upstream, as plotted in the middle iso-surface. At stage 3, the core heat source is situated downstream of the fuel jet, and

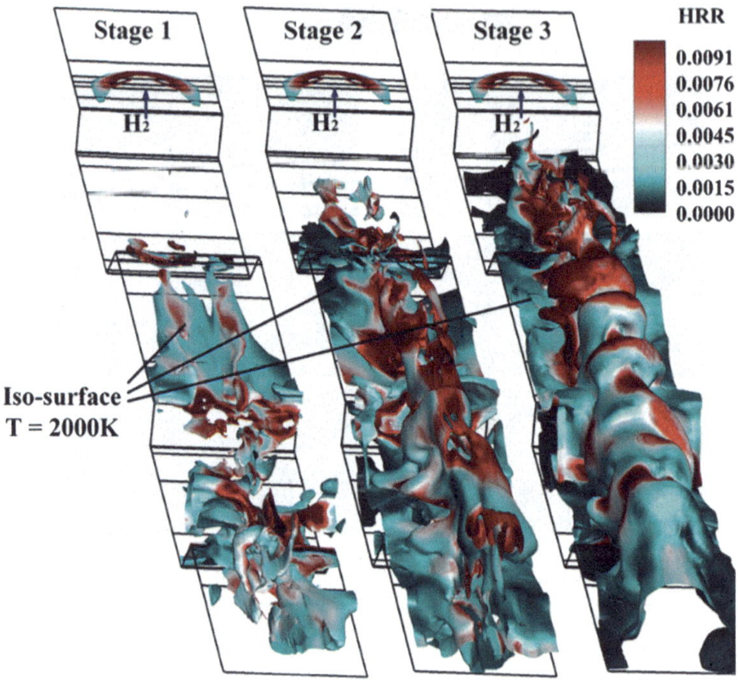

Fig. 4.28 Temporal growth of 2000 K iso-surface where the colour represents HRR [15]

stabilized by the upstream cavity. Figure 4.28 agrees well with flame luminosity images obtained by high-speed cameras.

Figure 4.29 shows HRR flux along the streamwise direction throughout the computational combustor. For stage 1, HRR flux has large values in $100 < x/d < 120$, near the rear wall of the downstream cavity. At stage 2, HRR flux then concentrates in $50 < x/d < 80$, between the two cavities. When stage 3 is reached, HRR flux peaks around $x/d = 65$. The results reveal where the strongest combustion exists and how it is established in the transient process lasting about 1 ms.

A plausible mechanism is illustrated in Fig. 4.30 to explain phenomena observed in the transient process. In the downstream cavity, initial flame generates a heated region with increasing pressure around it. This hot and active region combines with small flame packets anchored at the rear wall of the upstream cavity to produce larger flow separation between the two cavities. The separation zone has more appropriate conditions for combustion, the fuel ignition delay in the zone being reduced. As more hydrogen is consumed in the downstream cavity, the flame begins to propagate upstream, then merges with the flame in the upstream cavity, finally forming a robust flame. Though flame exists in the upstream cavity, the entire flame development is dominated by the combustion in the downstream cavity. In the separation zone the appropriate conditions, possibly ascribed to heat and radical transfer from the two cavities and the pressure rise, accounts for the fast flame motion.

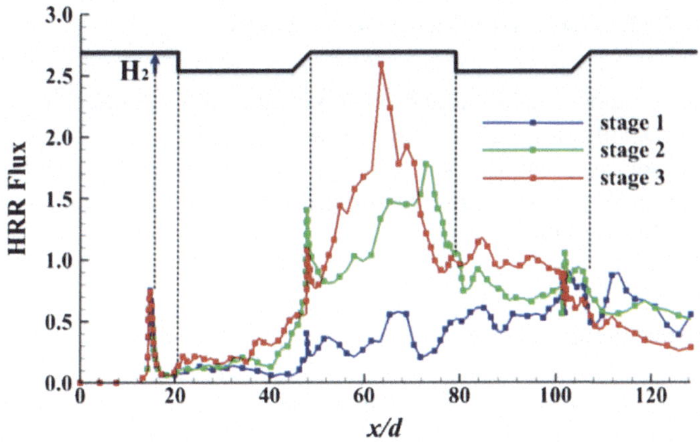

Fig. 4.29 Streamwise HRR flux distribution of three stages in the transient process of the tandem dual-cavity [15]

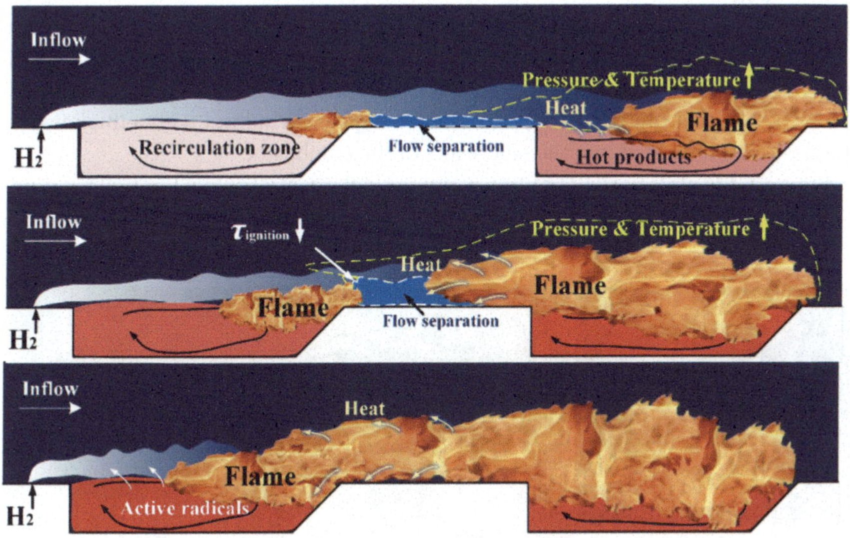

Fig. 4.30 Schematic of mechanism of the flame transients in the tandem dual-cavity [15]

4.3 Ignition Mechanism Analysis

The main purpose of studying the ignition process in a supersonic flow is to comprehensively reveal the ignition mechanism explaining diverse ignition behaviors. In this section, the ignition mechanism is clarified by introducing ignition stages as well as ignition modes.

4.3.1 Experimental and Numerical Setups

The supersonic inflow conditions and experimental facility have been introduced in Sect. 4.2.1.1. Details of the test facility can be found in Refs. [20, 24]. Information about the injectors are referred to Ref. [25] The experiments conducted in Sects. 4.3.2 and 4.3.3 are based on LIP ignition and share the same cavity geometry, but with different imaging systems.

4.3.1.1 Experimental Setups

A photo for the LIP ignition experiments in Sect. 4.3.2 is presented in Fig. 4.31. The combustor was equipped with a typical rear-wall-expansion cavity. As shown in Fig. 4.32a, the leading edge depth, rear wall depth, floor length and aft ramp angle of the cavity were 20 mm, 10 mm, 90 mm and 45°, respectively. The upstream wall of the cavity was in parallel with the downstream wall. Two porthole injectors with a diameter of 2 mm were situated 10-mm upstream of the cavity, dividing the spanwise plane equally. Ethylene fuel was injected vertically to the inflow at the room temperature. The global equivalence ratio was respectively set to 0.16, 0.30 and 0.48, corresponding to the fueling stagnation pressures of 0.8 MPa, 1.5 MPa and 2.4 MPa. The fuel injection lasted for 1 s. Two quartz windows with a thickness of 20 mm were embedded into the side walls of the combustor.

Figure 4.32a plots the LIP ignition positions in the central plane. Ignition positions *a*-1, *b*-1, *c*-1, *d*-1, *e*-1, and *f*-1 have different horizontal distances to the leading edge,

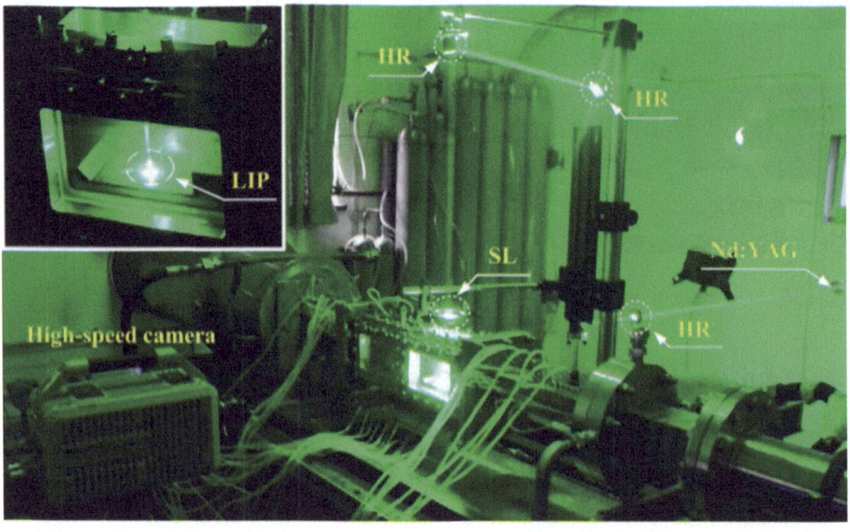

Fig. 4.31 Photo of the LIP ignition experiment facility [26]. The upper-left photo shows the combustor with incident laser

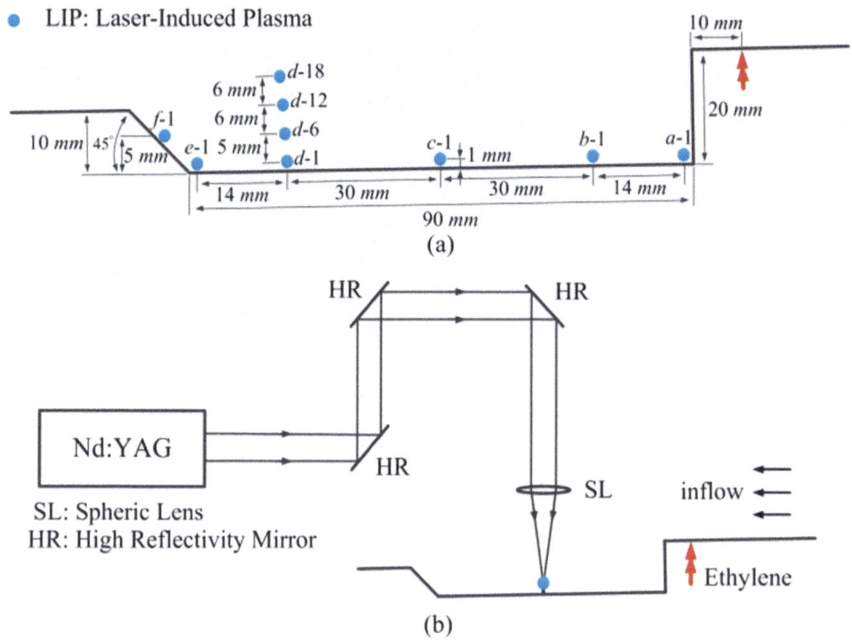

Fig. 4.32 Schematic of **a** the LIP ignition positions and the cavity, and **b** optical arrangement of LIP ignition experiments [26]

but their vertical distances to the cavity floor are all 1 mm. Ignition positions *d*-1, *d*-6, *d*-12 and *d*-18 are aligned vertically with different heights. Two fuel injectors were placed symmetrically about the central plane, and the distance of each injector to the central plane equals 8.4 mm.

The optical arrangement for the LIP ignition is shown in Fig. 4.32b. A frequency doubling Nd:YAG laser output 532-nm laser pluses with a pulse duration of 10 ns for ignition. After guided by several high-reflectivity (HR) mirrors, the laser beams were focused by a spherical lens (SL, f = 130 mm) in the central plane and above the cavity. Since the major laser energy is absorbed by the mixture, the energy reflected by the bottom wall is negligible. The laser pulse energy was kept at 300 mJ, and the absorbed energy approximately equalled 250 mJ.

The ignition and flame propagation processes were visualized by a high-speed camera (FASTCAM SA-X2) with a f/1.4 Nikkor lens. The camera worked at 50 kHz with a resolution of 768 × 328 pixels and an exposure time of 18 μs. A 10-nm bandpass filter centered at 431 nm was mounted before the camera to collect CH* chemiluminescence.The laser system and the camera were synchronized via a digital delay generator (DG645).

The LIP ignition process was also characterized by a schlieren system composed of a high-speed camera (FASTCAM SA-X2) equipped with a Nikkor lens (f = 200 mm, f/4). Both the two high-speed cameras share the same settings. Additionally, the wall

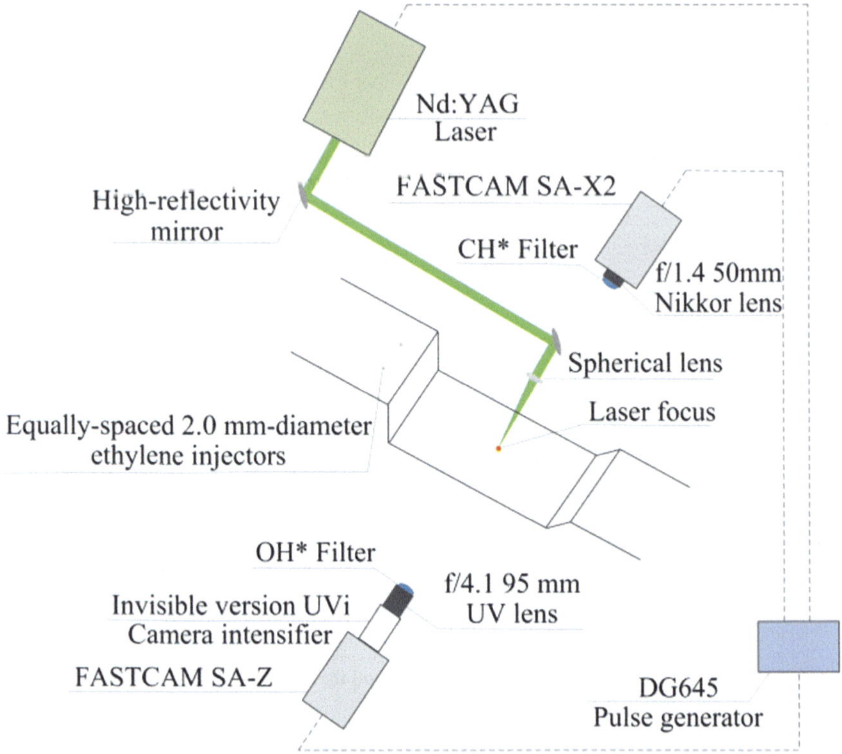

Fig. 4.33 Schematic of the experimental setup [28]

pressure along the combustor floor was measured through 32 static pressure taps with a sampling frequency of 100 Hz. The pressure transducers had an uncertainty of ±0.5% FS (full scale) [27].

The experimental setups in Sect. 4.3.3 are illustrated in Fig. 4.33. Configurations of the laser system and the optical paths were similar to that in Sect. 4.3.2. The laser beams were focused 1 mm above the cavity floor. The laser pulse energy was set as 200 mJ/pulse or 300 mJ/pulse. The efficiency of laser energy absorption was approximately 80%. It is noteworthy that two high-speed cameras simultaneously and respectively captured CH* and OH* chemiluminescence from two opposite sides of the combustor. The first camera (FASTCAM SA-X2) was equipped with a Nikkor lens (f = 50 mm, f/1.4) and a bandpass filter (centered at 431 nm, with a 10-nm FWHM). Another camera (FASTCAM SA-Z) utilized a camera intensifier (Invisible version UVi), a UV lens (f = 95 mm, f/4.1) and a bandpass filter (centered at 311 nm, with a 10-nm FWHM) to capture the OH* chemiluminescence. Both the cameras had an image frequency of 10 kHz, a resolution of 1024 × 512 pixels and an exposure time of 98 μs. A digital delay generator (DG645) synchronized the laser system with the two cameras to make the LIP ignition process be recorded precisely.

Schlieren imaging were widely employed to reveal the flow structures in supersonic flowfield [29, 30]. In Sect. 4.3.3, the flow structures during LIP ignition process were then visualized via schlieren imaging. The schlieren system consisted of a high-speed camera (FASTCAM SA-X2) with a Nikkor lens (f = 200 mm, f/4). The camera operated at a frame rate of 50 kHz, with a resolution of 768 × 304 pixels and an exposure time of 18 μs.

4.3.1.2 Numerical Setups

In Sect. 4.3.2, the fuel/air mixing and the flow structures before LIP ignition were studied through the large eddy simulation (LES). The LES equations were solved by Scramjet Foam [10], a solver modified from Open FOAM [31]. The solver utilized the Euler backward scheme for temporal integration and a second-order Gauss filtered linear scheme to discretize the diffusion terms. In the solver, the convective terms were discretized using the Kurganov and Tadmor scheme with the Van Leer Limiter. Thus, second order accuracy in both time and space could be achieved. The normalized grid resolution in the cavity where the mixing took place was $\Delta x^+ \approx$ 1–60, $\Delta y^+ \approx$ 1–40 and $\Delta z^+ \approx$ 20–60.

A 3D structured grid with a size of 11,690,775 was used. The cavity region where mixing took place had normalized grid resolutions of $\Delta x^+ \approx 1 \sim 60$, $\Delta y^+ \approx 1 \sim 40$ and $\Delta z^+ \approx 20 \sim 60$. The inflow condition was in accordance with that in the experiments. The ethylene injection had a stagnation pressure of 1.5 MPa, corresponding to a global equivalence ratio of 0.3. References [10, 11, 22, 32] detailed the numerical methods.

4.3.2 Four-Stages Dominated Ignition Process

Spark ignition and flame motions in the scramjet combustor with a rear-wall-expansion geometry has been studied both experimentally and numerically [12, 13, 32, 33]. In this section, experiments were performed to investigate LIP ignition processes in a rear-wall-expansion cavity model scramjet combustor with an inflow of Ma = 2.92. The LIP ignition and related combustion dynamics were investigated using CH* chemiluminescence imaging at 50 kHz. The results indicate that the ignition and flame stabilization are sensitive to the LIP location. To further understand the LIP ignition process in the combustor, large eddy simulations were conducted to characterize the mixing field and the flow dynamics. On basis of the experiments and the simulation, the LIP-assisted ignition in a cavity-based scramjet combustor is explained fundamentally.

4.3.2.1 Ignition Process in the Cavity

First, the initial flame propagation process within the rear-wall-expansion cavity is examined. The LIP ignition experiment was executed two times to observe the repeatability in each ignition position. The ignition results are presented in Fig. 4.34, where the red dots indicate the locations at which successful ignition and sustained combustion are achieved, and the back dots denote ignition failure. In *f*-1 and *d*-18, the ignition failed at all global equivalence ratios (Φ), but in ignition positions *a*-1 and *b*-1 the ignition was unsuccessful only at $\Phi = 0.16$. Successful ignition was achieved at all equivalence ratios for *c*-1, *d*-1, *e*-1, *d*-6 and *d*-12.

Processes of ignition and initial flame propagation were visualized through the high-speed imaging of CH* chemiluminescence and schlieren. CH* has been shown as a good marker of local reaction zones [34]. The high-speed imaging of CH* chemiluminescence shows the flame-front and heat-release regions on a time-resolved basis without spatial blurring, and the schlieren images indicate the density gradient.

Figure 4.35 gives the experimental observation with LIP ignition at *e*-1. The LIP ignition (with a laser pulse of 10 ns) event happens at t = 0 μs. It is seen that 20 μs after the LIP ignition, a rather large region of high CH* chemiluminescence appears around the ignition position where the schlieren image shows a white zone with saturated pixel intensities. The LIP energy focused on the ignition position results in an ultra-high temperature region (e.g. with a temperature possibly as high as 4000–16,000 K [35]) where the gas molecules undergo complex processes such as dissociation, atomization, ionization and excitation. Due to the large temperature and concentration gradients across the boundary of LIP ignition region, the high

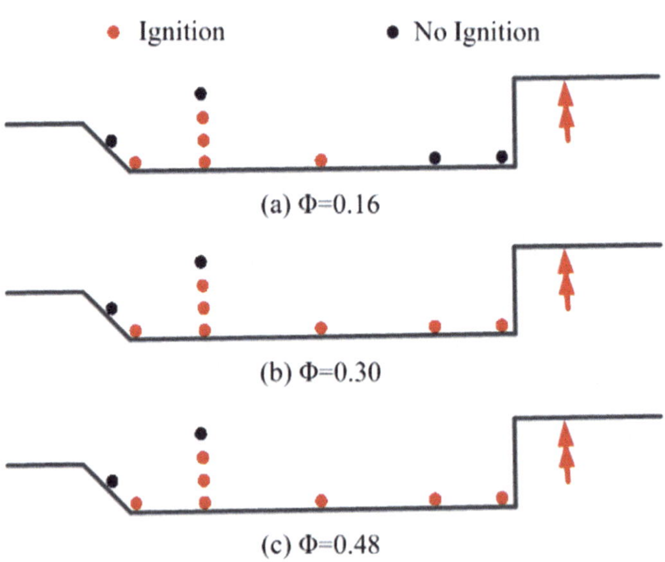

Fig. 4.34 Schematic of the LIP ignition results [26]

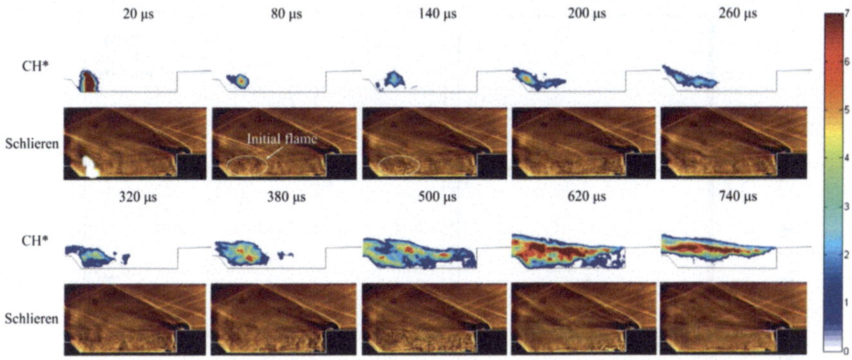

Fig. 4.35 Images of CH* chemiluminescence and schlieren after LIP ignition at *e*-1 with $\Phi = 0.30$ [26]

temperature gas around the region diffuses into the ambient colder fuel/air mixture. This diffusion cools the region and reduces the intensity of CH* chemiluminescence, as shown at $t = 80\ \mu s$. The region enclosed by the white ellipse shows finer structures with variation of density gradients than that the ambient, which denotes in this region significant heat and mass transfer. Moreover, this region is much larger than the region where exists observable CH* chemiluminescence.

It is expected that the temperature in the surrounding gas could initiate the chemical reactions, which is confirmed in the schlieren and CH* chemiluminescence images at 140 μs that some signals are detected outside the main signal region. As shown in the subfigures from 140 to 500 μs, the region of CH* radicals develops inside the cavity toward the injector. At 740 μs after LIP ignition a flame is then well stabilized. This observation is consistent with the results in Ref. [8], where the initial ethylene flame propagated first towards the leading edge and then spread downstream, regardless of the LIP ignition position in the cavity.

As depicted in Fig. 4.35, during 500 and 620 μs the CH* signal is found in nearly the entire cavity, which means the fuel in the cavity is consumed quickly. Thereafter, the main combustion occurs in the shear layer. As a result, the fuel entrained into the cavity becomes much less. The cavity then acts as a reservoir filled with hot products and radicals (e.g. OH) that stabilizes the flame in the shear layer [22]. This is confirmed by the signal distribution at 740 μs.

Figure 4.36 plots the integration of signals over each CH* chemiluminescence image during the ignition, with respect to different fuelling rates. The integrated signals are normalized by the mean value obtained at the stable flame stage. Very high integrated signal appears before 50 μs, corresponding the LIP ignition event. When $\Phi = 0.30$, the integrated signal increases from 50 μs and then peaks at 600 μs. In this period, as shown in Fig. 4.35, the flame propagates into and fills the entire cavity. Thereafter, the integrated signal initially decreases and then oscillates around the mean value after 740 μs, since a flame has been stabilized in the shear layer. With different fuelling rates, the integrated signal presents similar trends. However,

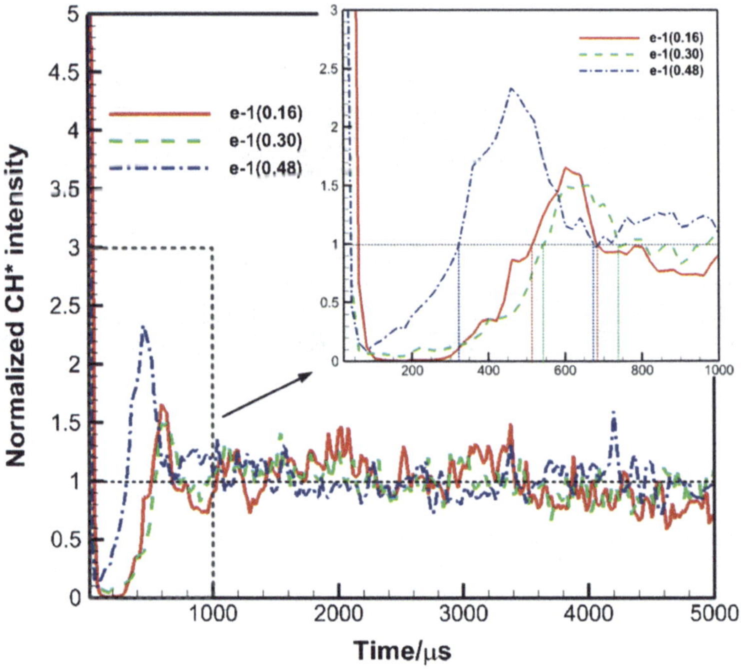

Fig. 4.36 Integrated CH* chemiluminescence intensity against time after LIP ignition at *e*-1 with $\Phi = 0.16$, 0.30 and 0.48 [26]

the flame development with a higher fuelling rate ($\Phi = 0.48$) takes a shorter time (from 50 to 420 μs), possibly because the fuel-richer mixture in the cavity is more in favor of ignition process. Another perceptible difference in the ignition process for the three fuelling rates is shown in Fig. 4.37. With a lower fuelling rate ($\Phi = 0.16$), there is no CH* chemiluminescence observed in the bottom right of the cavity. This phenomenon is mainly due to the low local equivalence ratios, which will be discussed in next section.

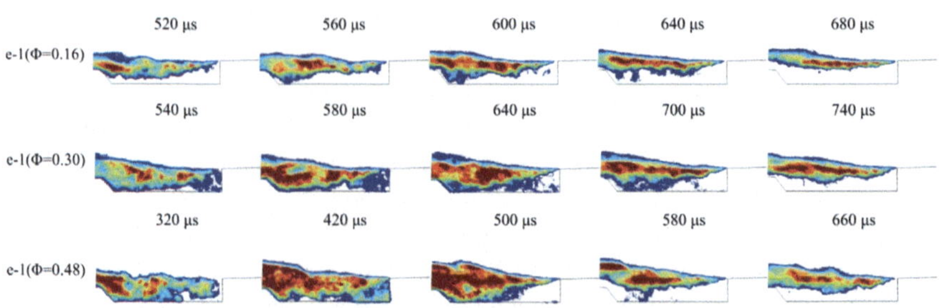

Fig. 4.37 CH* chemiluminescence images after LIP ignition at *e*-1 with $\Phi = 0.16$, 0.30 and 0.48 [26]

4.3.2.2 Mechanism of Cavity Ignition

Exploring the mixing field and the flow structure in the cavity helps to understand key factors leading to ignition of mixture in the cavity. Figure 4.38 shows contours of the velocity, the Mach number and the local equivalence ratio in the cavity and the shear layer prior to the LIP ignition event. The contours come from LES of the case $\Phi = 0.30$. Detailed LES results will be given in the next section, where the numerical accuracy and comparison with experiments are presented, as well as the three-dimensional mixing field and enstrophy field.

According to Fig. 4.38, the flow decelerates rapidly across the shear layer. The flow velocity decreases from Mach numbers greater than 1 in the free stream outside the cavity to rather low Mach numbers inside the cavity. Meanwhile the flow inside the cavity forms a large recirculation zone. Partial fuel/air mixture in the shear layer entrances into the cavity along the rear wall, and then flows towards the leading edge, and finally moves upward to the free stream. The recirculating flow inside the cavity convects the plasma/gas mixture from the LIP ignition position *e*-1 (illustrated in Fig. 4.32) to the leading edge. The hot mixture acts as a heat and radical source that eventually ignites the flammable mixture inside the cavity. Once the mixture inside the cavity is sufficiently burned, the cavity functions as a flame holder that continuously ignites the fuel/air mixture in the shear layer.

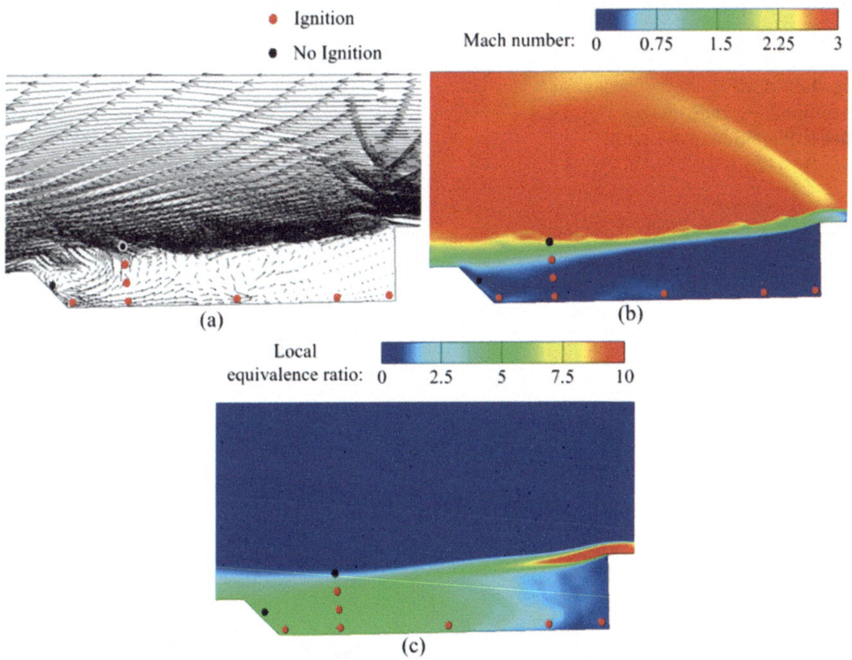

Fig. 4.38 Distributions of **a** velocity vector, **b** Mach number, and **c** local equivalence ratio. The results are from LES with $\Phi = 0.30$ [26]

Since the LIP positions *d*-18 and *f*-1 are located outside the recirculation zone, the generated plasma/gas mixture is convected directly downstream and then quenched due to the locally high Mach number. This explains the ignition failures for LIP ignition at *d*-18 and *f*-1. The LIP positions *d*-6 and *d*-12 are in the recirculation zone and thereby a stable flame is formed. Although the LIP positions *a*-1 and *b*-1 are inside the recirculation zone, the plasma/gas mixture formed in the two positions has a much shorter resident time in the cavity than the other cases, since the positions are close to the leading edge. Such short resident time limits the heat released by interaction of the plasma, the fuel and the air in the cavity, especially when the local equivalence ratio is low. This might explain why the LIP ignition at *a*-1 and *b*-1 with $\Phi = 0.16$ fails, since the small amount of heat release is unable to ignite the shear layer.

To confirm the above discussion, the CH* chemiluminescence images with LIP ignition at *a*-1 and $\Phi = 0.30$ are shown in Fig. 4.39. The initial signal from the LIP around *a*-1 lasts for 260 μs. From 320 to 580 μs, the CH* chemiluminescence quenched when the hot plasma/air mixture is convected to the shear layer along the leading edge. From 720 to 860 μs, the mixture inside the cavity is re-ignited. It is inferred that the radicals and heat form LIP is partially re-circulated into the cavity, which eventually initiates the auto-ignition in the cavity. A similar initial flame development for the LIP ignition at *c*-1 is given in Fig. 4.40. After a LIP ignition at *c*-1, the initial flame quenches at 140 μs and it takes a longer time (from 140 to 820 μs) for the mixture in the cavity to be re-ignited.

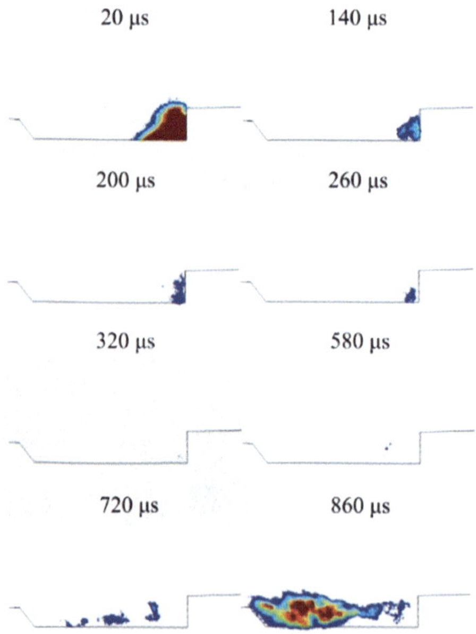

Fig. 4.39 CH* chemiluminescence images after LIP ignition at *a*-1 with $\Phi = 0.30$ [26]

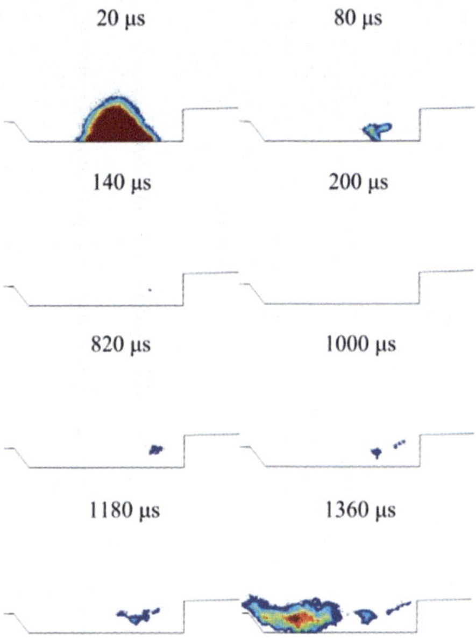

Fig. 4.40 CH* chemiluminescence images after LIP ignition at *c*-1 with $\Phi = 0.30$ [26]

Figure 4.41 shows the integration of intensities over CH* chemiluminescence images obtained in the cases with $\Phi = 0.30$. From Figs. 4.35–4.37 and Figs. 4.39–4.41, it is clear that the ignition process in the cavity at various LIP positions can be summarized as four stages. The first stage is the 'plasma ignition' stage (I), during which the residual thermal and plasma effects of the laser are important. The second stage is the 'plasma quenching' stage (II), during which the cooling from the ambient cold gas results in quenching of the chemical reactions introduced by the plasma. A lower CH* chemiluminescence signal indicates low chemical reaction rates. The third stage is the 're-ignition' stage (III), during which the residual gas from the LIP initiates auto-ignition of the mixture. This stage can be characterized with the onset of CH* chemiluminescence. The final stage is the 'stable flame' stage (IV), during which the ignition process in the cavity has completed and a stable flame has been established in the shear layer.

The ignition time may refer to the time interval from the LIP ignition event to the moment at which the integrated signal of CH* chemiluminescence image reaches its maximum (at the end of stage III). As plotted in Fig. 4.41, the ignition time is sensitive to the LIP ignition positions. The ignition time first increases when the LIP position varies from *a*-1 to *c*-1, then decreases when the LIP position changes from *d*-1 to *e*-1. Furthermore, the ignition time increases when moving the LIP position vertically from *d*-1 to *d*-12. Therefore, *e*-1 is preferred as the LIP ignition position since it has the shortest ignition time.

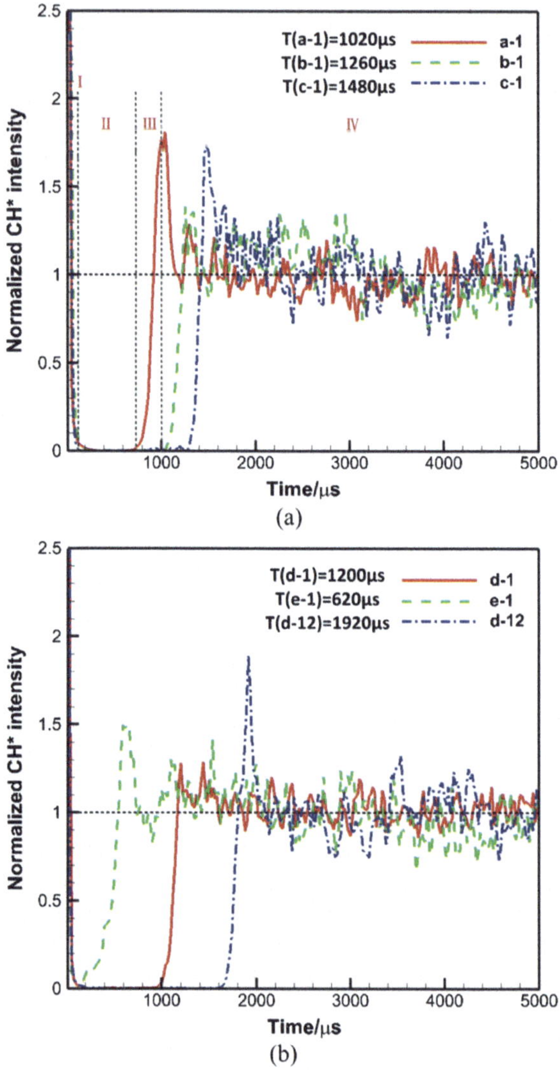

Fig. 4.41 Integrated CH* chemiluminescence against time after LIP ignition at different positions with $\Phi = 0.30$ [26]

4.3.2.3 Fluid Flow and Mixing in the Cavity

To further explore the LIP ignition process in the cavity, LES of the mixing process prior to the LIP ignition is carried out. The grid independency analysis adopts two grids: a moderate grid with 11,690,775 cells and a refined grid with 17,536,163 cells. Figure 4.42 plots the wall-pressure distributions from the numerical simulations and experiments when $\Phi = 0.30$. Here, $x = 0$ denotes the combustor inlet. The cavity is located in $x = 410$–520 mm. It can be seen that the numerical results match the experimental data reasonably well, except that around x = 600 mm the wall pressure

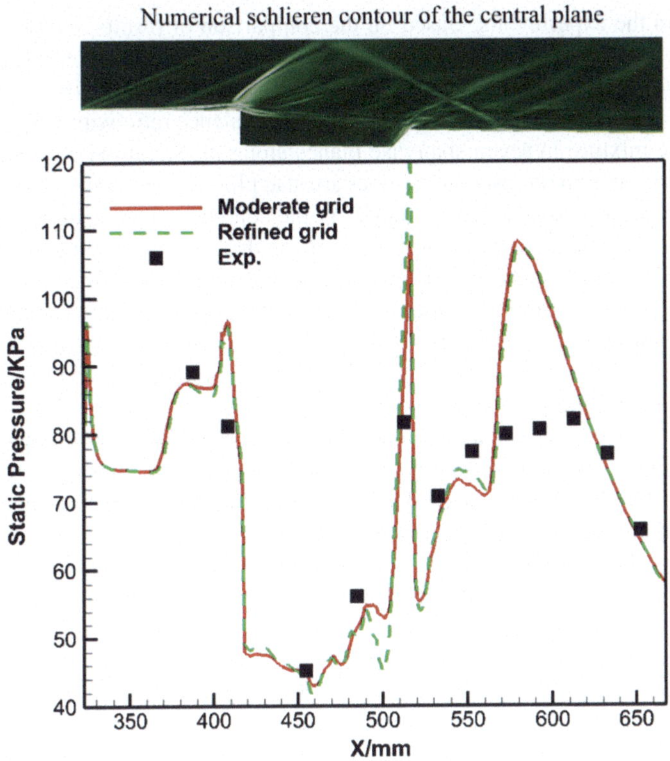

Fig. 4.42 Wall-pressure distribution along the combustor floor with $\Phi = 0.30$ from experiment (Exp.) and LES using two different grids, a moderately fine grid and a refined grid. A two-dimensional distribution of the density gradient (numerical schlieren) is also shown at the top [26]

is overestimated. Additionally, the refined grid leads to only a slight difference in the wall-pressure distribution. Thus the moderate grid with is chosen in the simulation.

There are two sources of errors in the simulations, namely numerical error and systematic error. The numerical error can be caused by the inaccuracy of temporal integration and spatial discretization, the spatial resolution, and the time duration of the numerical simulation to achieve a statistically stationary flow. The systematic error is often caused by the inaccuracy in the physical models such as the turbulent model and the boundary conditions. In the LES presented in this section, the time-averaged flow properties such as wall-pressure distribution shown in Fig. 4.42 are converged. So the numerical error is considered not significant. Considering the boundary-layer separation zone revealed in the numerical schlieren contour around $x = 600$ mm, the discrepancies in Fig. 4.42 are most likely ascribed to the slip wall boundary condition applied to the combustor top wall and the inaccuracy of inflow turbulent boundary conditions. The difference is nevertheless acceptable as the pressure distribution along the wall from the simulations is in a good agreement with that in the experiments. Based on the comparison of results and other related validation works [10, 11, 22, 32], it can be concluded that the present numerical solver can capture the non-reacting supersonic flow-field successfully.

Figure 4.43 shows the time-averaged local equivalence ratio with isolines of stoichiometric mixture in seven spanwise planes along the streamwise direction. The velocity fields in two representative slices are also plotted. From Figs. 4.43, 4.38 and 4.42, the mixing field and flow structures in the combustor can be understood. The fuel injections generate a pair of roll-up vortices that are convected downstream in the cross flow. Owing to the expansion of the combustor geometry the fuel is directed toward the cavity. A rather uniform distribution of equivalence ratio can be found in the four downstream spanwise planes. There exists a strong shockwave in the leading edge of the cavity, which is deflected at the top wall. Another weaker shockwave is formed at the trailing edge of the cavity.

It is obvious that fuel entrainment depends on the shear layer of the cavity and the recirculation zone structure. As plotted Fig. 4.38, the fuel stream is recirculated into the rearward of the cavity, in accordance with Ref. [36].The local equivalence ratios in the cavity rearward reveal a fuel-rich environment formed there. We have found in Sect. 4.3.2.2 that a shorter ignition time is obtained when implementing LIP ignition at *e*-1 and *d*-1. Then it is deduced that local fuel-rich environment contributes to successful ignition inside the cavity.

Figure 4.43 reveals that the stoichiometric lines are mainly above the fuel plume and in the outer edge of the shear layer. Inside the cavity, the stoichiometric lines exist in the bottom right corner where the equivalence ratios are low. This local fuel-lean environment agrees well with the discussion about failures of LIP ignition at *a*-1 and *b*-1 when $\Phi = 0.16$.

Figure 4.44 presents the distribution of enstrophy (Ω^2, where Ω is the magnitude of the vorticity) near the cavity. Enstrophy represents the production of turbulence, and turbulent kinetic energy reflects the fuel/air mixing rate. As shown in Fig. 4.44a, two large-scale fuel-plumes are convected downstream during which the plumes continuously break into smaller fuel-rich pockets. It is observed that the pair of roll-up vortices transports partial fuel into the cavity. The high enstrophy around the fuel plumes denotes strong turbulence and turbulent mixing in the shear layer. As plotted in Figs. 4.43b and 4.45, the enstrophy, scalar dissipation rate and turbulent kinetic energy tend to rise along the streamwise direction. This means as the plumes cross the cavity, the mixing rate increases and then more fuel would be entrained into the cavity, which explains the equivalence ratio distribution in the cavity shown in Figs. 4.38c and 4.43a.

According to the sonic lines plotted in Fig. 4.45, all LIP positions except *d*-18 are situated in the subsonic region. Obviously, the LIP position *d*-18 is inappropriate for ignition owing to the flow conditions. The present results show that LIP ignition site inside the subsonic zones is necessary to achieve a successful ignition in a scramjet combustor. Figure 4.45 also reveals the turbulent conditions in the cavity reward. The failure at *f*-1 can be seen as a combination of strong dissipation rate and turbulent kinetic energy. Considering the flow direction, the turbulent conditions and the equivalence ratios around *e*-1, it is in favor of the re-ignition stage (stage

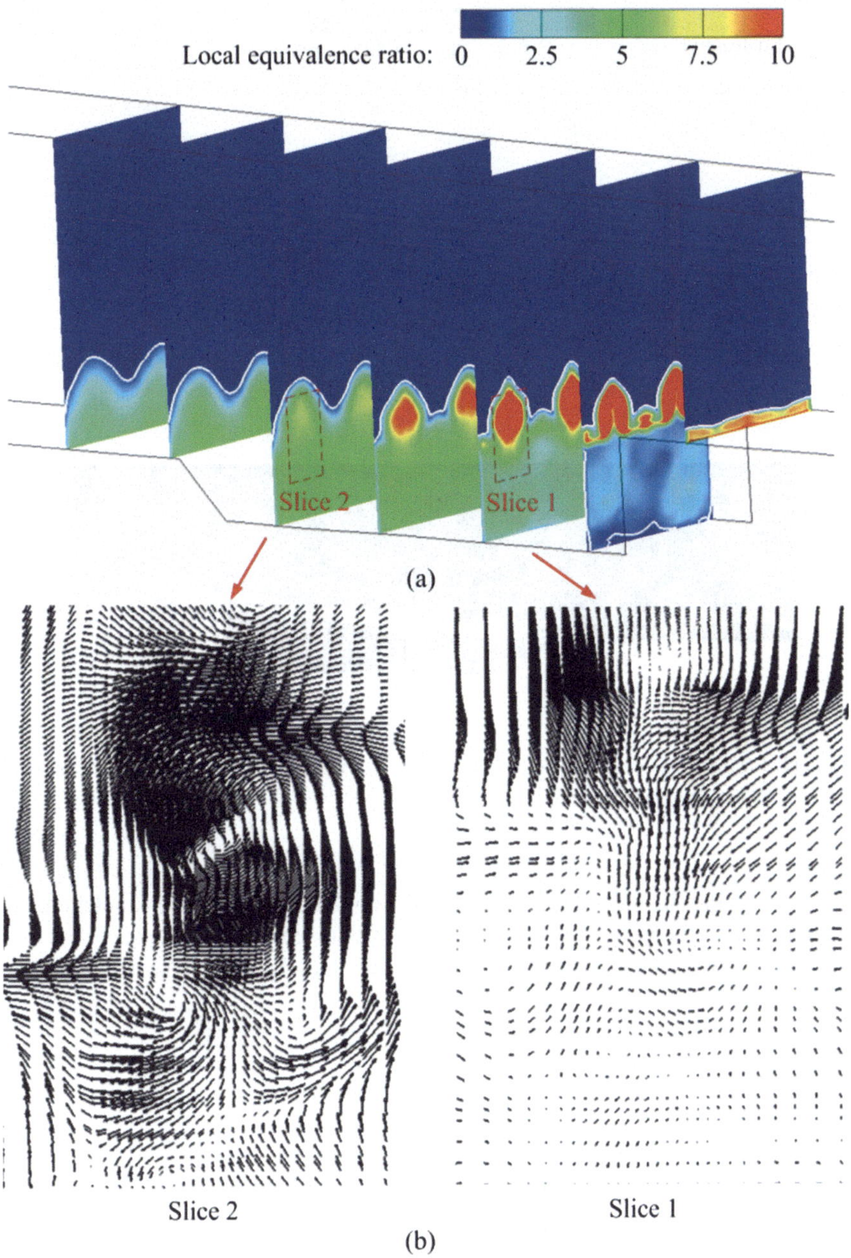

Fig. 4.43 **a** Time-averaged local equivalence ratio distribution and **b** velocity fields in different streamwise planes. The results are from LES with $\Phi = 0.30$ and the moderate grid [26]

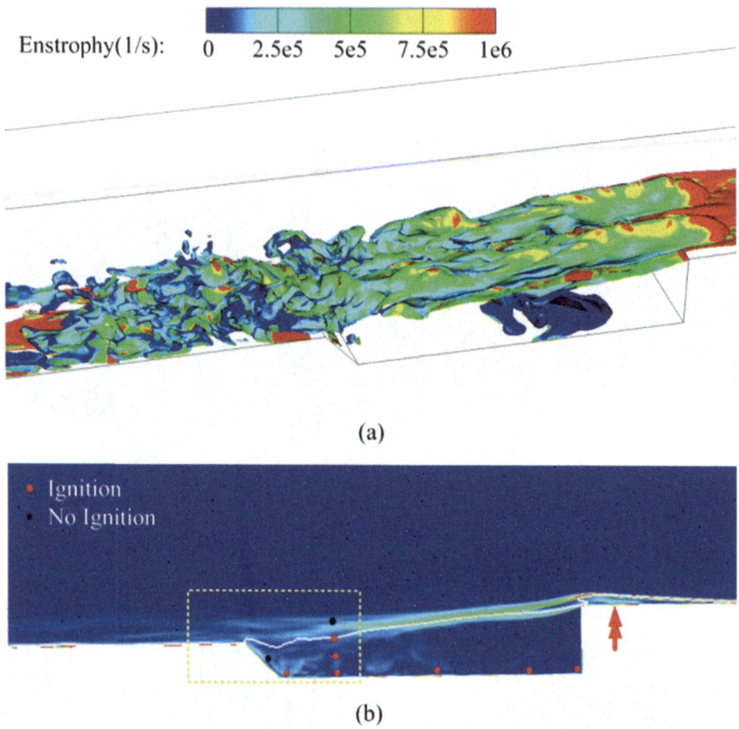

Fig. 4.44 Enstrophy distributions from LES with the moderate grid and $\Phi = 0.30$ [26]. **a** Representative instantaneous iso-surface with ethylene mass fraction being 0.15. **b** Time-averaged enstrophy field with sonic lines in the central plane

III). On the other hand, both the scalar dissipation rate and turbulent kinetic energy are higher at LIP ignition site *f*-1, which will enhance the plasma quenching stage (stage II) and less favorable to re-ignition stage (stage III), and possibly leading to LIP ignition failure.

4.3.3 Ignition Modes

As listed in Table 4.4, four ignition cases were conducted and each case was executed at least five times. In the table, P_i represents the fuel injection pressure and Φ the global equivalence ratio. The cavity fuelling rate is changed by altering the injection pressure. LIP ignitions in all the cases are successful. When the equivalence ratio is less than 0.15, however, successful LIP ignition could not be achieved. The global equivalence ratio in Case 1 is just above the lean ignition limit in the present supersonic combustor.

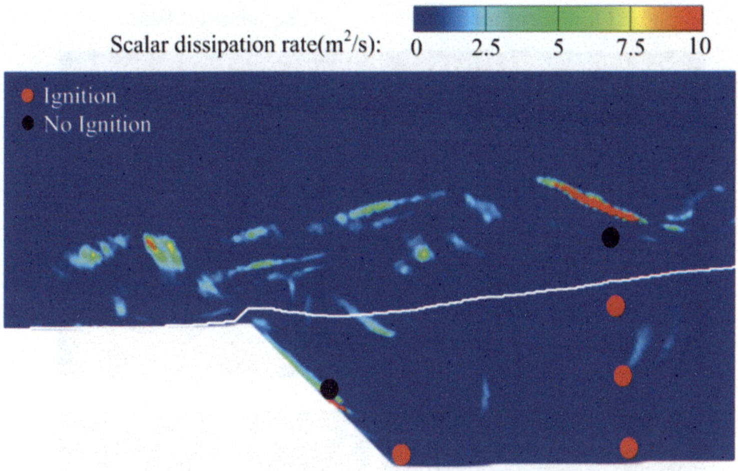

(a)

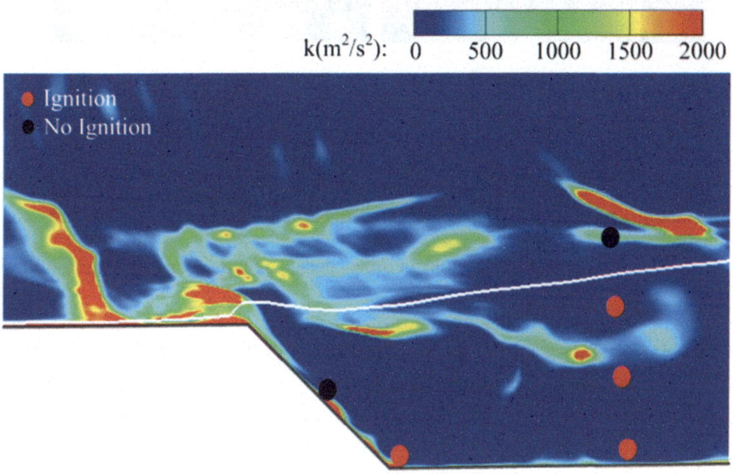

(b)

Fig. 4.45 Instantaneous local scalar dissipation rate and turbulent kinetic energy in the rearward of the cavity, obtained from LES with the moderate grid with $\Phi = 0.30$ [26]. **a** Representative instantaneous scalar dissipation rate with sonic lines. **b** Representative instantaneous turbulent kinetic energy with sonic lines

Table 4.4 Experimental arrangements of all test cases

Test case	P_i (MPa)	ϕ	Laser energy (mJ/pulse)
Case 1	0.75	0.15	200
Case 2	1.14	0.23	200
Case 3	1.15	0.23	300
Case 4	2.00	0.40	200

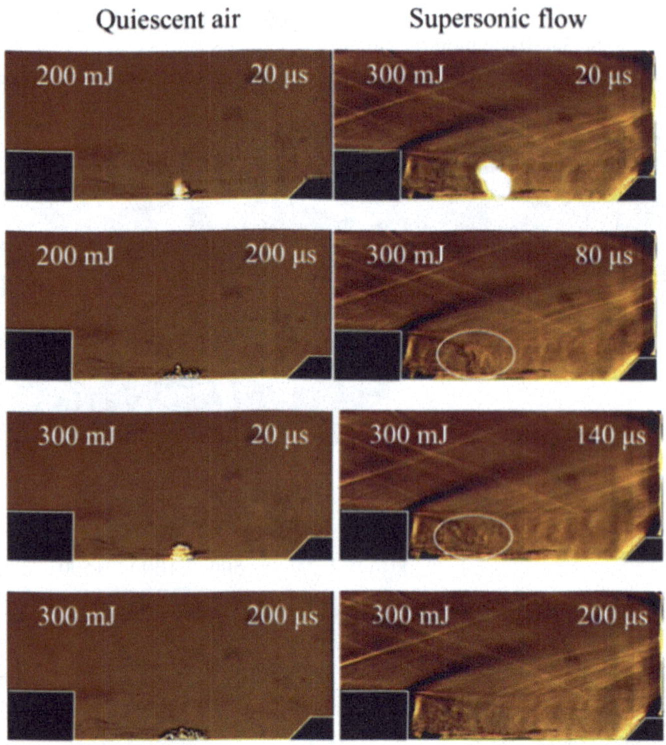

Fig. 4.46 Schlieren images of LIP motions in quiescent (left) and supersonic (right) air flow [28]

Figure 4.46 presents a series of Schlieren images for the laser excitation in quiescent and supersonic air flows. The LIP regions, with noticeable density gradients caused by thermal effects of the LIP, are clearly seen in the images. In the quiescent air flow the thermal region grows as the laser energy increases, which is plotted in the left column. In the supersonic air flow, the LIP region formed by a laser energy of 300 mJ disappears within a time span of 0.2 ms, demonstrating the strong turbulent heat transfer in the cavity. The CH* has been proved as a good reaction zone marker which could trace regions of flame front and heat release [34]. In addition, the high-temperature product zones can be traced by the OH* [37]. Figure 4.47 presents consecutive images of CH* and OH* chemiluminescence, and the corresponding Schlieren images for case 1. The LIP event begins at t = 0 ms.

The transition from LIP excitation to the establishment of a stable flame can be described as two main regimes. The first is the LIP initiation regime shown in Fig. 4.47 when t < 0.8 ms. The initial signals of CH* and OH* chemiluminescence are from the LIP excitation. Between t = 0.1 ms and t = 0.8 ms, the chemiluminescence intensities decrease rapidly because of the turbulent heat transfer shown in Fig. 4.46.

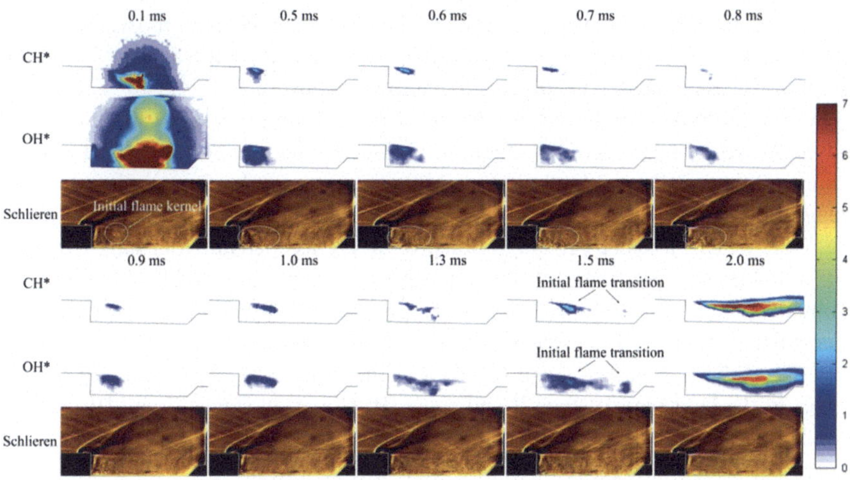

Fig. 4.47 Simultaneous CH*/OH* chemiluminescence and Schlieren images during the LIP ignition process for case 1 [28]

The second is the transient ignition reaction regime. After t = 0.8 ms, the chemiluminescence intensities start to increase, denoting the onset of the ignition. Before t = 2 ms, the CH* mainly lies in a narrow region in the shear layer near the front wall of the cavity. The OH* signals are found in a larger area than the CH*, since the life time of OH* is much longer in reactive flows [38]. It appears that the thermal effect of the LIP provides favorable thermal and chemical environment for the ignition process. When a stable flame is established at t = 2.0 ms, the CH* and OH* regions overlap each other with the OH* distribution slightly deeper into the cavity.

Figure 4.48 shows the CH* and OH* chemiluminescence images during the LIP ignition process for case 4. The two-regime ignition process discussed earlier for case 1 can be identified in the figure as well. It takes a shorter time to establish a stable flame due to the higher equivalence ratio in case 4 than in case 1. The first regime (the LIP initiation) ends at t = 0.4 ms, during which the intensities of CH* and OH* chemiluminescence attenuate because of the turbulent heat transfer. The onset of the ignition can be observed at t = 0.5 ms when the chemiluminescence starts to increase. Compared with case 1, in case 4 the transient ignition reaction regime is

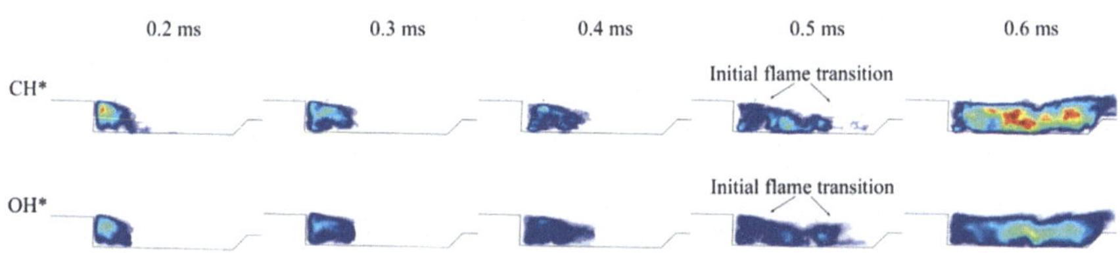

Fig. 4.48 Simultaneous CH*/OH* chemiluminescence images of the LIP ignition process for case 4 [28]

shorter and the CH* region is much wider. The CH* and OH* regions overlap rather well in case 4. It is clear that the equivalence ratio has a significant impact on the ignition process in the cavity-based supersonic combustor.

In order to characterize the two-regime LIP ignition process of the four cases, the intensities of CH* and OH* chemiluminescence images are integrated in each frame and the results are shown in Fig. 4.49. The integrated intensities in each case are normalized by its mean intensity at the stable flame period [39]. As discussed in Fig. 4.41, the LIP-assisted cavity ignition process is classified into four stages, involving 'initial plasma ignition', 'plasma quenching', 're-ignition' and 'stable flame'. It is obvious that the two-regime LIP ignition process discussed in Figs. 4.47 and 4.48 is consistent with the four stages. The first laser initiation regime corresponds to stage I and the first part of stage II, whereas the transient ignition reaction regime relates to the later part of stage II and stage III. However, stages II and III in cases 3 and 4 are much shorter than in cases 1 and 2, because case 3 has a higher laser energy and case 4 a larger fuelling rate (equivalence ratio).

The impacts of the fuelling rate and the laser energy on the cavity ignition process can be explained using the two-regime ignition mechanism. With a low fuelling rate (e.g., the global equivalence ratio is lower than 0.15), the heat release and radicals [40] formed from the LIP initiation regime are not sufficient to survive in the turbulent heat transfer, which results in the ignition failure. As the fuelling rate increases, the heat release in the upstream half cavity is larger, and then the reactions would not be quenched, the mixture in the rearward of the cavity therefore being ignited. With further increase in the fuelling rate or the laser energy, the onset of the ignition in the cavity rearward becomes faster and a stable flame can be established in a shorter time.

Figure 4.50 is plotted to further characterize the ignition process in the cavity. For each case the CH* chemiluminescence intensities in a frame are integrated respectively in the half cavity (from the front wall to the middle of the cavity) and the whole cavity. Then the integrated intensities are normalized by the time-averaged intensity of the whole cavity. The two normalized intensities are nearly the same during the LIP initiation regime, since the hot LIP gas is mainly found in the upstream half of the cavity. In the final stable flame stage, nevertheless, the normalized intensity in the upstream half of the cavity is much lower, showing that the chemical reactions chiefly occur in the downstream half.

Inspecting the differences of chemiluminescence images among diverse cases from Figs. 4.47, 4.48 and 4.50, it is evident that for the LIP ignition process there are two modes. In the weak ignition mode (case 1), the initial flame kernel anchored near the cavity shear layer is too weak to directly ignite the fuel in the rearward of the cavity. As in the whole cavity the environment is fuel-lean, the initial flame transits gradually from the upstream cavity to the rearward. This mode exists in the fuel-lean cavity. In the intense ignition model (case 4), a strong flame is formed immediately after the laser excitation owing to the favorable environment. Then following the recirculation flow the initial flame spreads quickly to the upstream cavity corner.

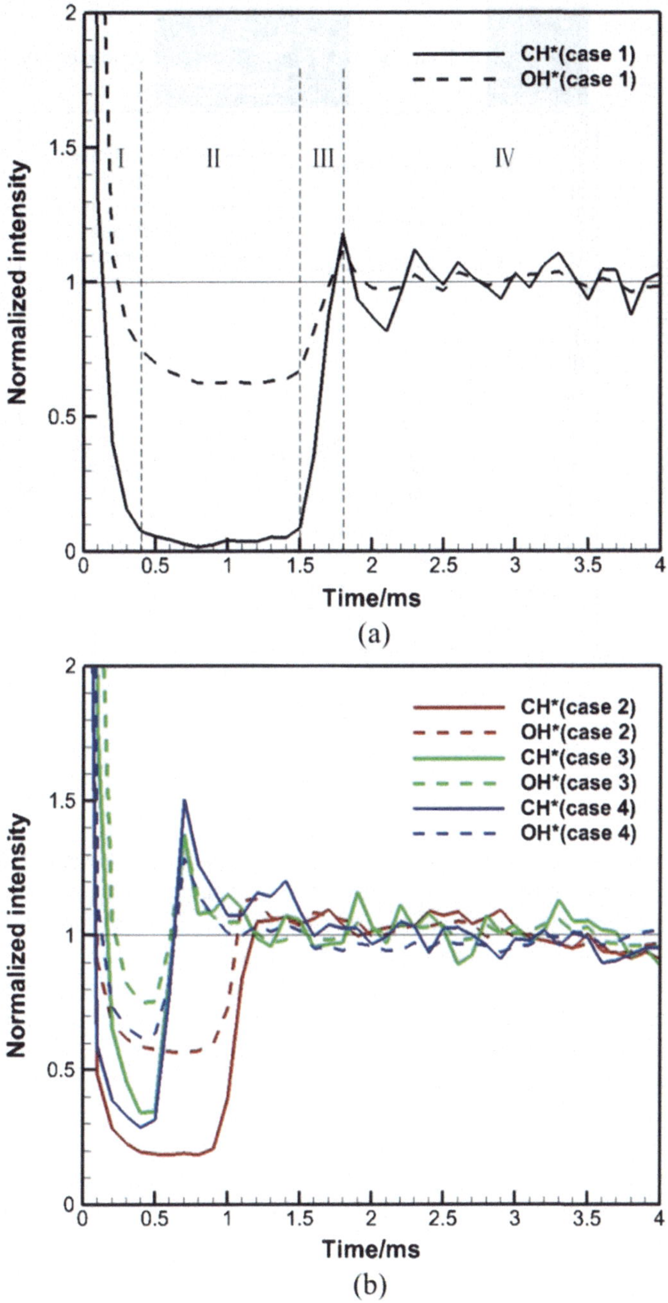

Fig. 4.49 Normalized CH* and OH* chemiluminescence intensity against time for different cases [28]. Each normalized intensity is a time-averaged result of repeated experiments for the same case

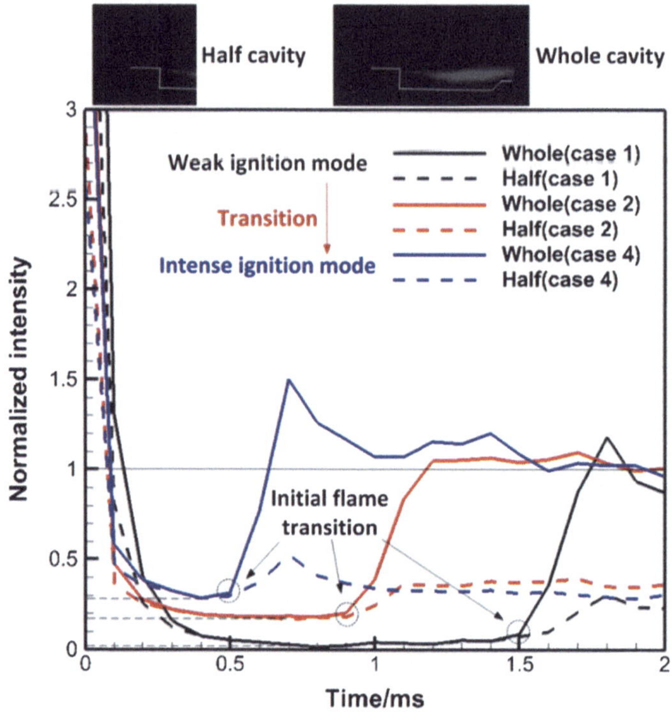

Fig. 4.50 Normalized CH* chemiluminescence intensity against time for different ignition modes [28]

Since the initial flame is robust enough to ignite the fuel in the rearward of the cavity, it only takes a rather short residence time before a stable flame is achieved. It is observed that there also exist transitional behaviors of the initial flame between the weak and the intense ignition modes, e.g. cases 2 and 3.

Figure 4.51 illustrates the schematic of the flowfield, fuel/air mixing and reaction zone structure based on the previous works in Refs. [10, 41, 42]. It is inferred that the ignition process is mainly assisted by the cavity recirculation zone in the weak ignition mode. But in the intense ignition mode, the initial flame directly leads the ignition process. For the transition between the two modes, the ignition process is dominated both by the cavity recirculation zone and the initial flame. It should be noted that in the intense mode the occurrence of the cavity recirculation zone is still favorable, because it also provides the flow field and mixing environments friendly for ignition.

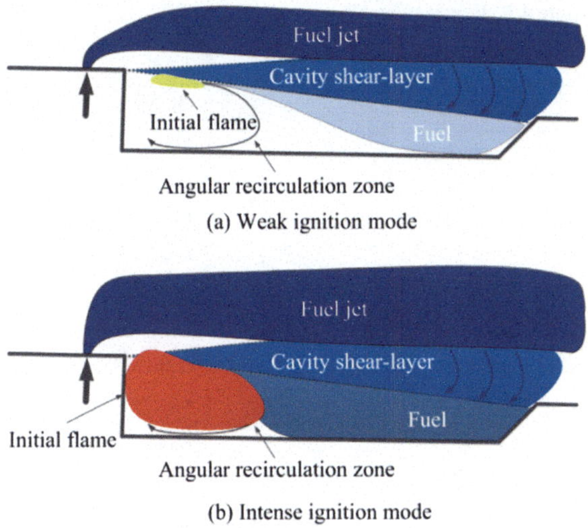

Fig. 4.51 Schematic of the reaction zone structure of the strong and weak ignition modes [28]

4.4 Auto-Ignition Effects

As the stagnation temperature increases, the effects of auto-ignition on combustion characteristics in the cavity-based scramjet combustor need more consideration. In this section, hydrogen was transversely injected upstream of the cavity. The effects of auto-ignition were investigated experimentally under stagnation temperature of 1400 K and 1600 K, respectively imitating the flight Mach numbers 5.5 and 6.

4.4.1 Experimental Setup

The experiments were carried out in a recently developed direct-connect test facility. Pure ethanol and oxygen were burned continuously to heat the air, providing a vitiated air stream of Mach 2.52 with the stagnation temperatures of 1400 K and 1600 K. The fuel injection last 1.5 s, and during this period the fueling rate was steady according to the mass flowmeter. Table 4.5 reports the operation conditions.

Schematic of the test section is plotted in Fig. 4.52. The combustor had a width of 50 mm and a height of 40 mm. Installed on the bottom wall was a cavity with the depth D = 8 mm, the length-to-depth ratio L/D = 7 and the rear wall angle A = 45°. A spark ignition plug was mounted 10-mm downstream of the leading edge of the cavity. An injector with an orifice diameter of 2 mm was fixed 160-mm upstream of the cavity. High-speed imaging of flame luminosity and schlieren were introduced in the test section. The frame rate was set to 15000 frame/s with an exposure time of 0.067 ms. The pressure distribution along the centrelines of the top and bottom walls in the test section was measured by a pressure scanner with a 100-Hz acquisition

Table 4.5 Experimental conditions

Parameter		Scheme 1	Scheme 2
Air	T_0, K	1600	1400
	P_0, MPa	1.65	1.55
	Ma	2.52	2.52
	Y_{O_2}, %	23.38	23.38
	Y_{H_2O}, %	7.13	5.92
	Y_{CO_2}, %	11.67	9.68
	Y_{N_2}, %	57.82	61.0
H_2	P_{0jet}, MPa	3.5–5.0	
	Equivalence ratio (Φ)	0.22–0.32	
	Y_{H_2}, %	99.5–99.8	
	Y_{other}, %	0.2–0.5	

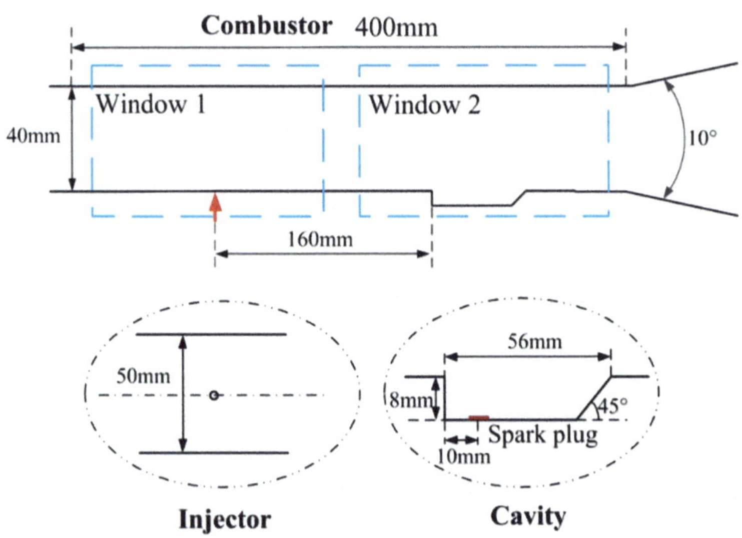

Fig. 4.52 Schematic of the test section (top) and zoomed views around the injector and the cavity (bottom) [43]. The red arrow denotes the fuel injection location

frequency and 0.25% full scale accuracy. More details about the experimental facility can be obtained from the previous works [12, 44].

There is a concern on the OH concentration in the vitiated air produced by the ethanol/O_2 combustion, since OH radicals have a significant role in the H_2/O_2 auto-ignition. The OH-PLIF results, however, showed that the OH concentration was actually low and can be negligible, as discussed in Ref. [45].

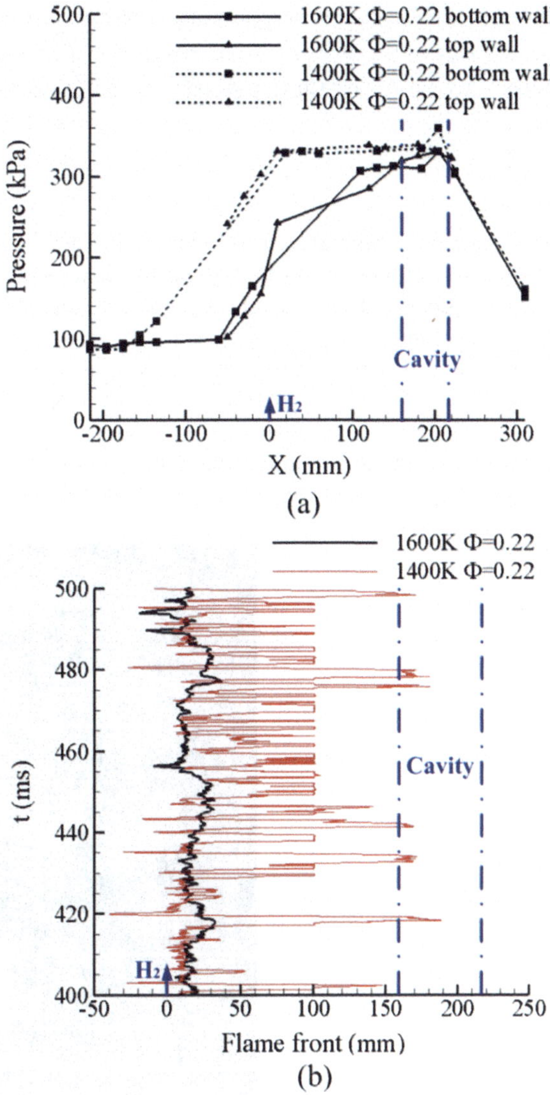

Fig. 4.53 **a** Comparison of the time-averaged pressure along the centerlines of bottom wall and top wall in the combustor, and **b** the dynamics of the flame front from 400 to 500 ms after the fuel injection begins [43]

4.4.2 Auto-Ignition in the Ignition Process

Table 4.6 lists the ignition schemes and combustion stabilization modes at the equivalence ratios from 0.22 to 0.32 under the stagnation temperature T_0 of 1400 K and 1600 K, respectively. When $T_0 = 1400$ K, the fuel could be ignited successfully only by the spark ignition. When $T_0 = 1600$ K, however, auto-ignition was achieved in the combustor. Obviously, auto-ignition would be more significant with increasement of

Table 4.6 Ignition schemes and combustion stabilization modes

T_0, K	P_{0jet}, MPa	Φ	Ignition scheme	Combustion stabilization mode
1400	3.5–5.0	0.22–0.32	Spark	Jet-wake stabilized mode combined with cavity stabilized mode
1600	3.5–5.0	0.22–0.32	Auto-ignition	Jet-wake stabilized mode

the stagnation temperature. It provides some new ideas for designing the ignition schemes for scramjet under high flight Mach numbers.

As reported in Table 4.6, the jet-wake stabilized mode is obtained under both stagnation temperatures. Figure 4.54a shows the time-averaged pressures along the combustor walls; Fig. 4.54b presents the dynamics of flame fronts against times. The pressure when $T_0 = 1600$ K is lower from X = 0 mm to X = 100 mm than in $T_0 = 1400$ K, which suggests that the time-averaged flame in the jet is less intense in the former. On the contrary, the flame front is more stable in the jet wake with

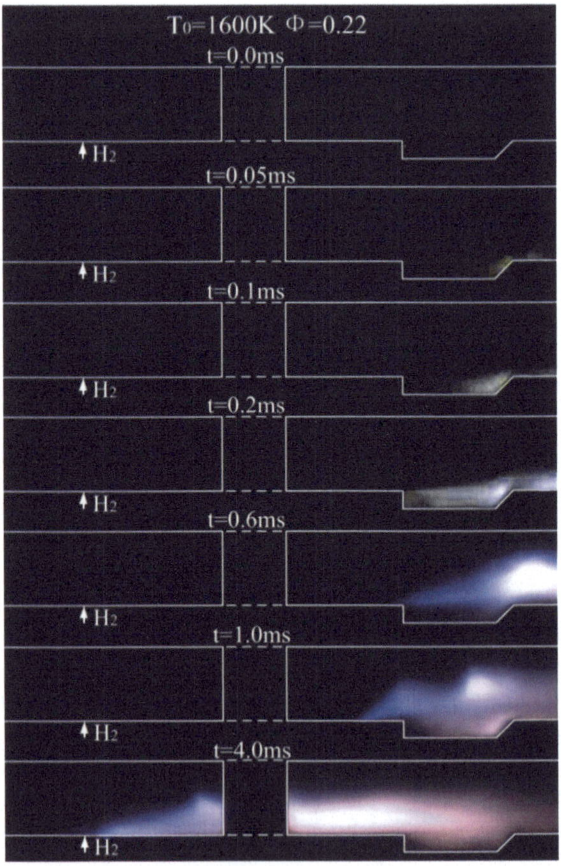

Fig. 4.54 High-speed flame luminosity images in the initial phase of the combustion with auto-ignition [43]

$T_0 = 1600$ K. With $T_0 = 1400$ K, the flame front fluctuates between $X = -50$ mm (upstream of the injector) and $X = 180$ mm (in the cavity). It is then concluded that the combustion is accompanied by intermittent blow-off and restabilization in the jet wake. During the unsteady process an interesting phenomenon was repeatedly observed that an individual flame appeared in the jet upstream of the original flame in the cavity. This phenomenon took place under different equivalence ratios from 0.22 to 0.32.

Figure 4.54 presents the instantaneous flame luminosity images with $T_0 = 1600$ K and $\Phi = 0.22$. The initial flame core initially appears close to the rear wall of the cavity because of auto-ignition. Then the flame core develops in the recirculation zone of the cavity. Following that the jet around the cavity is ignited, and then the flame propagates upstream. Finally, a jet-wake stabilized flame is formed, as shown in the bottom subfigure.

Figure 4.55 shows the typical loop from blow-off to restabilization under the stagnation temperature of 1400 K. The loop begins at t = 477 ms and ends at t = 481 ms, as plotted in Fig. 4.53b. As shown in Fig. 4.55, the flame is first stabilized in the jet wake with the flame front close to the injector. The precombustion shock train is located upstream of the injector. Then, the flame is blown downstream along with the

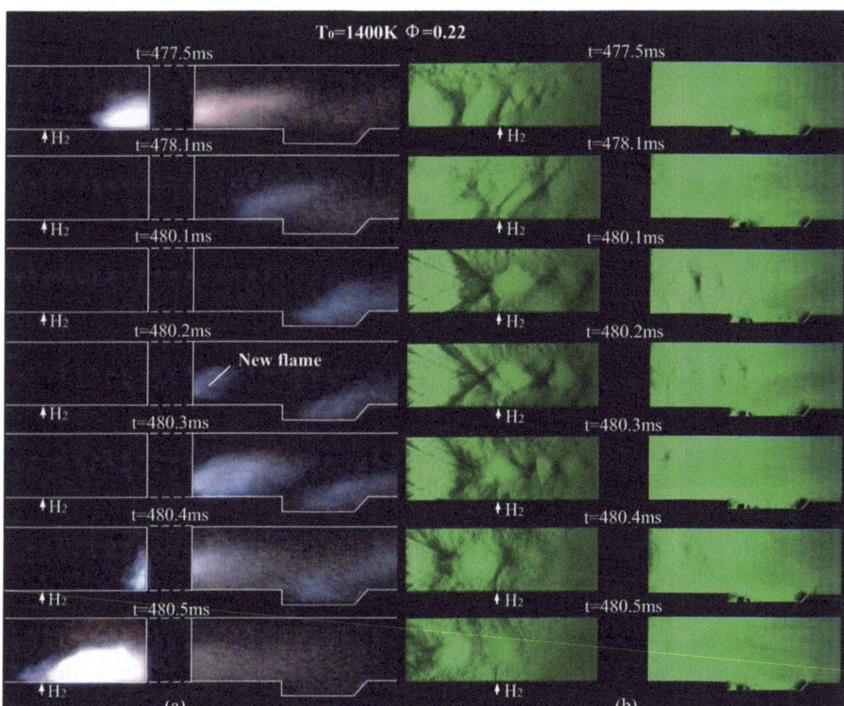

Fig. 4.55 Typical loop from blow-off to restabilization with $T_0 = 1400$ K [43]. **a** Instantaneous high-speed flame luminosity images. **b** Simultaneously instantaneous high speed schlieren images

shock train. When the weak flame reaches the cavity, the shock train remains upstream of the cavity, compressing the mixed gas and increasing the static temperature. At t = 480.1 ms, there exists no flame observed in the jet wake and the flame is completely stabilized in the shear layer of the cavity. At t = 480.2 ms, a new flame core suddenly occurs and then develops into an individual flame in the upstream of the cavity. The new flame alters the shock structures and then combines with the original cavity-stabilized flame. The combined flame further propagates upstream, eventually the flame being re-stabilized in the jet wake.

Figure 4.56a shows dynamics of the flame front in the typical loop from blow-off to restabilization. The time interval between adjacent subfigures is 0.067 ms whereby the flame-front-propagation speed is calculated, as plotted in Fig. 4.56b. At t = 480.1 ms when the new flame appears, the instantaneous speed would be over 1000 m/s if we assume that the new flame is ascribed to the flame propagation from the cavity. This flame speed is far beyond the turbulent flame propagation limits. Therefore, the new flame is caused by auto-ignition. The different flame behaviors and flame-stabilized modes with $T_0 = 1400$ K and $T_0 = 1600$ K demonstrate the auto-ignition has significant influences on ignition process and combustion stabilization, especially when the stagnation temperature is high.

4.5 Summary

Cavities are often used in practical scramjets owing to their minimal system complexity, small total pressure losses and the good performance in flame holding. The cavities could hold the flame because they provide subsonic regions where recirculation zones form and successively furnish the core flow with heat and radicals vital for combustion. Considering the inflow conditions, ignition of cavity-based scramjets aims to initialize the combustion in the cavities, since in other places the turbulent dissipation is usually too severe to maintain the initial flame.

Ignition is to inject a certain amount of energy into the fuel/air mixture, with the purpose of triggering oxidation reactions which could further propagate in the combustor. Variables concerned in the energy injection process includes how to inject energy, the ignition energy amount, the ignition position, and the inflow conditions. These variables and their influences are usually the intents of studies about cavity ignition. Since cavity ignition is intrinsically and partially stochastic, it is of significance that the dynamics during the ignition process. Both experimental and numerical investigations show that the cavity ignition process is a strong coupling of subsonic flow, supersonic flow, heat and molecular transfer, and chemical reactions.

There are various ignition techniques corresponding to different ways to inject the ignition energy. Spark ignition, piloted ignition, GAD ignition and LIP ignition are introduced. Except for the piloted ignition, the other three techniques are all based on plasmas. In the spark ignition and the LIP ignition, equilibrium plasmas are produced, thus the mixture being ignited mainly by the heat effects of the plasmas. Nonetheless, the GAD generates nonequilibrium plasmas which could selectively stimulate

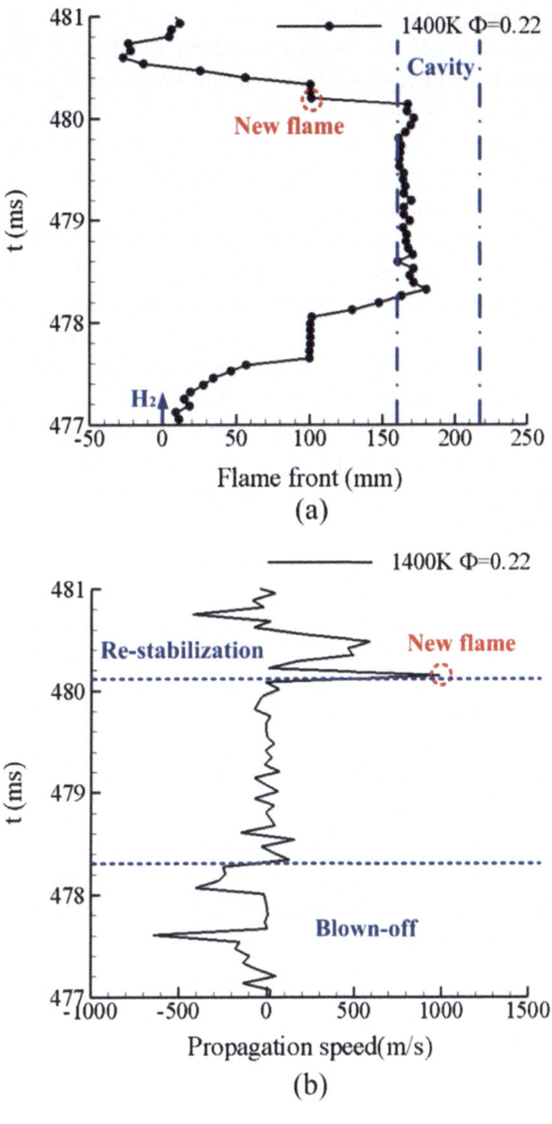

Fig. 4.56 **a** The dynamic of the flame front in the typical loop from blow-off to restabilization; **b** the propagation speed of the flame front during the loop [43]

nonequilibrium chemical reactions important for the ignition process. Instead of plasmas, the piloted ignition is based on the combustion of piloted fuel (often Hydrogen) and air. Further, the generated heat and radicals are distributed in both in and out the cavity, very different from the other ignition techniques. Among the four techniques, the piloted ignition could provide a large amount of ignition energy just by increasing the mass flow rate of piloted fuel; however, the LIP ignition is most suitable for studying the flame kernel formation because this technique has the highest accuracy in manipulating the ignition energy, position and timing.

After the ignition energy is injected into the cavity at a specific region and initial chemical reactions are triggered there, the flame kernel formation is primarily dominated by confrontation between the chemical reactions and the turbulent dissipation. A success cavity ignition requires that the chemical reactions in the cavity produce enough heat and radicals which ignite the fuel/air mixture in the shear layer. According to the flow direction and the fuel distribution in the cavity, the ignition position matters. Ignition near the rear wall of the cavity is preferred: first the equivalence ratio is high there; second the hot plasmas have enough time to fully contact with the fuel/air mixture in the cavity. If the ignition position is close to the front wall, the hot plasma would be quickly transferred into the shear layer and then quenched. Then the reduced heat release and radical generation might fail to ignite the fuel/air mixture in the shear layer. Increasement of the ignition energy enhances the chemical reactions in the cavity and thus promotes the cavity ignition.

The whole ignition process can be divided into four stages. The first stage begins with the ignition event and ends when the plasmas quench, meaning that the energy in the plasmas is transferred to the mixtures in the cavity. The second stage corresponds to an accumulation of heat and radical in the cavity. After a period of time the fuel/air mixture in the cavity would starts to combust and the mixture in the shear layer would be ignited, which is the third stage. In the final stage the flame is mainly in the shear layer, and partial high-temperature products are transported from the rear wall of the cavity to the front wall, and finally into the shear layer. As duration of the second stage decreases, the cavity ignition becomes stronger.

It is noteworthy that the fuel/air mixture might auto-ignite. If the stagnation temperature is improved to a critical value, there is no need for forced ignition. The oxidation reactions happen spontaneously when the fuel/air mixture passes through the isolator and the combustor, and then the heat and radicals accumulate in the cavity. After some accumulation there would be auto-ignition in the cavity. Slightly lowering the stagnation temperature and utilizing forced ignition in the cavity, the auto-ignition could occur upstream of the cavity. A possible position of auto-ignition is the downstream separation zone induced by the fuel jet. This kind of auto-ignition is able to bring about combustion unsteadiness.

References

1. Sun, M., Gong, C., Zhang, S., Liang, J., Liu, W., & Wang, Z. (2012). Spark ignition process in a scramjet combustor fueled by hydrogen and equipped with multi-cavities at Mach 4 flight condition. *Experimental Thermal and Fluid Science, 43,* 90–96.
2. Xi, W., Wang, Z., Sun, M., Liu, W., & Li, Q. (2014). Experimental investigation of ignition transient phase in model supersonic combustor. *Journal of Aerospace Engineering, 27,* 1–9.
3. Feng, R., Li, J., Wu, Y., Zhu, J., Song, X., & Li, X. (2018). Experimental investigation on gliding arc discharge plasma ignition and flame stabilization in scramjet combustor. *Aerospace Science and Technology, 79,* 145–153.

4. Huang, S., Wu, Y., Song, H., Zhu, J., Zhang, Z., Song, X., et al. (2018). Experimental investigation of multichannel plasma igniter in a supersonic model combustor. *Experimental Thermal and Fluid Science, 99,* 315–323.
5. Phuoc, T. X. (2006). Laser-induced spark ignition fundamental and applications. *Optics and Lasers in Engineering, 44,* 351–397.
6. Li, X., Liu, W., Pan, Y., Yang, L., & An, B. (2017). Experimental investigation on laser-induced plasma ignition of hydrocarbon fuel in scramjet engine at takeover flight conditions. *Acta Astronautica, 138,* 79–84.
7. Li, X., Liu, W., Pan, Y., Yang, L., An, B., & Zhu, J. (2018). Characterization of ignition transient processes in kerosene-fueled model scramjet engine by dual-pulse laser-induced plasma. *Acta Astronautica, 144,* 23–29.
8. An, B., Wang, Z., Yang, L., Li, X., & Zhu, J. (2017). Experimental investigation on the impacts of ignition energy and position on ignition processes in supersonic flows by laser induced plasma. *Acta Astronautica, 137,* 444–449.
9. Cai, Z., Wang, Z., Sun, M., Wang, H., & Liang, J. (2014). Investigation of the spark ignition enhancement in a supersonic flow. *Modern Physics Letters B, 28,* 1–8.
10. Cai, Z., Liu, X., Gong, C., Sun, M., Wang, Z., & Bai, X.-S. (2016). Large Eddy simulation of the fuel transport and mixing process in a scramjet combustor with rearwall-expansion cavity. *Acta Astronautica, 126,* 375–381.
11. Cai, Z., Wang, Z., Sun, M., Bai, X.-S. (2016). Effect of combustor geometry and fuel injection scheme on the combustion process in a supersonic flow. *Acta Astronautica, 129,* 44–51.
12. Cai, Z., Yang, Y., Sun, M., & Wang, Z. (2016). Experimental investigation on ignition schemes of a supersonic combustor with the rearwall-expansion cavity. *Acta Astronautica, 123,* 181–187.
13. Cai, Z., Zhu, J., Sun, M., & Wang, Z. (2018). Spark-enhanced ignition and flame stabilization in an ethylene-fueled scramjet combustor with a rear-wall-expansion geometry. *Experimental Thermal and Fluid Science, 92,* 306–313.
14. Cai, Z., Zhu, X., Sun, M., Wang, Z. (2018). Experimental study on the combustion process in a scramjet combustor with a rear-wall-expansion geometry. *Journal of Aerospace Engineering, 31*(5), 04018077.
15. Yang, Y., Wang, Z., Sun, M., Wang, H., & Li, L. (2015). Numerical and experimental study on flame structure characteristics in a supersonic combustor with dual-cavity. *Acta Astronautica, 117,* 376–389.
16. Jiang, G., & Shu, C. W. (1996). Efficient implementation of weighted ENO schemes. *Journal of Computational Physics, 126,* 917–923.
17. Wang, H., Qin, N., Sun, M., & Wang, Z. (2012). A dynamic pressure-sink method for improving large eddy simulation and hybrid Reynolds-averaged Navier-Stokes/large eddy simulation of wall-bounded flows. *Proceedings of the Institution of Mechanical Engineers, Part G: Journal of Aerospace Engineering, 226,* 1107–1120.
18. Jachimowski, C. J. (1988). An analytical study of the hydrogen-air reaction mechanism with application to scramjet combustion, NASA Technical Paper 2791.
19. Wang, H., Qin, N., Sun, M., Wu, H., & Wang, Z. (2011). A hybrid LES (Large Eddy Simulation)/assumed sub-grid PDF (Probability Density Function) model for supersonic turbulent combustion. *Science China Technological Sciences, 54,* 2694–2707.
20. Wang, H., Wang, Z., Sun, M., & Qin, N. (2013). Combustion characteristics in a supersonic combustor with hydrogen injection upstream of cavity flameholder. *Proceedings of the Combustion Institute, 34,* 2073–2082.
21. Wang, H., Wang, Z., Sun, M., & Qin, N. (2015). Large eddy simulation of a hydrogen-fueled scramjet combustor with dual cavity. *Acta Astronautica, 108,* 119–128.
22. Wang, Z., Cai, Z., Sun, M., Wang, H., & Zhang, Y. (2016). Large Eddy simulation of the flame stabilization process in a scramjet combustor with rearwall-expansion cavity. *International Journal of Hydrogen Energy, 41,* 19278–19288.
23. Kouchi, T., Masuya, G., Mitani, T., & Tomioka, S. (2012). Mechanism and control of combustion-mode transition in a scramjet engine. *Journal of Propulsion and Power, 28,* 106–112.

24. Wang, Z., Sun, M., Wang, H., Yu, J., Liang, J., & Zhuang, F. (2015). Mixing-related low frequency oscillation of combustion in an ethylene-fueled supersonic combustor. *Proceedings of the Combustion Institute, 35,* 2137–2144.
25. Chong, D., Hu, M., Chen, W., Wang, J., Liu, J., & Yan, J. (2014). Experimental and numerical analysis of supersonic air ejector. *Applied Energy, 130,* 679–684.
26. Cai, Z., Zhu, J., Sun, M., Wang, Z., & Bai, X.-S. (2018). Laser-induced plasma ignition in a cavity-based scramjet combustor. *AIAA Journal, 56,* 4884–4892.
27. Gao, T., Liang, J., Sun, M., & Zhao, Y. (2017). Analysis of separation modes variation in a scramjet combustor with single-side expansion. *AIAA Journal, 55,* 1307–1317.
28. Cai, Z., Zhu, J., Sun, M., Wang, Z., & Bai, X.-S. (2018). Ignition processes and modes excited by laser-induced plasma in a cavity-based supersonic combustor. *Applied Energy, 228,* 1777–1782.
29. Geerts, J. S., & Yu, K. H. (2016). Shock train/boundary-layer interaction in rectangular isolators. *AIAA Journal, 54,* 3450–3464.
30. Geerts, J. S., & Yu, K. H. (2017). Systematic application of background-oriented schlieren for isolator shock train visualization. *AIAA Journal, 55,* 1105–1117.
31. Greenshields, C. J., Weller, H. G., Gasparini, L., & Reese, J. M. (2010). Implementation of semi-discrete, non-staggered central schemes in a colocated, polyhedral, finite volume framework, for high-speed viscous flow. *International Journal for Numerical Methods in Fluids, 63,* 1–21.
32. Cai, Z., Wang, Z., Sun, M., Bai, X.-S. (2017). Large Eddy simulation of the flame propagation process in an ethylene fueled scramjet combustor in a supersonic flow. In *21st AIAA International Space Planes and Hypersonics Technologies Conference* (pp. 2017–2148). Xiamen, China: AIAA.
33. Cai, Z., Zhu, X., Sun, M., & Wang, Z. (2017). Experiments on flame stabilization in a scramjet combustor with a rear-wall-expansion cavity. *International Journal of Hydrogen Energy, 42,* 26752–26761.
34. Allison, P. M., Frederickson, K., Kirik, J. W., Rockwell, R. D., Lempert, W. R., & Sutton, J. A. (2017). Investigation of supersonic combustion dynamics via 50 kHz CH chemiluminescence imaging. *Proceedings of the Combustion Institute, 36,* 2849–2856.
35. Zhang, S., Wang, X., He, M., Jiang, Y., Zhang, B., Hang, W., et al. (2014). Laser-induced plasma temperature. *Spectrochimica Acta Part B, 97,* 13–33.
36. Wang, H., Li, P., Sun, M., & Wei, J. (2017). Entrainment characteristics of cavity shear layers in supersonic flow. *Acta Astronautica, 137,* 214–221.
37. Wang, H., Wang, Z., Sun, M., & Wu, H. (2013). Combustion modes of hydrogen jet combustion in a cavity-based supersonic combustor. *International Journal of Hydrogen Energy, 38,* 12078–12089.
38. Rasmussen, C. C., Dhanuka, S. K., & Driscoll, J. F. (2007). Visualization of flameholding mechanisms in a supersonic combustor using PLIF. *Proceedings of the Combustion Institute, 31,* 2505–2512.
39. Ombrello, T. M., Carter, C. D., Tamb, C.-J., & Hsuc, K.-Y. (2015). Cavity ignition in supersonic flow by spark discharge and pulse detonation. *Proceedings of the Combustion Institute, 35,* 2101–2108.
40. Brieschenk, S., O'Byrne, S., & Kleine, H. (2013). Laser-induced plasma ignition studies in a model scramjet engine. *Combustion and Flame, 160,* 145–148.
41. Gruber, M. R., Donbar, J. M., Carter, C. D., & Hsu, K.-Y. (2004). Mixing and combustion studies using cavity-based flameholders in a supersonic flow. *Journal of Propulsion and Power, 20,* 769–778.
42. Sun, M., Zhong, Z., Liang, J., & Wang, Z. (2014). Experimental investigation of supersonic model combustor with distributed injection of supercritical kerosene. *Journal of Propulsion and Power, 30,* 1537–1542.
43. Wang, Y., Wang, Z., Sun, M., & Wang, H. (2018). Effects of auto-ignition on combustion characteristics in a hydrogen-fueled dual-mode scramjet combustor. *Acta Astronautica, 153,* 154–158.

44. Sun, M. B., Cui, X. D., Wang, H. B., & Bychkov, B. V. (2015). Flame flashback in a supersonic combustor fueled by ethylene with cavity flameholder. *Journal of Propulsion and Power, 31,* 976–980.
45. Wang, H. B., Wang, Z. G., & Sun, M. B. (2013). Experimental study of oscillations in a scramjet combustor with cavity flameholders. *Experimental Thermal and Fluid Science, 45,* 259–263.

Chapter 5
Flame Flashback in Supersonic Flows

Non-stationary flame propagation against the incoming supersonic flow in scramjet engines has been in focus of numerous experimental and computational studies. In this chapter, we give the research work of the flame flashback phenomenon in scramjet combustor under a condition of flight Mach 4 and Mach 5.5.

5.1 Flame Flashback Phenomenon in a Flight Mach 4 Condition

This part experimentally investigates the flame flashback in scramjet combustor with a cavity flameholder under a condition of flight Mach 4 adopted to gain additional insights into the driving mechanism of combustion instability in this special condition. Section 5.1.1 presents experimental investigation of flame flashback phenomenon and the calculated flashback flame speed. Section 5.1.2 presents the injection parametric study of flame flashback which includes fuel-equivalence ratio and injection schemes.

5.1.1 Flashback Flame in a Single-Side Expansion Scramjet Combustor

In scramjet combustors, Reynold's number is huge and the boundary layer is very thin. The high compressibility of supersonic flows suppresses the developments of instabilities. Under such conditions, the mechanism of flame propagation is still not clear, which concerns especially flame flashback against the supersonic flow. Wang et al. [1] have observed flame propagation against the incoming supersonic flow under certain mixing state with a fixed fuel equivalence ratio. In this part, Sun et al. [2] obtain strong variations the flame flashback parameters at various equivalence

M. Sun et al., *Unsteady Supersonic Combustion*,
https://doi.org/10.1007/978-981-15-3595-6_5

ratio in an ethylene-fueled model scramjet combustor with inflow $M = 2.1$ and stagnation temperature $T_0 = 846$ K, which simulates Mach 4 flight condition. The results obtained allow obtaining more efficient heat release in the scramjet.

5.1.1.1 Experimental Description

The model combustor shown in Fig. 5.1 has a total length of 2200 mm and consisted of one constant area section and three divergent sections with the expansion angles of 2.5°, 3.5°, and 4°, respectively. The entry cross section of the combustor is 54.5 mm in height and 75 mm in width. There is one cavity (denoted as 'T1') installation in the top wall of the test section. The distance from the cavity leading edge to the combustor inlet is 544 mm. The cavity depth is $D = 15$ mm, length to depth ratio $L/D = 7$, and the aft wall angle $A = 45°$. Figure 5.1 shows also the optional fuel injection locations: I1, I2 denote the injection locations set upstream with a distance of 250 mm and 260 mm to T1 cavity leading edge, respectively. The injector configuration (orifice number × diameter, 3 × 2.0 mm, the distance between the orifices is 15 mm) is conducted on a module which could be uninstalled. A high-frequency voltage signal corresponding to wall pressure change was acquired by a water-cooled high-frequency sensor (PCB model 112A05) at the point on the side wall where the location is 240 mm, 120 mm, 0 mm and −115 mm upstream of the T1 cavity leading edge respectively and 36 mm above the bottom wall (shown in Fig. 5.1). The acquisition frequency of the pressure signal is 50 kHz. The flowfield is visualized through two quartz window for high speed imaging camera with 8000 frames per second. The Schlieren system uses a laser light and the shutter time is 1/120,000 s. In the present work, the test facility has been operated twice for each scheme in Table 5.1 to allow comparison between the flame photos and the Schlieren images. The flow conditions of the scramjet combustor entry are $M = 2.1$ and stagnation temperature $T_0 = 846$ K, which simulates Mach 4 flight condition.

Table 5.1 shows the detailed injection schemes, where $\dot{m}_f$ represents the fuel mass flow rate and P_{0i} is the injection stagnation pressure. Schemes 1–5 were employed to compare the combustion instability features for injection with various fuelling rates and demonstrate different flame dynamics regimes. Ethylene was injected at the condition of stagnation temperature $T_{0i} = 300$ K. For Scheme 1, the flame quenches

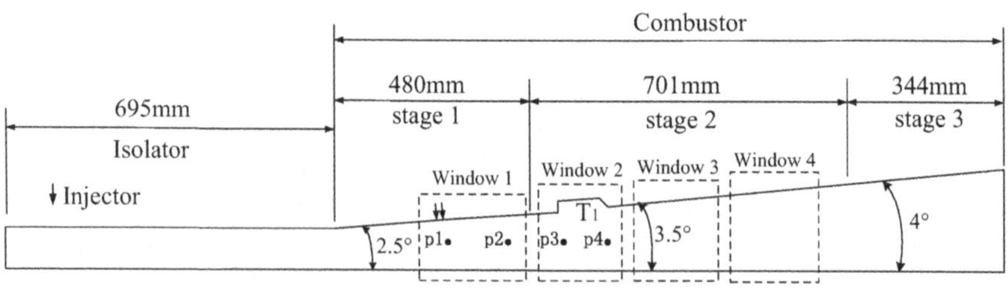

Fig. 5.1 Schematic of test section and cavity installation scheme [2]

Table 5.1 Injection schemes and flame dynamics phenomena [2]

	Injectors	P_{0i} (MPa)	$\dot{m}_f$ (g/s)	Flame dynamics
Scheme 1	$I_1 + I_2$	4.0	65.3	Extinction
Scheme 2	$I_1 + I_2$	3.0	55.0	Oscillating
Scheme 3	$I_1 + I_2$	2.0	39.5	Oscillating
Scheme 4	I_1	3.0	28.2	Oscillating
Scheme 5	I_1	2.0	19.1	Weak flame

immediately after forced-ignition is turned off; quenching happens due to the higher injection penetration, which causes little fuel entrained into the cavity shear layer and the flame cannot be stabilized. For Schemes 2–4, oscillating or fluctuating flames are obtained. For Scheme 5, we observe a weak flame in the T1 cavity. It is inferred from Table 5.1 that $\dot{m}_f$ plays an important role in the flame dynamics.

5.1.1.2 Results and Discussion

Figure 5.2 shows high-speed flame luminosity images of Scheme 2. The images demonstrate the details of the flame flashback (from T1 cavity to injection location) and flame blow-off. It is seen from Fig. 5.2a–c that, with support of the pilot flame in the cavity shear layer, explosive combustion develops the in supersonic flow downstream of the cavity. During this process the flame in the main flow becomes bright, which indicates considerable increase of the combustion intensity. At this stage the flame propagates against the incoming supersonic flow and spreads from the boundary layer to the bulk flow. In Fig. 5.2d–f, the flame base moves upstream of the cavity into the wake of the fuel jet very quickly until it goes through the jet injection location. In Fig. 5.2g–j, the flame quenches and is blown off downstream in a short time to the cavity location where flame is sustained again in the cavity shear layer. In Fig. 5.2k–i, burning stabilized in the cavity shear layer acts as a pilot flame, and reignites the bulk of the fuel/air mixture after certain time (about 1.9 ms averagely). From the Schlieren images in the first window, it is found that the pre-combustion shock train is pushed forward and the I1 and I2 injection has a higher penetration height during the flame flash-back. When a transient thermal throat is formed near the injection location, the fuel-air mixing is strongly enhanced and the combustion region transversely expands to the whole flow.

For Schemes 3 and 4, the flame stabilized in T1 cavity shear layer can flash back, while the leading flame is mostly stabilized in the I1 jet wakes and fluctuates intermittently within a short distance downstream of the I1 injector. The flame is not able to flash back through the jet injection location. For Scheme 5, a weak flame is always stabilized in the cavity shear layer and no flame flashback occurs in the whole procedure.

Figure 5.3 shows the measured pressure histories of different PCB sensors and the calculated flashback flame speed. The pressure signal of Scheme 2 has an intense

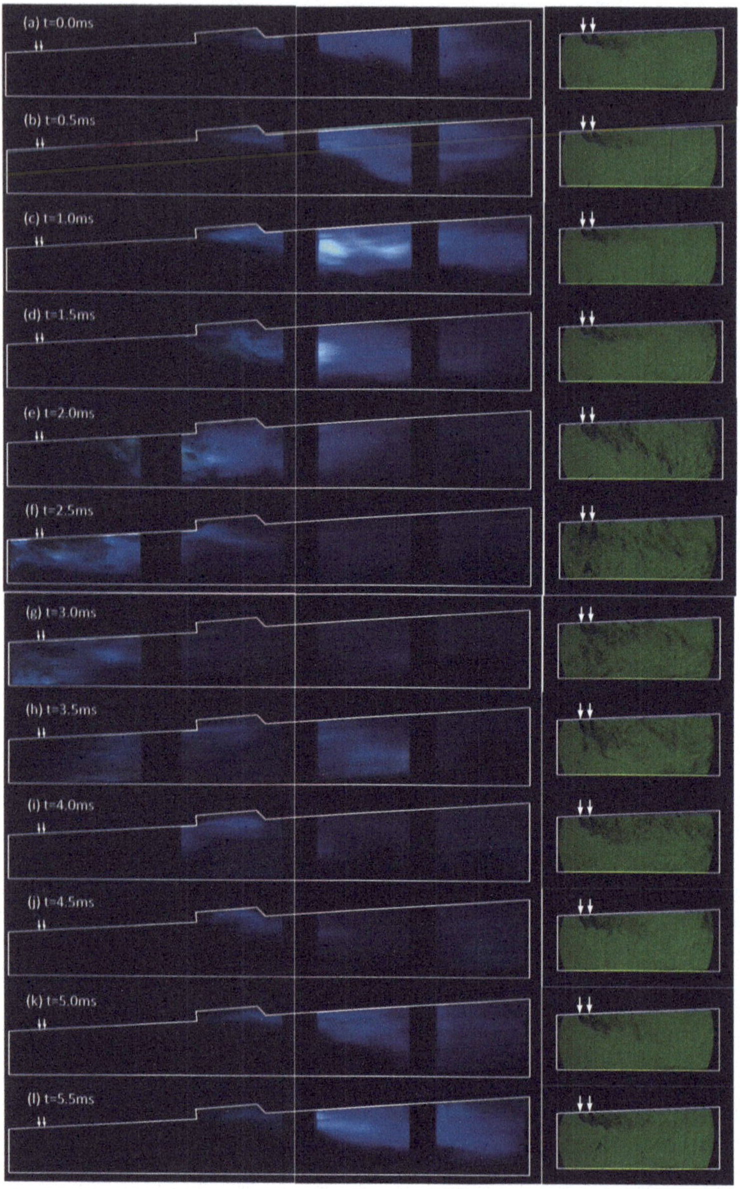

Fig. 5.2 Typical luminosity and schlieren movies of flame flash-back and blow-off event for scheme 2 (the left side are luminosity images and the right side are schlieren images) [2]

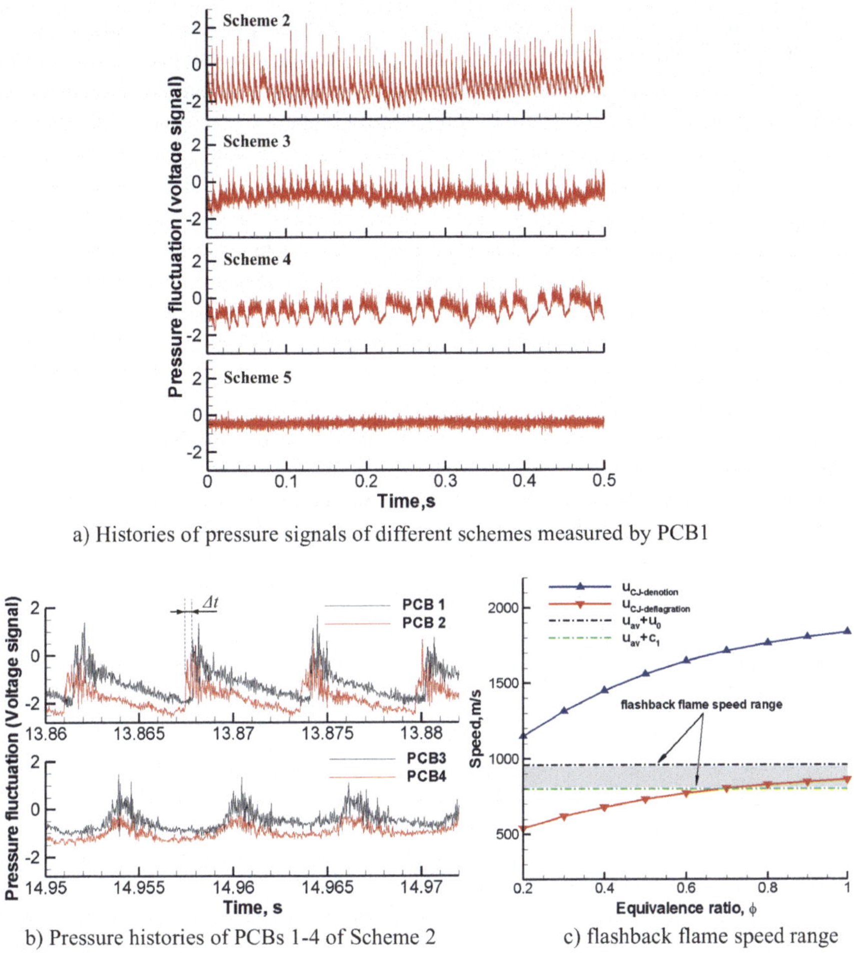

Fig. 5.3 Pressure histories inside the combustor and the calculated flashback flame speed [2]

magnitude with transition between the pressure peak and the minimum point happening very quickly. For Scheme 4, an intermittent character of pressure pulsations is observed with much lower pulsation amplitudes as compared to Scheme 2. For Scheme 3, the oscillations develop in a hybrid mode of Schemes 2 and 4; in that case leading flame is stabilized in the I1 jet wakes at most of time, while intermittent pressure drops sporadically occur. Scheme 5 has the lowest pressure fluctuations among Schemes 2–5. During the flame flashback in Scheme 2, the shock movement ahead of the flame leads to a specific shape of the pressure signal in the oscillation period, with extremely sharp start and subsequent relaxation as shown in Fig. 5.3a. We believe that, as the equivalence ratio increases, the flame in the cavity T1 ignites the fuel/air mixture in the main stream and produces an intense heat release, which leads to explosive combustion. The unstable flame propagates through the premixed region, which increases considerably energy release in the process and, in turn, promotes fast flame propagation. For Scheme 4, the mixing equivalence ratio is lower and the ignition cannot produce an explosive combustion. Even if the flashbacks occur, the flame propagation cannot produce sufficient energy release in the process and the amplitude of flame oscillations remains moderate. In that case the flame front cannot go through the premixed region sufficiently fast, and the flame base remains stabilized in the low speed region of the jet-wake with a proper local equivalence ratio. In Scheme 4, the pressure drops are presumably related to large enough flow fluctuations able to extinguish the flame, which is then convected downstream to the cavity stabilized location. As the fuel equivalence ratio increases, the energy of the instantaneous heat release is increased too, and hence the amplitude of the flame flashback oscillations becomes larger.

As shown in Fig. 5.3b, the pressure peaks are recorded in sequence when the flame front or the leading shock passes by the PCB transducers. No obvious phase difference is observed for the peak pressure for PCB 3 and 4. However, the pressure peak of PCB 1 and 2 give obvious time shifts Δt, which means acceleration of the flamefront as well as increase in the propagation speed and amplitude of the leading shock. Here we remind that PCBs 3–4 are located in the downstream region and PCBs 1–2 are placed upstream of the cavity T1.

Since the flame is stabilized/propagates in the supersonic flow, the flame propagation speed with respect to the flow during the flash-back is comparable to the speed behind the oblique shock wave. Then the average flame speed relative to a fixed observer or walls is calculated as two instantaneous pressure peak signals, $u_{av} = \frac{1}{n}\sum_i^n \frac{\Delta L}{\Delta t_i}$, where n is the number of pressure peaks, ΔL is the distance between the two PCB transducers, and Δt_i is the time interval of two step signals shown in Fig. 5.3b. For Schemes 2–4, the flame speed relative to the walls is approximately 318.3 m/s, 218.2 m/s and 207.7 m/s, respectively. The shocks ahead of the flame are oblique instead of normal, and the flow velocity behind oblique shocks is somewhat higher than the local sound speed. Here we take the local sound speed c_1 as the lower limit to evaluate the flame speed, that is $u_{av} + c_1$. The upper limit of the flame speed relative to the walls is taken as $u_{av} + u_0$, where u_0 is the normal shock velocity relative to the incoming flow. Figure 5.3b shows the flame speed range and the respective speed for the Chapman–Jouguet (CJ) detonation and deflagration versus the ethylene-air equivalence ratio. The CJ deflagration speed relative to the walls is evaluated as [3, 4]

$$(u_{CJ})_{deflagration} \approx \frac{\gamma(\gamma-1)+2(\gamma+1)}{2(\gamma+1)^2}\sqrt{2(\gamma^2-1)Q} \tag{5.1}$$

where Q is the chemical energy release in the ethylene-air mixture and γ is the specific heat ratio.

As shown in Fig. 5.3c, the estimated flame speed range is well located in between the CJ detonation and deflagration speed, with much better correlation to the CJ deflagration regime. This correlation indicates that the observed flame flashback process in Schemes 2–4 has much in common with the final stages of deflagration-to-detonation transition (DDT) in channels, for which the quasi-stationary CJ deflagration accelerates due to explosive run-away just ahead of the front and goes over to CJ detonation [3, 4]. We infer that the observed flame flashback demonstrates also similar features with strong flame acceleration (FA) in a tube filled with premixed gas [5]. During the flame flashing back to the I1/I2 injectors, the energy release rate appears to be sufficiently high and generates explosive compression, resulting in fast flames propagating at speed exceeding the CJ deflagration speed relative to the incoming flow between PCB 1 and PCB 2, see Fig. 5.3c. Though flame acceleration is the initial and the most important part of DDT [3, 4], it does not necessarily mean that flame acceleration always leads to DDT. In the present experimental conditions, the DDT process is presumably moderated and interrupted when flame reach to the injectors by the insufficient mixing. For the present conditions, the distance between the I1/I2 injectors and the T1 cavity is not long enough to establish the run-up distance for detonation. The flame front cannot propagate upstream of the injectors, which hence leads to flame extinction and blow-off. Still, for Scheme 2, quasi-detonation may be expected with higher equivalence ratio.

For Scheme 3, the weak flame sustained in the cavity shear layer corresponds to the case of weak FA [5]. Due to the limited heat release, the flame is not able to generate strong compression and propagate backwards. Compared to Schemes 3–5, it can be inferred that there is a critical fuel mass flow rate between 19.1 and 28.2 g/s, for which the flame flash-back can occur.

5.1.2 Injection Parametric Study in a Single-Side Expansion Scramjet Combustor

In addition to observe the detailed transient process of the flame flashback in a single-side expansion scramjet combustor. Ouyang et al. [6] had carried out a series of comparative experiments on different conditions for the injection parametric study of combustion oscillation in a single-side expansion scramjet combustor with inflow $M = 2.1$ and stagnation temperature $T_0 = 947$ K, which simulates Mach 4 flight condition.

5.1.2.1 Experimental Description

The facility is composed of air heater, supersonic nozzle and scramjet combustor. The model scramjet combustor is directly mounted downstream the supersonic nozzle of the air heater which heats the air by means of air/ethanol/O_2 combustion. A weight sensor mounted on the forehead of the air heater was used to measure the facility thrust changes during the experiments [7]. This system yielded a maximum force reading of 10,000 N with an uncertainty of 0.5%. The flow conditions of the supersonic nozzle exit, that is the scramjet combustor entry, are listed in Table 5.2

Table 5.2 Flow conditions at the scramjet combustor entry [6]

Ma	*P* (kPa)	*T* (K)	P_0 (MPa)	T_0 (K)	Y_{O2}
2.1	71	528	0.65	947	23.3%

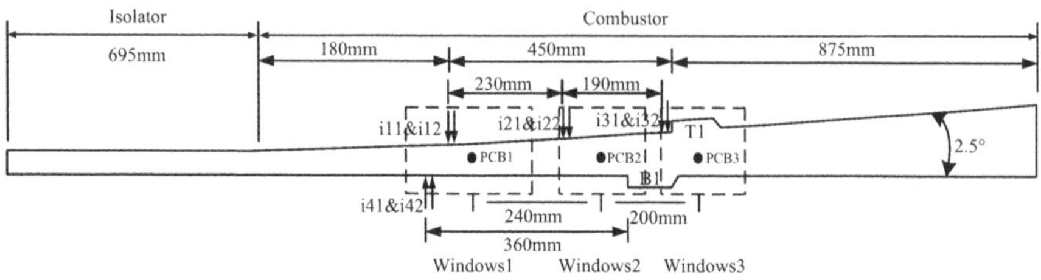

Fig. 5.4 Schematic diagram of the scramjet combustor model [6]

As Fig. 5.4 shows, the model scramjet combustor consists of a constant cross-section isolator and a single-side expansion combustor. The entry cross section of the combustor is 54.5 mm in height and 75 mm in width. The combustor has an expansion angle of 2.5° on the upside wall. The T1 cavity is arranged in the expansion-side wall (upside) and the B1 cavity is arranged in the straight-side wall (downside). The parameters of the cavities are set to depth $D = 15$ mm, length to depth ratio $L/D = 7$, and the aft wall angle $A = 45°$. Figure 5.1 also shows the fuel injection locations. Four group injectors named i1, i2, i3 and i4 respectively, will be used in the experiments. Every group contains two rows injectors (orifice number × diameter, 3 × 2.0 mm in every row, the distance between the two rows is 15 mm), named im1 and im2, m denotes group name. The distance between groups is shown in Fig. 5.4.

The ethylene flame behavior is captured by a high speed movie camera through the three quartz windows on the model scramjet combustor, for which 4000 fps (frames per second) is chosen with a resolution of 1024 × 512 pixels and a shutter time of 1/5000 s. The flow structure is obtained with the schlieren system using a laser light and the shutter time is set 1/148,000 s with a frequency of 9300 fps. The pressures of combustor along the centerline of the upside wall in the test section are measured by a series of strain-gauge pressure transducers through taps with the diameter of 0.5 mm distributed on the upside wall. Furthermore, a high-frequency voltage signal corresponding to wall pressure change is acquired by a water-cooled high-frequency sensor (PCB model 112A05) at the points on the sidewall (shown in Fig. 5.4). The acquisition frequencies of the pressure signals are 50 kHz.

5.1.2.2 Effect of Fuel-Equivalence Ratio

In order to investigate the effect of equivalence ratio on the combustion oscillation in scramjet combustor,two group comparative experiments have been carried out which are listed in Table 5.3. The first group experiments are conducted when the

Table 5.3 Comparison experiments about effect of varying equivalence ratio [6]

Group no.	Scheme no.	Fuel	Injector	Cavity	Φ	Net thrust
1	Scheme 1	Ethylene	i11 + i12	T1	0.266	303
	Scheme 2				0.4	408
	Scheme 3				0.533	515
2	Scheme 4		i41 + i42	B1	0.266	Extinction
	Scheme 5				0.4	361
	Scheme 6				0.533	571

injectors and flame-holding cavity are mounted on the expansion-side wall (upside), and the second group on the straight-side wall (downside). The thrust increment of different schemes are also listed in Table 5.3, which is calculated from only the air heater working status to the scramjet engine plus air the heater working status [7]. The thrust increment can be used as one target parameter for the combustor performance assessment. Figure 5.5 shows the high-frequency pressure-time history of all available PCB transducers and for a detailed analysis of periodic characteristics, the frequency spectrum of all available PCB transducers is obtained using FFT (Fast Fourier Transform). The corresponding FFT transformed results are also displayed in Fig. 5.5.

Firstly, the results of first group experiments will be discussed. It can be found that when the global equivalence ratio is 0.266 (scheme 1), the irregularity of the pressure peak period is obvious, the combustion behaves as an intermittent oscillation. When the global equivalence ratio increases to 0.4 (scheme 2), however, the surprising change happens. The PCB results become regular and periodic, and it is easy to identify a clear principal frequency according to the frequency spectrum, which is about 126 Hz. The results when the global equivalence ratio is 0.533 (scheme 3) are analogous to scheme 2, its principal frequency is about 118 Hz. As the equivalence ratio is raised, the combustion oscillation becomes regular and its intensity becomes higher, mainly attributed to an increase in the combustion heat release, which can be validated by the net thrust in Table 5.3.

As to the second group results, when the global equivalence ratio is 0.266 (scheme 4), the flame cannot be hold and extinct, and when the global equivalence ratio is 0.4 (scheme 5), the result is analogous to scheme 1 instead of scheme 2, when the global equivalence ratio increases to 0.533 (scheme 6) in the end, the obvious periodic combustion oscillation achieved in the first group experiments cannot yet be observed, but the combustion oscillation is more frequent than scheme 5.

Comparing the two group experiments, we can found some similar and different results. The similar results contain two sides. On the one hand, as the global equivalence ratio increases, the combustion oscillation becomes more regular and frequent. It implies that the local thermal choking should be responsible for the combustion oscillation process. As Fig. 5.6 shows, the cavity-hold flame will be strengthened and expand transversely to the main flow before upstream propagation, and the ensuing flame upstream propagation exists in the bulk of flowpath instead of the vicinity of

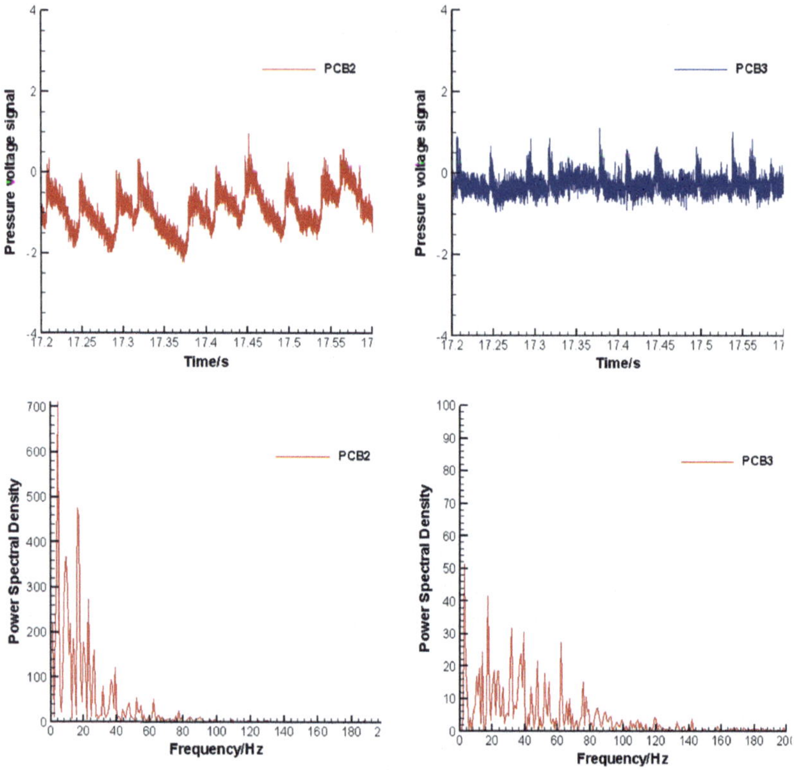

(a) scheme 1

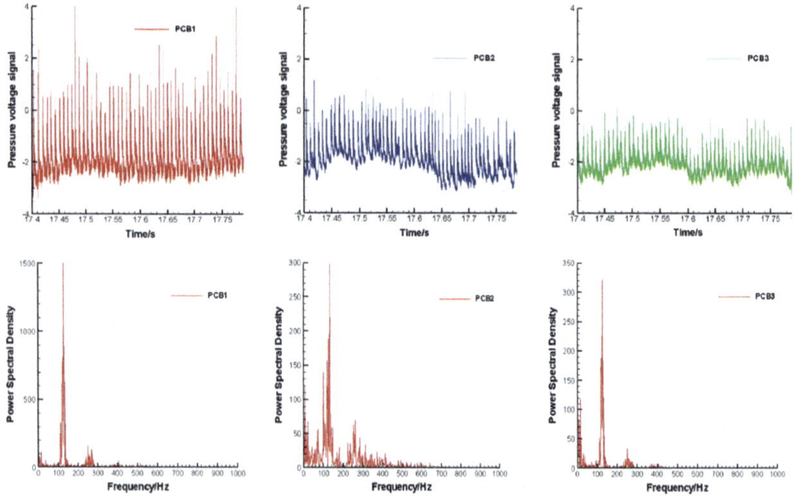

(b) scheme 2

Fig. 5.5 Histories and power spectra of high-frequency pressure signal of scheme 1–3, 5–6 [6]

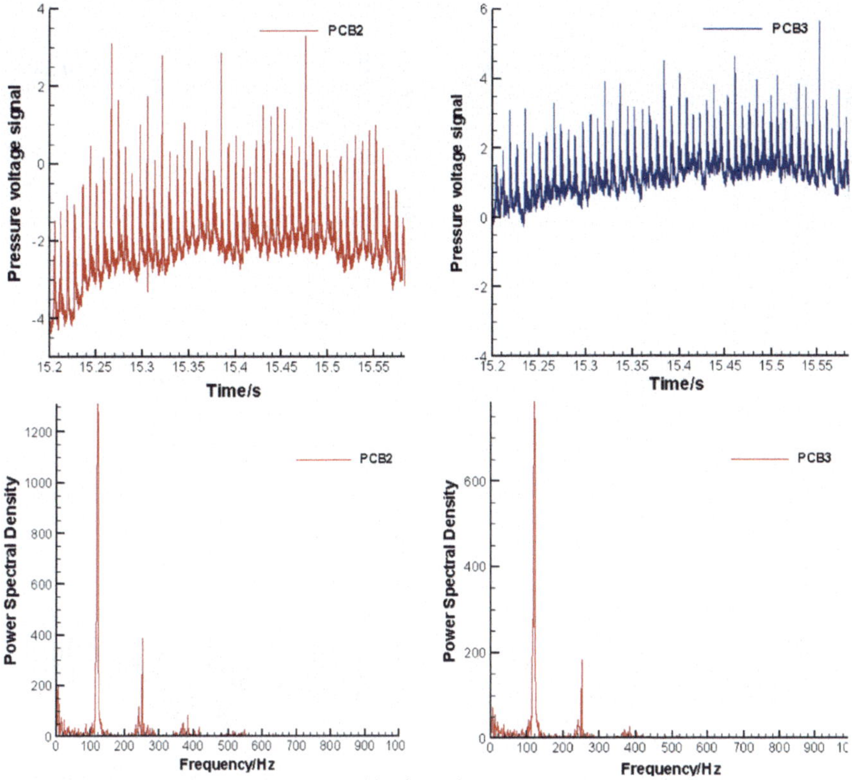

(c) scheme 3

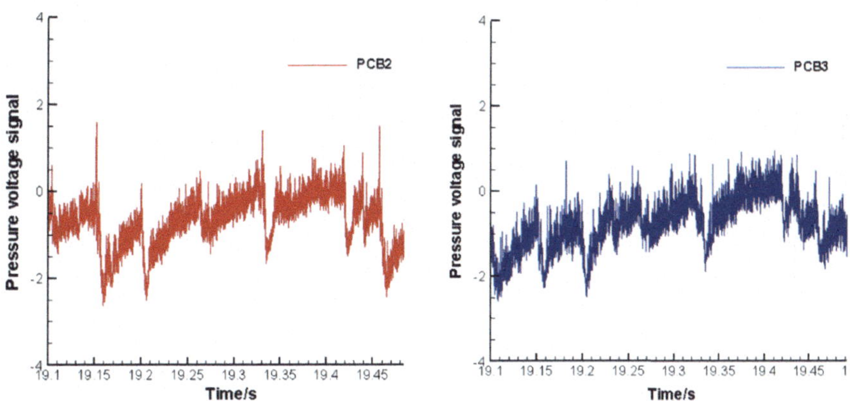

Fig. 5.5 (continued)

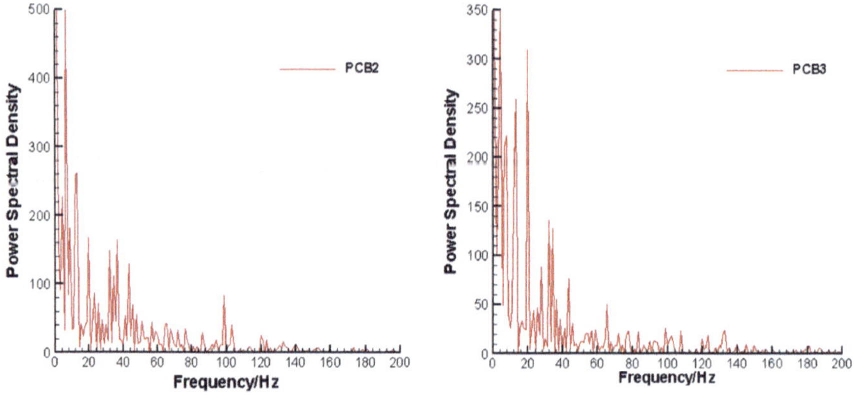

(d) scheme 5

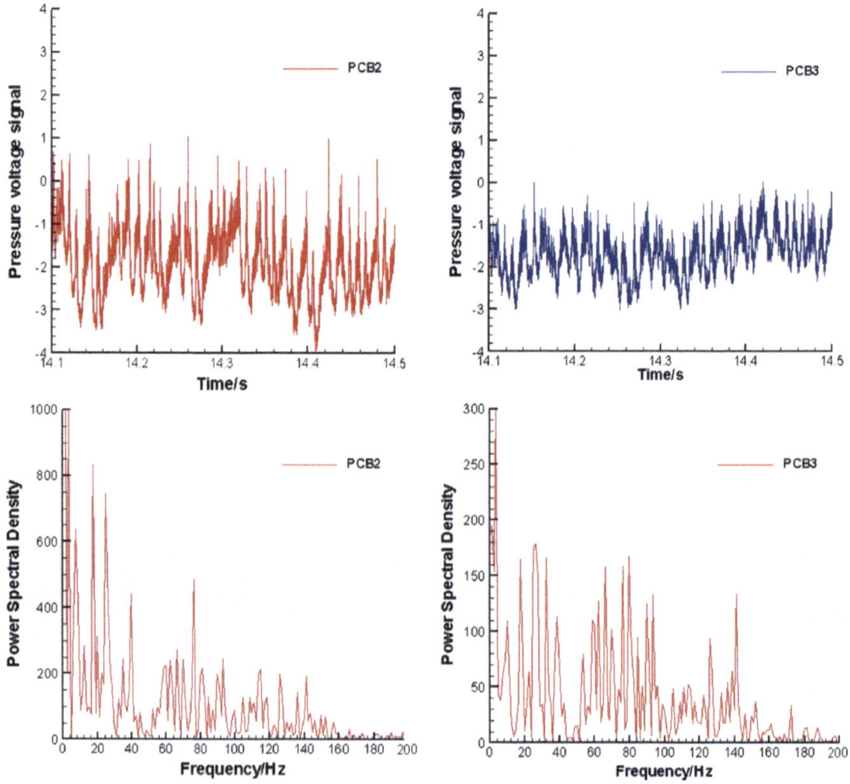

(e) scheme 6

Fig. 5.5 (continued)

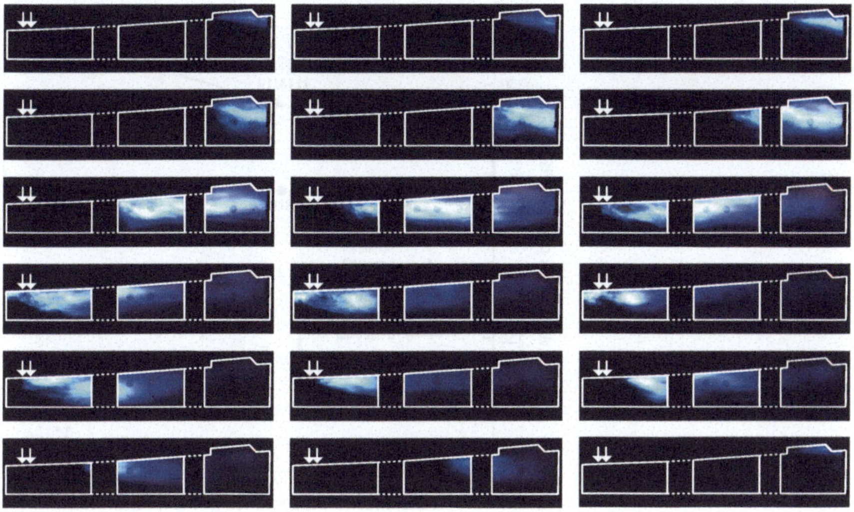

Fig. 5.6 Flame frames of one typical oscillation period in scheme 2. Time between images: 1/2000 s. Figure must be read from left to right and then, top to bottom [6]

wall. On the other hand, the combustion oscillation in the vicinity of the upstream PCB transducers is more intense at all global equivalence ratios which maybe be attributed to the intensive interaction between jet and combustion, which is shown as Fig. 5.7. The different result can be summarized that the regular and periodic

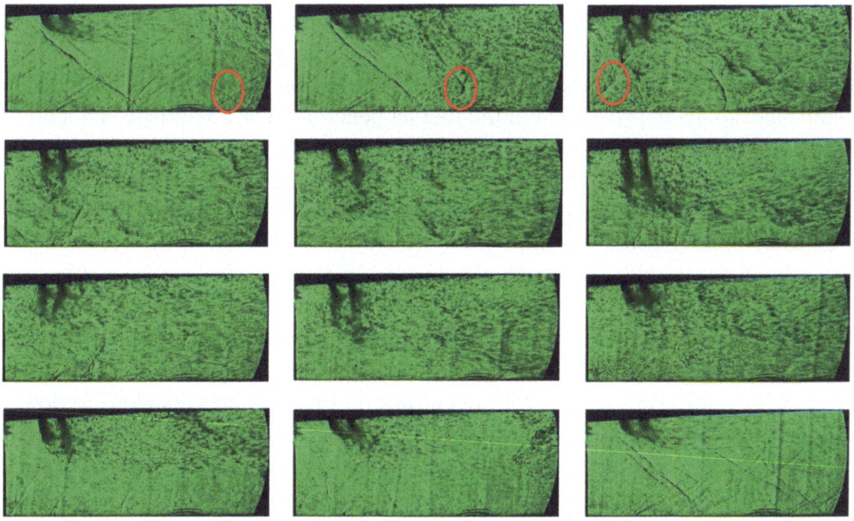

Fig. 5.7 Schlieren images of one typical oscillation period in scheme 2. Time between images: 3/3100 s. Figure must be read from left to right and then, top to bottom [6]

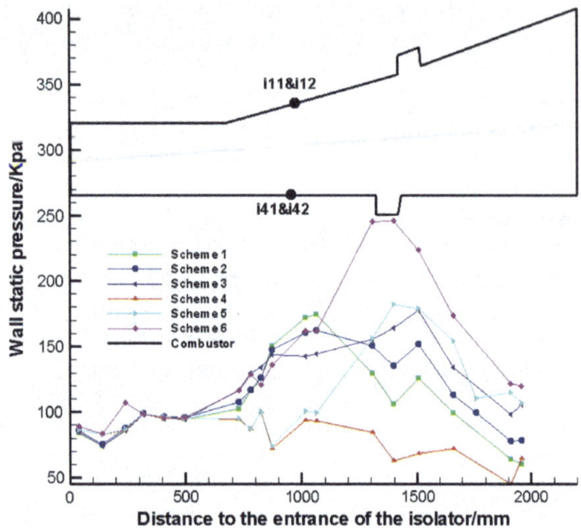

Fig. 5.8 Comparison of wall static pressure of scheme 1–6 [6]

combustion oscillation is more possible when the injectors and flame-holding cavity are mounted on the expansion-side wall (upside). This difference is related to the formation of the low-speed separation zone. Figure 5.8 shows the time-averaged upside static pressure of the six schemes. It can be found that when the injectors and flame-holding cavity are mounted on the expansion-side wall (upside), the main combustion heat zone is between injectors and cavity. However, when the injectors and flame-holding cavity are mounted on the straight-side wall (downside), the main combustion heat zone concentrates in the vicinity of cavity even the flame cannot be hold, which means the low-speed separation zone in front of cavity B1 has been suppressed, which can be deduced from Fig. 5.9. In Fig. 5.9, the injecting bow shock and its reflect shock exist most of the time, which indicates the flowfield maintain supersonic, so the improvement of the fuel jet penetration characteristic is not obvious. However, most of the flowfield in Fig. 5.7 is decelerated to subsonic when the flame upstream propagates to the vicinity of the fuel jet, which is identified by the disappearance of the injecting bow shock and its reflect shock. As Fig. 5.7 shows, it even can deflect towards incoming flow, which means that the flow or combustion instability reaches sufficient amplitude, and reversal of the flow occur which results in the low-speed separation zone spreads to the upstream of the injectors. Compared the shockwave structures in Figs. 5.7 and 5.9 carefully, some surprised difference can be find. In Fig. 5.7, as the flame propagates upstream, the pre-combustion shock wave (which has been marked with red ellipse in Fig. 5.7) appears close to downside, which will be strengthened and pushed upstream subsequently. This indicates that the flame upstream propagation should be attributed to the downstream thermal choking, which is in line with the standpoint of Laurence et al. [8–13]. As Fig. 5.9 shows, however, when the injectors and flame-holding cavity are mounted on the straight-side wall (downside), as flame propagate upstream, the boundary layer separation shock (which has been marked with white ellipse in Fig. 5.9) instead of the

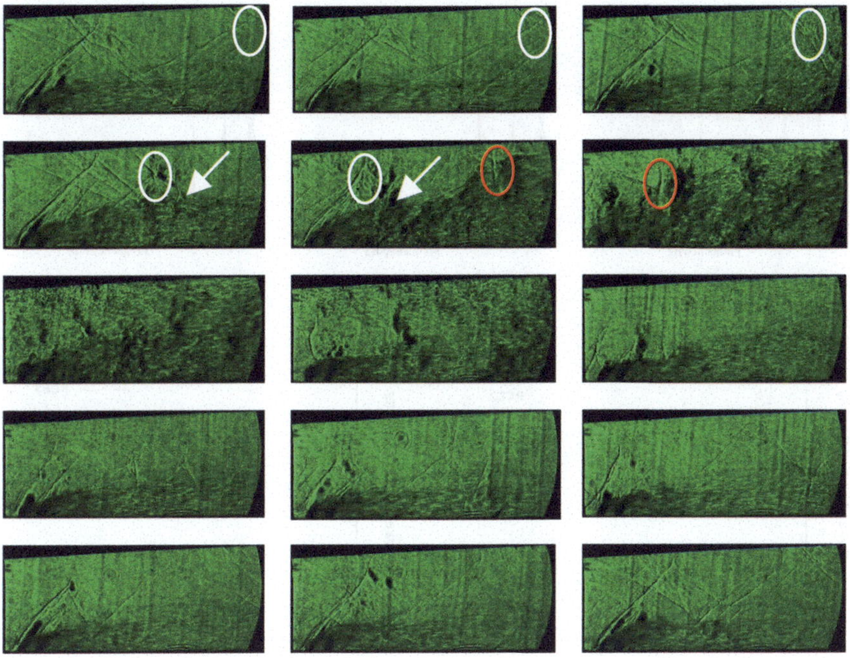

Fig. 5.9 Schlieren images of one typical oscillation period in scheme 5. Time between images: 3/3100 s. Figure must be read from left to right and then, top to bottom [6]

pre-combustion shock wave will appear close to upside, which will be strengthened and pushed to fuel jet subsequently. The strengthened separation shock will act on the fuel jet. The action will expand the jet wake, thus improve the penetration and mixing effect notably, which has been pointed out by the white arrowhead in Fig. 5.9. Subsequently, the improved mixing fuel will promote the downstream combustion and result in the pre-combustion shock wave (which has been marked with red ellipse in Fig. 5.9). The promoted combustion will only cause transient local thermal choking, which can be revealed by the successive schlieren images in Fig. 5.9. Unlike Fig. 5.7, in this operating condition, the flame upstream propagation should result from the boundary layer separation, which is corresponding to the perspective of Frost et al. [14–16]. The different mechanisms of flame upstream propagation lead to the different combustion oscillation characteristics, as the PCB results in Fig. 5.5 shows.

5.1.2.3 Effect of Injection Schemes

The former numerical study [17] has indicated that fuel injection scheme has important influence on ethylene reacting flowfiled. So it is necessary to investigate the influence of injection scheme on the combustion oscillation in scramjet combustor. Six comparative experiments have been considered in this investigation, as shown

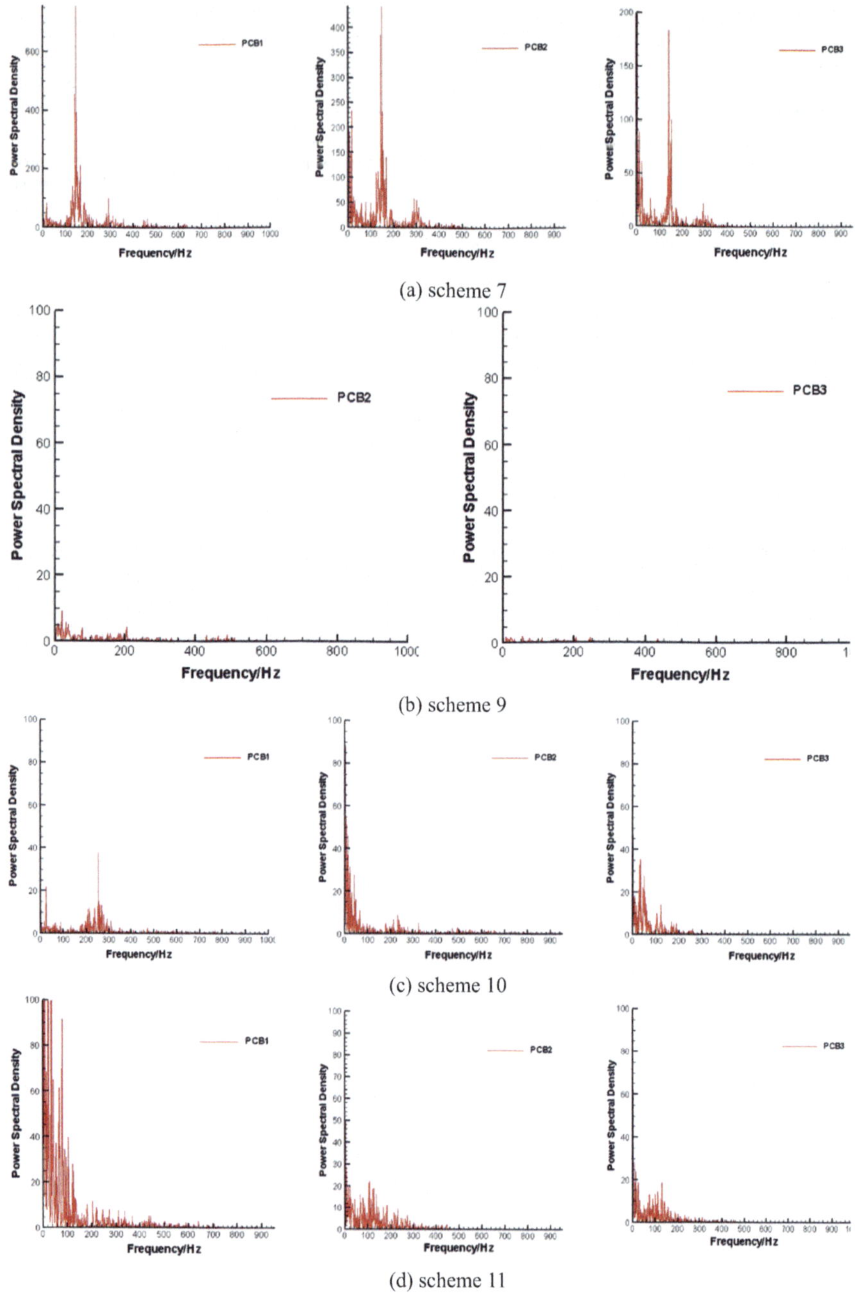

(a) scheme 7

(b) scheme 9

(c) scheme 10

(d) scheme 11

Fig. 5.10 Power spectra of pressure signal of scheme 7, 9–11 [6]

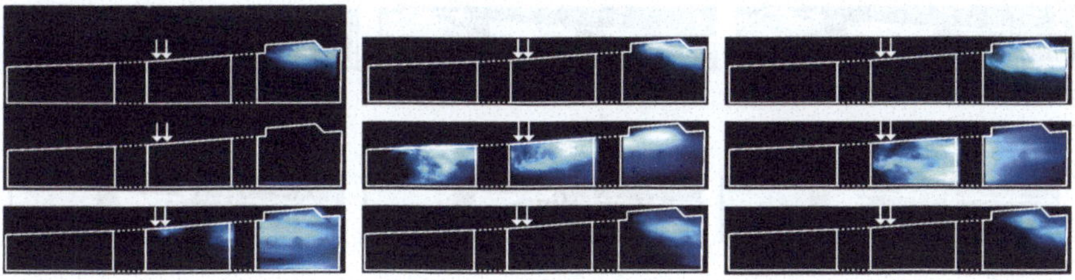

Fig. 5.11 Flame frames of one typical oscillation period in scheme 7. Time between images: 3/4000 s. Figure must be read from left to right and then, top to bottom [6]

Table 5.4 Comparison experiments about effect of varying injection schemes [6]

Group no.	Scheme no.	Fuel	Injectors	Cavity	Φ	Net thrust
Group 1	Scheme 2	Ethylene	i11 + i12	T1	0.4	408
	Scheme 7		i21 + i22			374
	Scheme 8		i31 + i32			Extinction
Group 2	Scheme 9		i11 + i31			286
	Scheme 10		i21 + i31			256
	Scheme 11		i11 + i21			394

in Table 5.4. The group 1 can be considered as centralized injection and group 2 as distributed injection. The thrust increment of different schemes is listed in Table 5.4, and for brevity, the PCB results are omitted in this section.

The result of group 1 shows that as the injectors are installed more closely to the inlet, the thrust increment become more remarkable. That is to say, increasing premixing distance will improve ethylene combustion characteristic. According to the FFT and flame results, however, we can found that the centralized injection will result in combustion oscillation in scramjet combustor. The principal frequency is about 150 Hz in scheme 7, which is higher than that of scheme 2, but its intensity of the FFT signal is lower. It indicates that although the longer premixing distance can increase thrust, it will increase the oscillation period and intensity. The extinction of scheme 8 may be not only because of the lacking of premixing but also the combustion oscillation, which reveals that the combustion oscillation in scramjet combustor can cause working uncertainty and instability. Thus, in order to avoid the combustion oscillation, the distributed injection in group 2 is considered. The intensity of the FFT signal of group 2 is lower than that of group 1 and the intensity of the FFT signal of case 9 is the lowest, which indicate that increasing the streamwise distance between injectors can relieve the combustion oscillation in scramjet combustor. Fig. 5.10 shows the comparison of combustion oscillations between the two injection schemes, and Fig. 5.11 shows one typical flame oscillation period of the distributed injection. As Fig. 5.12 shows, the separation of injectors will divide the flame into two parts, the upstream part and downstream part. It can be inferred that the insufficiently combustion hot products in upstream act as the plot flame of the downstream combustion, and thus the downstream combustion will result in sufficient adverse

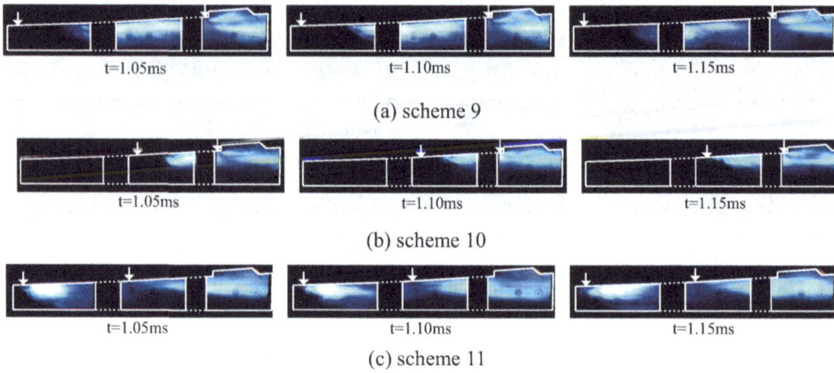

Fig. 5.12 Flame frames of scheme 9–11 [6]

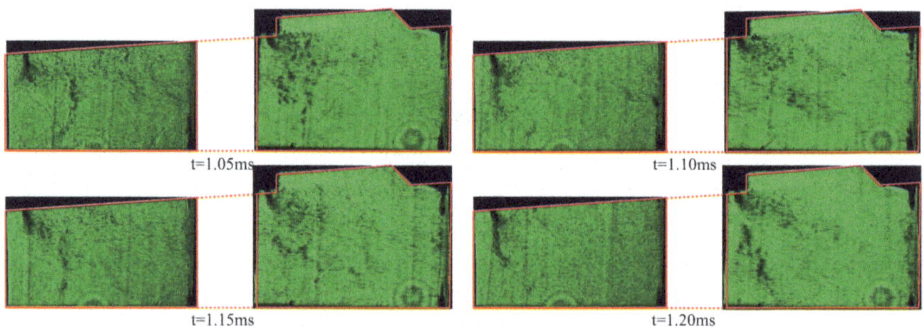

Fig. 5.13 Schlieren images of scheme 10 [6]

pressure gradient and cause local thermal choking, which can contribute to the stability of low-speed separation zone as Fig. 5.13 shows in which the injecting bow shock and its reflect shock is not visible through the observation windows. Thus, the stabilized divided flame has formed. Unlike the distributed injection, as Fig. 5.11 shows, the centralized injection will cause the generation and disappearance of low-speed separation zone periodically, which will result in combustion oscillation. According to the comparison between group 1 and group 2, it can be found that the distributed injection can avoid the combustion oscillation effectively as shown in Fig. 5.10.

5.2 Flame Flashback Phenomenon in a Flight Mach 5.5 Condition

This part experimentally investigates the combustion oscillation in scramjet combustor with a cavity flameholder under a condition of flight Mach 5.5. Especially, a combined numerical, and theoretical approach is adopted to gain additional insights into the driving mechanism of combustion instability in this special condition. The subchapter is organized as follows. Section 5.2.1 presents experimental investigation of flame flashback phenomenon. Section 5.2.2 presents numerical models and discusses the influencing factors of flame flashback, such as the boundary-layer effects, thermal disturbance, or local mixing degree. Section 5.2.3 introduces a simplified combustion opening system model to illustrate flame flashback mechanisms. In addition, a theoretical analysis model of the auto-ignition was established to investigate whether there is auto-ignition behaviour.

5.2.1 Experimental Investigations on Flame Flashback

Supersonic flame flashback is an important sub-process in the combustion oscillation process. Although flame flashback has been demonstrated in many experiments, the academic community has not yet reached a unified understanding of the exact factors that cause supersonic flame flashback. In order to improve the combustion efficiency of the scramjet engine while taking into account the reliability and robustness of the combustion chamber, this chapter uses the high-speed photography and schlieren technology, combined with the quantitative analysis method to simulate the flashback dynamic process and triggering factors in the combustion chamber of the scramjet engine. Systematic experimental studies, from injection parameter changes (total fuel equivalent ratio, fuel premixing distance, nozzle angle of incidence and number of nozzles) and cavity parameters (cavity length to depth ratio, cavity trailing edge inclination and cavity downstream air Throttling) two major aspects of discussion.

5.2.1.1 Experimental Description

The experiments are carried out in a direct-connected test facility, as detailed by Wang et al. [18], which is composed of an air heater, a supersonic nozzle, and a scramjet combustor. The air heater burns continuously pure ethyl alcohol and oxygen, and heats the gas to imitate flight Mach 5.5 conditions. As a result, we obtained a Mach 2.52 flow at the combustor entrance with a mass flux of about 1 kg/s. The detailed operation status at the nozzle and fuel injection are listed in Table 5.5. Ethylene is injected at the condition of stagnation temperature $T_0 = 300$ K and stagnation pressure $P_0 = 2.1, 2.4, 2.7$ MPa, respectively. The corresponding

Table 5.5 Experimental conditions

Parameter	T_0 (K)	P_0 (MPa)	Ma	Y_{O_2} (%)	Y_{H_2O} (%)	Y_{CO_2} (%)	Y_{N_2} (%)	$Y_{C_2H_4}$ (%)
Air	1480	1.6	2.52	23.38	6.22	10.16	60.24	0.0
f_1/f_2 jet	300	2.1/2.4/2.7	1.0	0.0	0.0	0.0	0.0	100.0

global equivalence ratio, $\phi = 0.28, 0.34, 0.37$. The fuel–air equivalence ratio of a system is defined as the ratio of the fuel-to-oxidizer ratio to the stoichiometric fuel-to-oxidizer ratio. Mathematically,

$$\phi = \frac{m_{C_2H_4}/m_{O_2}}{\left(m_{C_2H_4}/m_{O_2}\right)_{st}} \tag{5.2}$$

where, m represents the mass, suffix st stands for stoichiometric conditions. It is important to note that the local equivalence ratios in the field near the upper wall are higher since only the upper wall injection is conducted. Local equivalence ratios in the field near the upper wall are higher since only the upper wall injection is conducted.

Figure 5.14 shows a schematic of the test section. A constant 200 mm long isolator is directly connected to the 40 mm high and 50 mm wide nozzle exit, followed by the 680 mm long test section consisting of a constant area section and a diverging section with expansion angle of 10°. A cavity 'T_1' is located on the top wall 220 mm downstream of the combustor entrance. The cavity is D = 12 mm in depth, L = 56 mm in length, with the aft wall angle $\theta = 45°$. Two pair fuel injectors (f1/f2) are fixed $L_1 + L_2$ mm and L_2 mm upstream of the cavity leading edge, respectively. Unless specified, L_1 equals 20 mm and L_2 equals 150 mm; each of injectors has 3 orifices at an angle of φ to the wall; each orifice is 1.0 mm in diameter with 10 mm interval distance in the transverse direction. To maintain a total constant jet area, the equivalent injector diameter of dual-jet case is 1.7 mm. A spark plug is fixed on the cavity floor to assist ignition. Its supply voltage is 220 V; however, the instantaneous discharge voltage is nearly 2000 V. The spark rate is 100 Hz.

The flow field is visualized by high speed imaging camera and schlieren system, through the quartz window which is 140 mm × 60 mm for photograph camera and schlieren observation. The schlieren system utilizes the semiconductor continuous laser as light source. The wavelength of the laser is 532 nm. To eliminate the influence of combustion radiation, a 532 nm single pass filter is installed in front of the image recorder. To observe the flame dynamics, the high-speed imaging and schlieren image

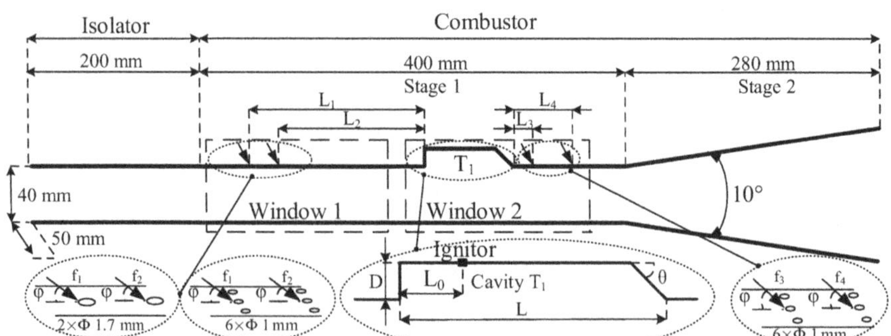

Fig. 5.14 Schematic diagram of test section and zoom diagram of injector and cavity

Table 5.6 Comparative experiments of the influence of injection parameters

Case	Total injection pressure/fuel-equivalence ratio	f_1 and f_2 injector	Outcome
1	$P_{jet} = 2.1$ MPa/$\Phi = 0.28$	$\varphi = 60°$ $L_2 = 150$ mm	Stable combustion
2	$P_{jet} = 2.4$ MPa/$\Phi = 0.34$	$\varphi = 60°$ $L_2 = 150$ mm	Periodic oscillation
3	$P_{jet} = 2.7$ MPa/$\Phi = 0.37$	$\varphi = 60°$ $L_2 = 150$ mm	Periodic oscillation with inlet unstart
4	$P_{jet} = 2.1$ MPa/$\Phi = 0.28$	$\varphi = 90°$ $L_2 = 150$ mm	Periodic oscillation
5	$P_{jet} = 2.1$ MPa/$\Phi = 0.28$	$\varphi = 90°$ $L_2 = 110$ mm	Periodic oscillation
6	$P_{jet} = 2.1$ MPa/$\Phi = 0.28$	$\varphi = 120°$ $L_2 = 110$ mm	Periodic oscillation
7	$P_{jet} = 2.7$ MPa/$\Phi = 0.37$	Dual-jet $\varphi = 60°$ $L_2 = 170$ mm	Stable combustion

cameras are set with 10000 fps (frames per second) 1024 × 256 pixels and shutter time of 1/25000 s. A pressure scanner is introduced to obtain the pressure along the combustor centerline of the top wall in the test section through taps with 1.0 mm diameter. The acquisition frequency of the pressure scanner is 100 Hz, with 0.25% full scale accuracy. The repeatability of the experimental procedure has been validated using numerous experiments [18–20].

5.2.1.2 Effect of Injection Parameters

The effects of equivalence-ratio, injection angle, and jets number on combustion oscillation phenomenon are investigated. The detailed flow conditions and outcomes are listed in Table 5.6.

Figure 5.15 shows the average pressure distribution along the model combustor upper wall. There are no combustion oscillation phenomena in cases 1 and 7, only stable flames anchored at the cavity. The combustions are concentrated mainly in the cavity and the boundary layer downstream of the cavity, with a peak pressure of about 240 kPa. In contrast, combustion oscillation occurs in other cases. As expected in case 3, the higher equivalence ratio causes the pressure profiles to increase. Violent combustion leads to increased pressure in isolator which leads to inlet unstart. But the pressure decreases in the cavity and downstream of the cavity, which indicates that the combustion is subsonic. For the rest cases, it is seen that the isolators almost keep the flow steady, and pressure fluctuations only start at the fourth measurement point. Due to the moderate combustion, the pressure decreases in the cavity and downstream of the cavity, with the peak pressure corresponding to about 350 kPa. Scram-ram transition occurs in types of periodic oscillation cases, when the pressure profile abruptly increases from stable combustion case (scram) to periodic oscillation with inlet unstart case (ram).

The effect of fuel-equivalence ratio

Firstly, the influence of injection pressure (or the global equivalence ratio) has been studied by the comparative experiments listed in Table 5.7. It should be noted that

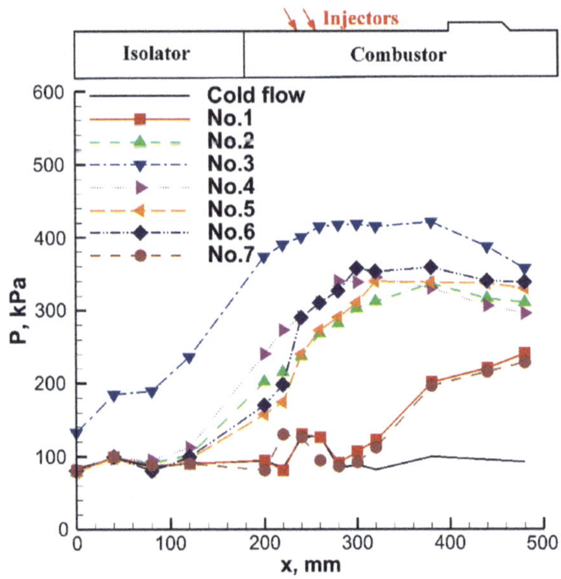

Fig. 5.15 Wall pressure distribution for different cases along the upper wall

Table 5.7 Comparative experiments of the influence of global equivalence-ratio

Case	Total injection pressure/fuel-equivalence ratio	f_1 and f_2 injector	Outcome
1	$P_{jet} = 2.1$ MPa/$\Phi = 0.28$	$\varphi = 60°$ $L_2 = 130$ mm	Stable combustion
2	$P_{jet} = 2.4$ MPa/$\Phi = 0.34$	$\varphi = 60°$ $L_2 = 130$ mm	Periodic oscillation
3	$P_{jet} = 2.7$ MPa/$\Phi = 0.37$	$\varphi = 60°$ $L_2 = 130$ mm	Periodic oscillation with inlet unstart

the dark shadow between two quartz glass windows is the supporting structure of test section. As Fig. 5.16 shows, the pressure of injection is too weak to yield a high global equivalence ratio, the lower equivalence ratio cannot produce an explosive combustion in the premixed region. Hence, there is no combustion oscillation. In addition, the injection bow shock and its reflect shock exist all the time, which indicates the flow field maintains supersonic. The separated boundary layer is too thin to hold flame. Hence, the flame is stable in cavity shear layer.

Figure 5.17 shows the combustion oscillation for case 2, it can be seen that the large-amplitude combustion oscillation can be divided into three distinctive stages consisting of flame flashback, flame blown off, and flame re-holding. Compared with case 1, the higher injection pressure leads to the much more fuel injection and higher global equivalence, and more intense combustion occurs in the cavity and downstream boundary layer. By accumulating the energy from the exothermic reactions, the combustion intensity in the separated boundary layer is gradually enhanced. When energy exceeds a certain threshold level, the separated boundary full of flames occupies the flow channel. Then, the flame begins to propagate from the leading edge of the

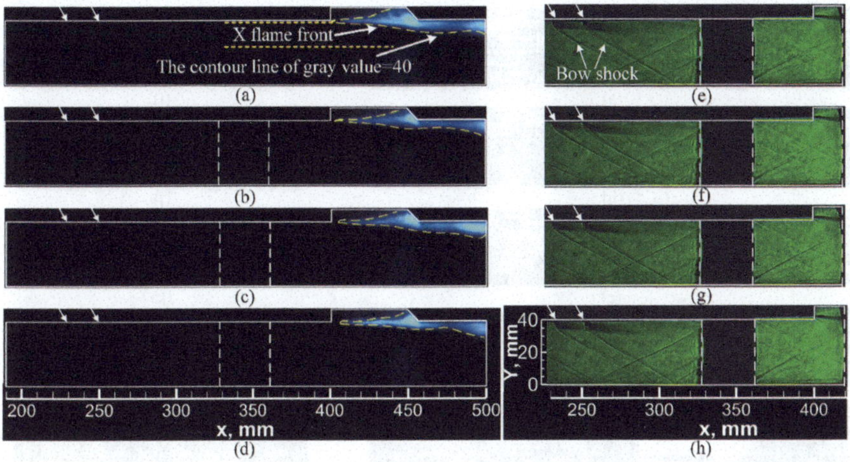

Fig. 5.16 Luminosity and schlieren visualizations of stable combustion for low equivalence-ratio. The time interval between two consecutive images is 1 ms [21]

cavity wall to the fuel injectors. As schlieren images show (Fig. 5.17i–l), the oblique shocks created by the impingement of the fuel injection are compressed and become a set of shock trains. The presence of shock trains indicates that thermal choking has occurred. Sharing the same standpoint with Laurence et al. [11, 13, 22], the thermal choking is the necessary condition of the flashback, and it results in large low-speed separation zone which plays a crucial role in the stability of flame combustion and in the enhancement of the combustion intensity. Along with more combustion heat-release going into heating the air flow, the pre-combustion shock trains near the fuel injector are gradually pushed forward, accompanying with enhancement of combustion. During the flame flashback process, the shock waves confine a high-pressure, high-temperature, low-speed region. The increased residence time promotes fuel mixing, thus enhancing combustion. It is also clearly seen in schlieren images that the low-speed promotes deeper jet penetration depth. All these effects lead to higher combustion intensity downstream of the cavity. The increased heat release rate forces the flame to propagate upstream, and the strengthened pre-combustion shock trains are subsequently pushed upstream. The flame core moves upstream very rapidly along the combustor wall, until it reaches the fuel injector locations.

After this stage, due to the lack of the fuel premixing effect, the local equivalence ratio is lost. Hence, the flame surrounding the injector cannot be sustained, and it is blown quickly back to the cavity. The flame then is re-stabilized in the rear of the cavity and the cavity shear layer, forming the cavity stabilized mode flame again. According to the corresponding schlieren images (Fig. 5.17m–p), the bulk flow is subsonic during most of this stage, and then recovers to supersonic, once the separation zone is pushed downstream owing to lack of heat release blockage. At the end of this stage, the bow shock (at the upper wall) re-appears upstream, and the flame is held in the cavity shear layer again, preparing for triggering of the next

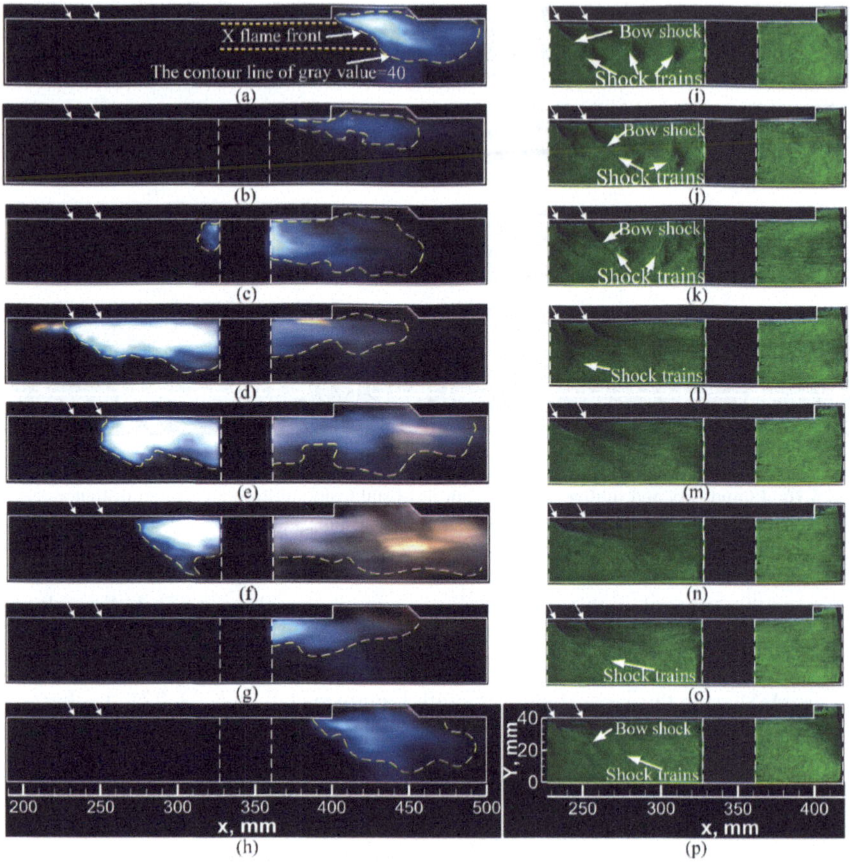

Fig. 5.17 Luminosity and schlieren visualizations of combustion oscillation for medium equivalence-ratio. The time interval between two consecutive images is 0.4 ms [21]

combustion oscillation period. Hence, a closed-loop of the flame stabilized mode is completed over one typical oscillation period.

Actually, the periodic oscillation period is a mutual transformation of scramjet and ram mode which is the same standpoint of Fotia and Driscoll [8]. Along with more combustion heat-release going into heating the air flow, the shock trains are pushed to propagate upstream as the flame propagates upstream and disappeared from the quartz windows. The combustor mode is changed from the scram mode to ram mode. Eventually, the shock trains are visualized again accompanying with the combustor mode changing from the ram to scram.

As Fig. 5.18 shows, the jet penetration is deeper because of the higher injection pressure. The higher equivalence ratio promotes more intense combustion leading to flame flashback to upstream of injector even to isolator, which can be proved by wall pressure distribution in Fig. 5.15. As schlieren images show, the shock trains cannot be visualized through the quartz windows, which is accompanied by the weak bow shock. It indicates that the flow has transitioned from a confined supersonic core flow to a pseudo shock accompanied subsonic flow, thus, combustor mode changing from the scram to ram.

The flame structures of the cases from 4 to 6 are analogous to case 2 and can also flashback to injectors. Hence, it only gives initial schlieren image in rest of the subsections for purpose of avoiding repetition.

In order to quantitatively investigate the experimental flame oscillation characteristics, Following the method of Micka [23], 42,000 total images were analyzed for 7 cases to characterize the combustion zone outline by iso-luminosity contours of gray value = 40. The averaged streamwise locations of the flame front are then calculated in range of y = 20–40 mm. The flame propagation speed can be obtained by the flame front positions and the corresponding time intervals. The probability density of flame propagation speed can be obtained by statistics of a number of discrete flame propagation speeds within the certain width interval of 50 m/s. The power spectrum of flame front oscillation frequency obtains by Fast Fourier Transform (FFT). Although the existence of error caused by the discontinuity and limitation of the quartz windows, the general trend should not change qualitatively.

Figure 5.19 is the global view of time history of flame front position. For the convenience of analyzing, only zoom views are given below. Some interesting trends are observed. For case 1 (as seen in Fig. 5.20), since there is only a stable combustion downstream of the cavity, the flame front is maintained at the rear of the cavity and the cavity shear layer. The propagation speed relative to the combustor (see Fig. 5.21) is concentrated within the 100 m/s region corresponding with the small spatial movements of the flame front. For cases 2 and 3, the increasing proportion of flame front surrounding the injector demonstrates that the flame can be held around

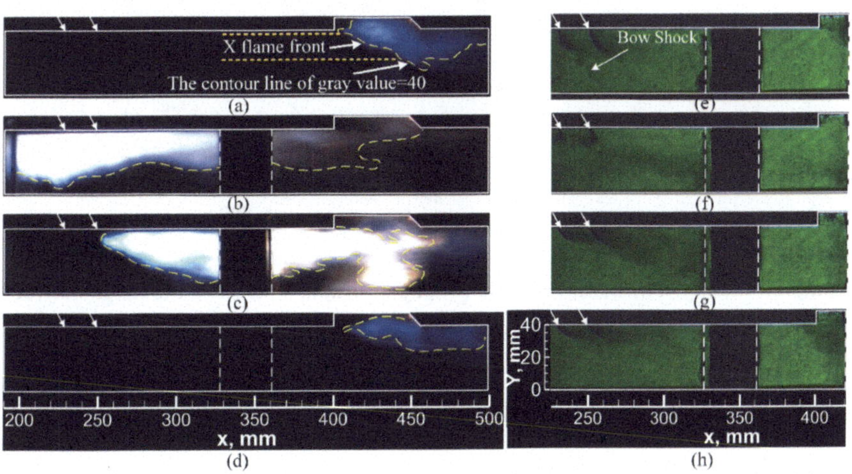

Fig. 5.18 Luminosity and schlieren visualizations of combustion oscillation for high equivalence-ratio. The time interval between two consecutive images is 1 ms [21]

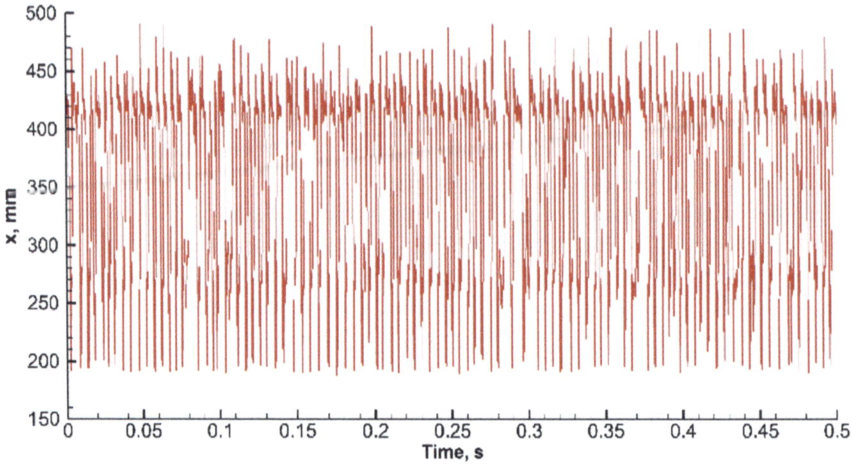

Fig. 5.19 Global view of time history of flame front position [21]

Fig. 5.20 Time history of flame front position in cases 1–3 [21]

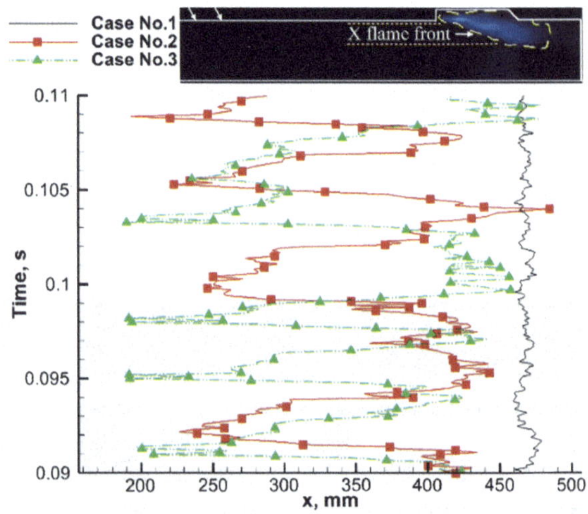

the injector easily. The main values of propagation speed focus on the range between −400 and 400 m/s. Although the speed histogram of cases 2 and 3 is similar, case 3 has slightly higher probability in range of −100 to 100 m/s. The same conclusion can be found in the propagation speed graph (Fig. 5.22), in which the distinct dominant frequency of case 3 is higher than case 2.

The effect of pre-mixing distance

In this part, effects of the different fuel mixing distances on combustion oscillation are investigated (Table 5.8). From schlieren visualizations (see Fig. 5.23), the position and structure of the bow shocks in the two cases exhibit huge differences

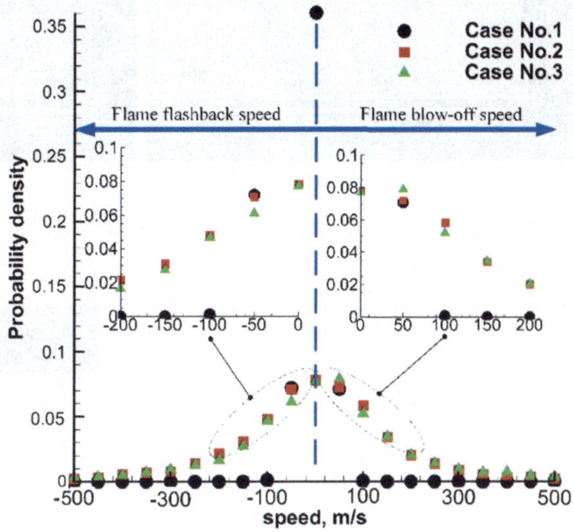

Fig. 5.21 Probability density of flame propagation speed in cases 1–3 [21]

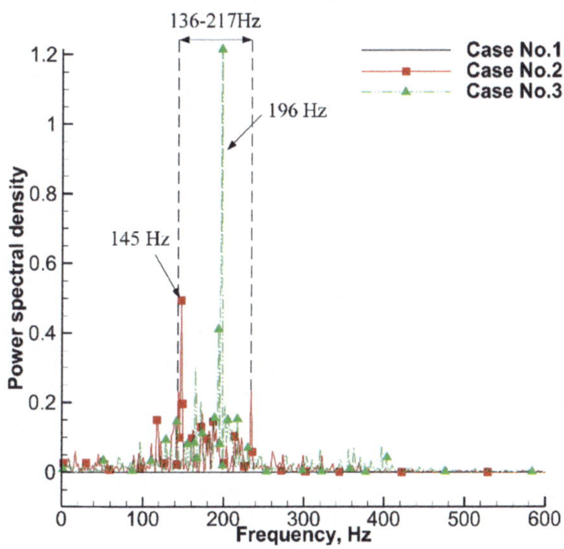

Fig. 5.22 Power spectral density of flame front oscillation in cases 1–3 [21]

Table 5.8 Comparative experiments of the influence of pre-mixing distance

Case	Total injection pressure/fuel-equivalence ratio	f_1 and f_2 injector	Outcome
4	$P_{jet} = 2.1$ MPa/$\Phi = 0.28$	$\varphi = 90°$	Periodic oscillation
5	$P_{jet} = 2.1$ MPa/$\Phi = 0.28$	$\varphi = 90°$ $L_2 = 110$ mm	Periodic oscillation

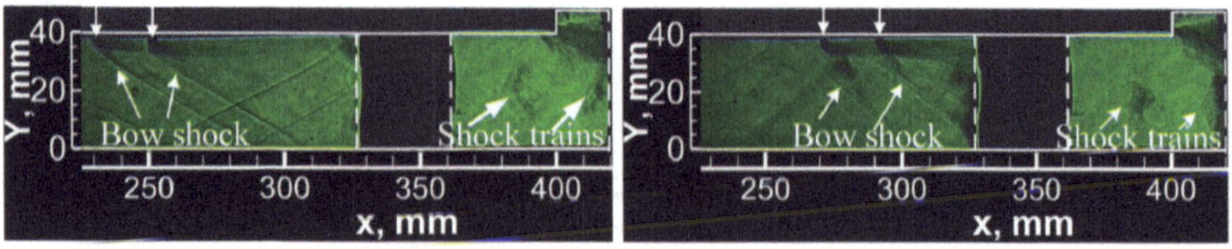

Fig. 5.23 Schlieren visualizations of combustion oscillation in cases 4 and 5 [21]

Fig. 5.24 Time history of flame front position in cases 4 and 5 [21]

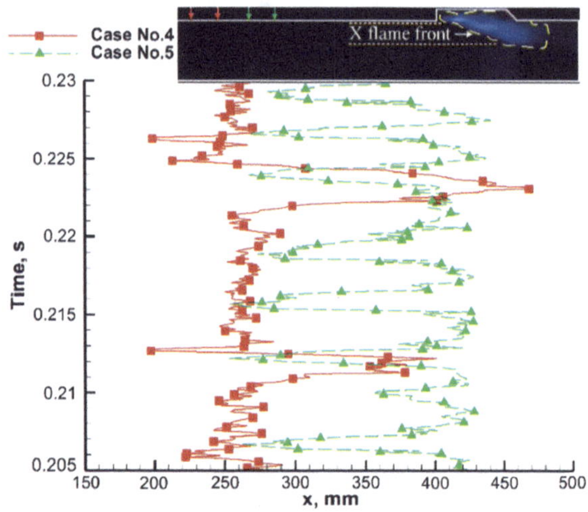

because of the different premixing distances. From the second quartz window, the pre-combustion shock trains exist in the upstream of cavity, which push the bow shock closer to the upstream in case 5. From Fig. 5.24, the reason for different flame front position may be explained as follows. Owing to longer premixing distance in case 4, the more effective fuel-mixing promotes the combustion which holds the flame around the injector for sufficiently long time. Hence, the flame propagation speed in case 4 is lower than case 5 (Fig. 5.25). The corresponding distinct dominant frequency can be obtained, which are about 64 and 335 Hz (see Fig. 5.26).

The effect of injection degree

In the present section, Table 5.9, Figs. 5.27, 5.28, 5.29 and 5.30 compare the effect of injection degree. As explained by Ref. [24], the recirculation upstream of the injectors is caused by the suction of injection. Comparing the schlieren images carefully, it can been find that increasing the injection angle leads to a stronger bow shock and a bigger separation region. The sharper injection angle promotes the larger recirculation region, and more injection mess enter the thick the boundary layer, resulting in higher fuel mixing. The recirculation region plays a vital role on mixing and consequently cases 4, 5, and 6 show better mixing performance. Under this effect, the combustion intensity in the cavity and separated boundary layer downstream of the

Fig. 5.25 Probability density of flame propagation speed in cases 4 and 5 [21]

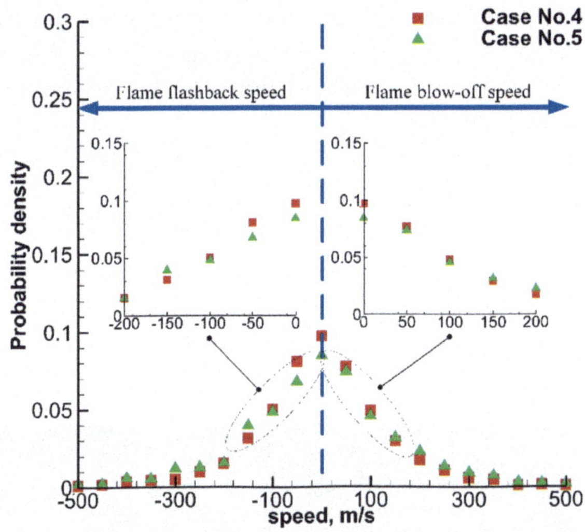

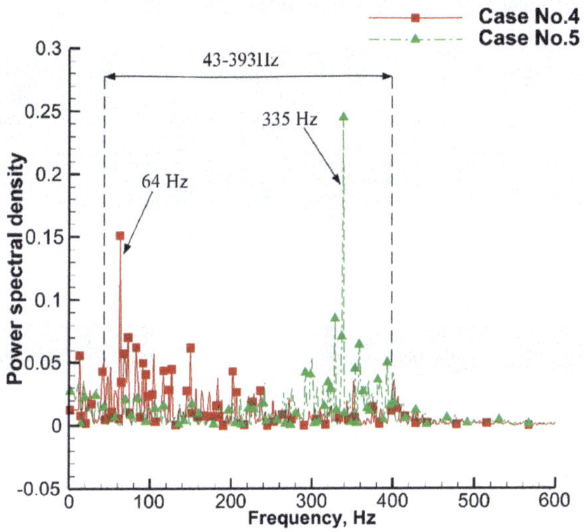

Fig. 5.26 Power spectral density of flame front oscillation in cases 4 and 5 [21]

cavity is gradually enhanced, forcing the pre-combustion shock trains and the bow shock to move forward. Under the interaction between combustion and separation region, the flame occupies the flow channel and forms a thermal throat, thus triggering flame flashback. It is worth emphasizing that case 6 not only generates the stronger bow shock, but also the compression to the bulk flow, where the fluctuation frequency and flame propagation speed are lower.

Table 5.9 Comparative experiments of the influence of injection degree

Group	Case	Total injection pressure/fuel-equivalence ratio	f_1 and f_2 injector	Outcome
1	1	$P_{jet} = 2.1$ MPa/$\Phi = 0.28$	$\varphi = 60°$	Stable combustion
	4	$P_{jet} = 2.1$ MPa/$\Phi = 0.28$	$\varphi = 90°$	Periodic oscillation
2	5	$P_{jet} = 2.1$ MPa/$\Phi = 0.28$	$\varphi = 90°$ $L_2 = 110$ mm	Periodic oscillation
	6	$P_{jet} = 2.1$ MPa/$\Phi = 0.28$	$\varphi = 120°$ $L_2 = 110$ mm	Periodic oscillation

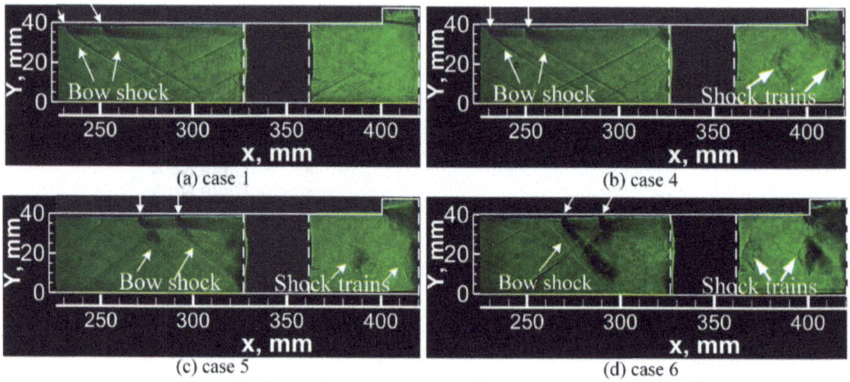

Fig. 5.27 Schlieren visualizations of combustion oscillation in cases 1, 4, 5, and 6

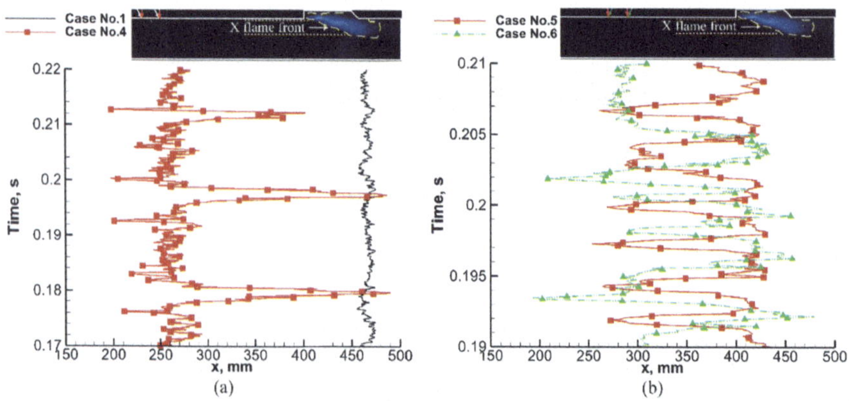

Fig. 5.28 Time history of flame front position in cases 1, 4, 5, and 6 [21]

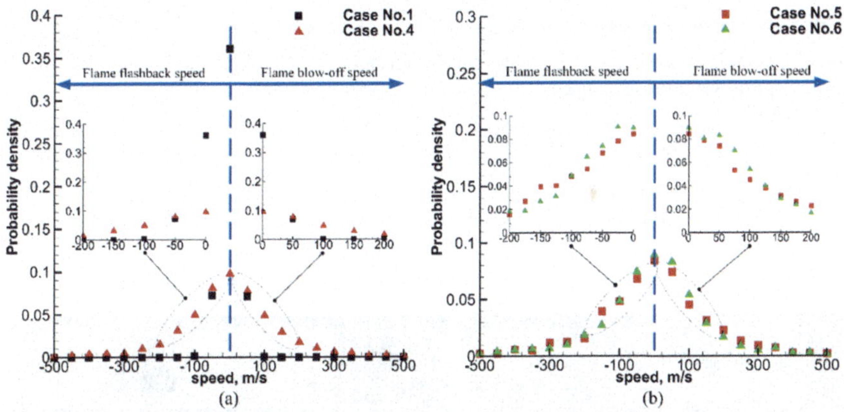

Fig. 5.29 Probability density of flame propagation speed in cases 1, 4, 5, and 6 [21]

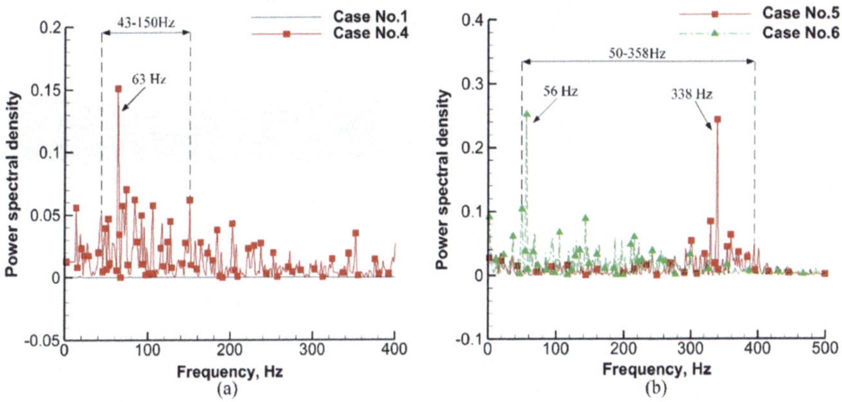

Fig. 5.30 Power spectral density of flame front oscillation in cases 1, 4, 5, and 6 [21]

The effect of jets number

In the present section, Table 5.10, Figs. 5.31, 5.32, 5.33 and 5.34 compare the effect of jets number. Compared with case 1, despite the higher fuel-equivalence ratio, the single-jet case only promotes the lower jet penetration depth, where the weaker bow shock exists. According to Pudsey and Boyce [25], the multi-jets act to enhance the near field mixing, thus improving overall mixing efficiency when compared with the dual-jet case. Compared with case 3, although the longer premix distance, case 7 only exhibits the stable combustion anchored in cavity layer, without interacting with boundary-layer. The weaker combustion could not generate pre-combustion shock trains in flow field.

Table 5.10 Comparative experiments of the influence of jets number

Case	Total injection pressure/fuel-equivalence ratio	f_1 and f_2 injector	Outcome
3	$P_{jet} = 2.7$ MPa/$\Phi = 0.37$	$\varphi = 60°$	Periodic oscillation with inlet unstart
7	$P_{jet} = 2.7$ MPa/$\Phi = 0.37$	Dual-jet $\varphi = 60°$ L_2 = 170 mm	Stable combustion

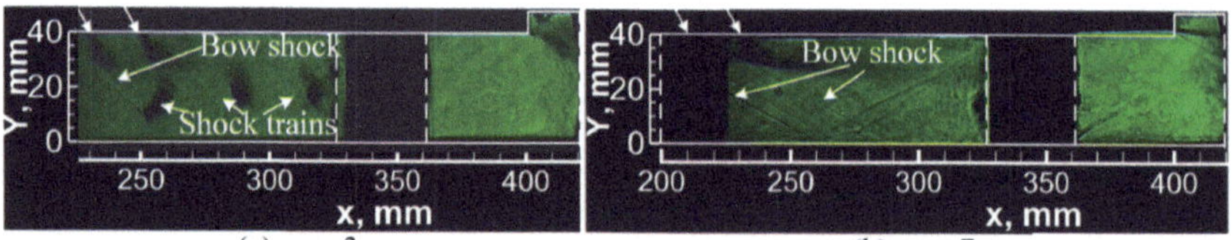

(a) case 3 (b) case 7

Fig. 5.31 Schlieren visualizations of combustion oscillation in cases 3 and 7 [21]

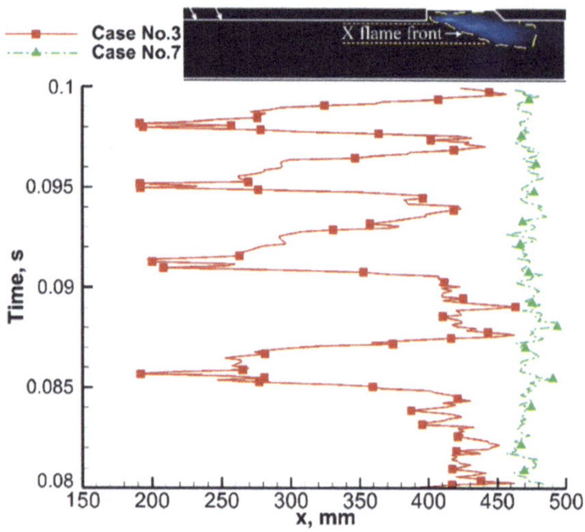

Fig. 5.32 Time history of flame front position in cases 3 and 7 [21]

5.2.1.3 Effect of Cavity Parameters

The detailed flow conditions and outcomes are listed in Table 5.11. Three groups of comparative cases respectively investigate the effects of cavity length-to-depth ratio, aft ramp angle of the cavity, and air throttling distance downstream of the cavity on combustion oscillation phenomenon.

In Fig. 5.35 there is no combustion oscillation for case 1, only the stable combustion is maintained downstream of the cavity. An expansion wave (as indicated in the schlieren photograph by its light colour) is observed at the separation corner of

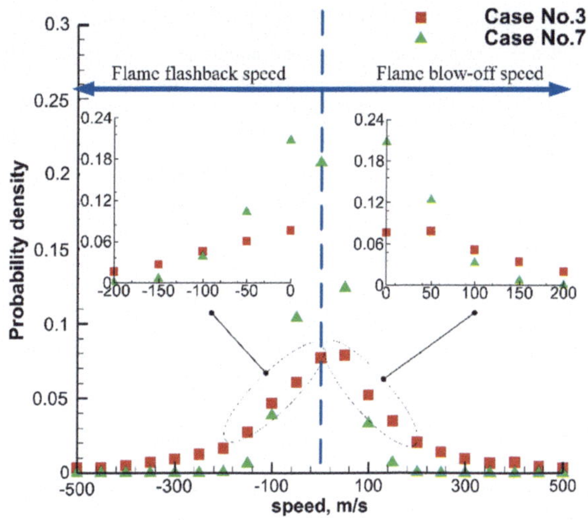

Fig. 5.33 Probability density of flame propagation speed in cases 3 and 7 [21]

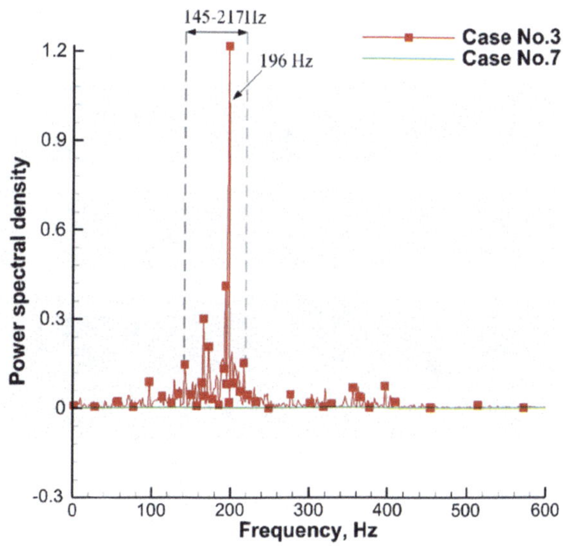

Fig. 5.34 Power spectral density of flame front oscillation in cases 3 and 7 [21]

leading edge of the cavity. In addition, the injection bow shock and its reflect shock exist all the time, which indicates the flow field maintains supersonic. From the separation of the boundary layer at the leading edge of the cavity, a shear layer is formed, which extends toward the cavity floor. The separated boundary layer upstream of the cavity is too thin to hold flames. Hence, the flames are stable in cavity shear layer. Note that the core flow direction in all photographs is from left to right. The shadow between the two quartz windows is the combustor supporting structure. Due to the limitation of experimental equipment, schlieren image only shows the region

Table 5.11 The cavity parameters of comparative cases

Group	Case	Total injection pressure/fuel-equivalence ratio	Cavity aft ramp angle/cavity length/N_2 injector distance	Outcome
1	1	$P_{jet} = 2.4$ MPa/$\phi = 0.34$	$L = 40$ mm	Stable combustion
	2	$P_{jet} = 2.4$ MPa/$\phi = 0.34$	$L = 56$ mm	Periodic oscillation
2	3	$P_{jet} = 2.1$ MPa/$\phi = 0.28$	$\theta = 45°$	Stable combustion
	4	$P_{jet} = 2.1$ MPa/$\phi = 0.28$	$\theta = 90°$	Periodic oscillation
3	5	$P_{jet} = 2.1$ MPa/$\phi = 0.28$	$P_{N2} = 0.5$ MPa/$L_4 = 30$ mm	Stable combustion
	6	$P_{jet} = 2.1$ MPa/$\phi = 0.28$	$P_{N2} = 0.5$ MPa/$L_3 = 10$ mm	Periodic oscillation

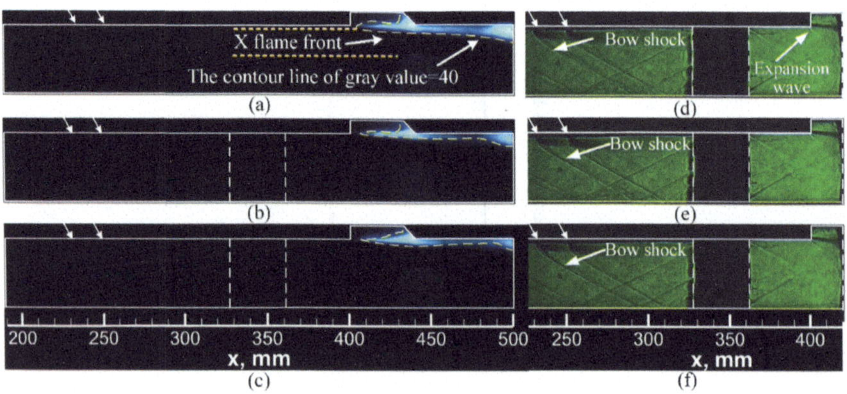

Fig. 5.35 Luminosity and schlieren visualizations of stable combustion for case 1. The time interval between two consecutive images is 2 ms [27]

between fuel injectors and leading edge of cavity. The flame frames of case 3 and case 5 are analogous to case 1, so they have been omitted to avoid repetition.

Figure 5.36 shows the large-amplitude combustion oscillation which can be divided into three distinctive stages such as flame flashback (Fig. 5.36a–c, i–k, q–s), flame blown off (Fig. 5.36d–g, l–o, t–w) and flame re-holding (Fig. 5.36h, p, x). At the beginning, the flames are maintained in the cavity and downstream of the cavity. For some reasons (which will be detailed investigated in following subsections), the intense combustion occurs in the cavity and boundary layer downstream of the

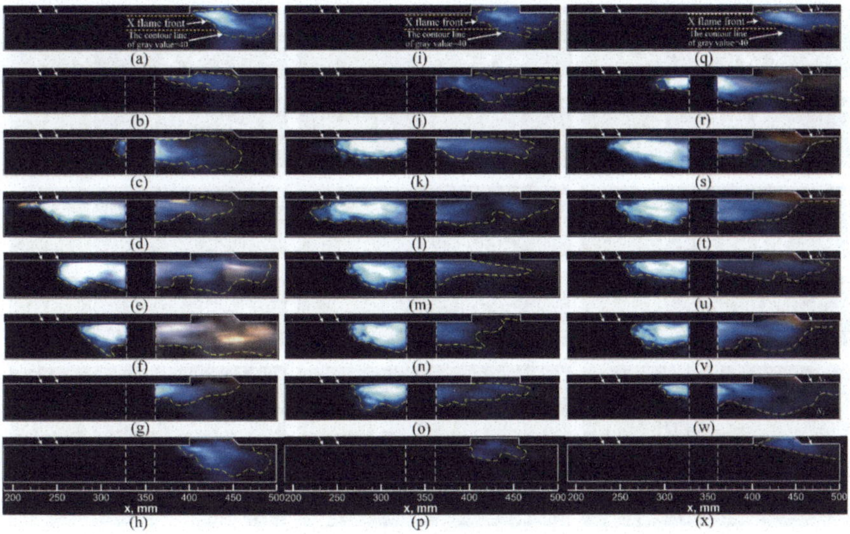

Fig. 5.36 Luminosity visualizations of combustion oscillation for cases 2, 4 and 6 (Each column). The time intervals between two consecutive images for each case are 0.4, 0.5 and 1 ms [26]

cavity. By accumulating the energy from the exothermic reactions, the combustion intensity in the separated boundary layer is gradually enhanced. The flames occupy the flow channel forming a thermal throat. Then, the flames begin to propagate from the leading edge of the cavity wall to the fuel injectors. The flame core rapidly moves upstream along the combustor wall, until reaches the fuel injectors location.

As schlieren images at the flame flashback stage (Fig. 5.37a–c, i–k, q–s) show, pre-combustion shock trains near the fuel injector are gradually pushed forward, and jet penetration depth are increased. During the flame flashback process, as we know, the shock waves confine a high-pressure, high-temperature, low-speed region. The increased fuel residence time promotes fuel mixing, leading to enhanced combustion intensity downstream of the cavity. The increased heat release rate leads to separated boundary layer enlarged. So far, the thermal throat is formed and forces the flames propagate upstream. After the flame flashback stage, due to the lack of the fuel premixing effect, the local equivalence ratio is lost. Hence, the flames surrounding the injector cannot be sustained, and they are blown quickly back to the cavity. The flames then are re-stabilized in the rear of the cavity and the cavity shear layer, forming the cavity stabilized mode flames again. According to the corresponding schlieren images (Fig. 5.37l–o, t–w), the core flows are supersonic during flame blown off stage. In contrast, Fig. 5.37d–g show a subsonic flow. At this stage, the separation zone is pushed downstream owing to lack of heat release blockage. At the flame re-holding stage, the flames are held in the cavity shear layer again, preparing for triggering of the next combustion oscillation period. Hence, a closed-loop of the flame stabilized mode is completed over one typical oscillation period. It can be concluded that, the boundary layer separation is the necessary condition of the

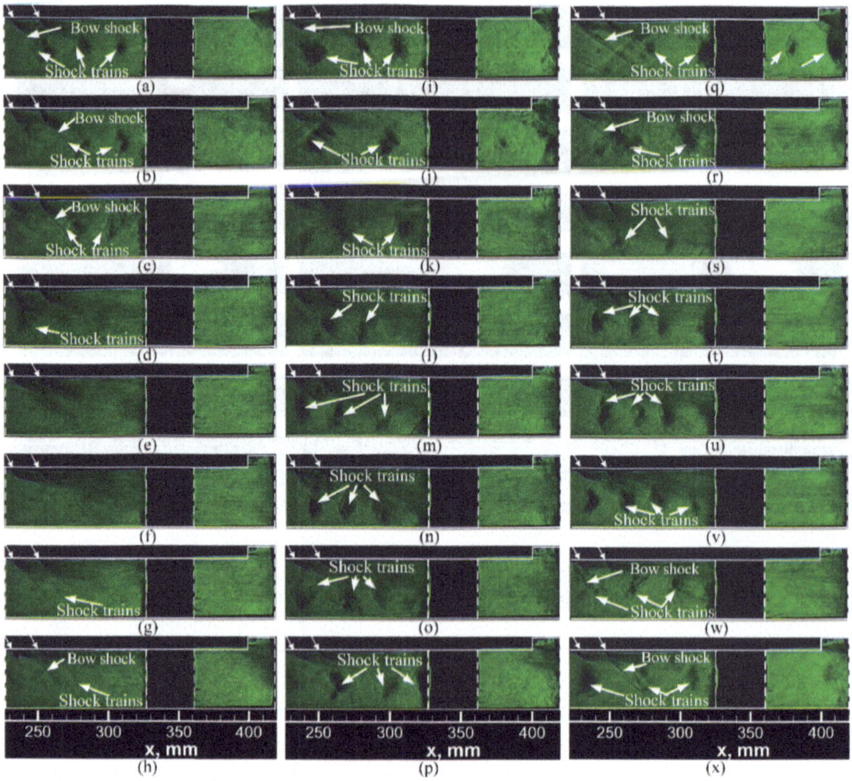

Fig. 5.37 Schlieren visualizations of combustion oscillation for cases 2, 4 and 6 (Each column). The time intervals between two consecutive images for each case are 0.4, 0.5 and 1 ms [26]

flashback, and it will result in thermal throat which plays a crucial role in the stability of combustion and in triggering flame flashback.

In order to quantitatively investigate the experimental combustion oscillation characteristics, following the method of Micka [23], 36,000 total images are analyzed for 6 runs to characterize the combustion zone outline by iso-luminosity contours of gray value = 40. The average streamwise location of the flame front is then calculated for y = 20–40 mm. In the following analysis, figures show the streamwise position of the flame front (Fig. 5.38) the streamwise propagation speed of the flame front calculated by flame front position and corresponding time interval (Fig. 5.39) and power spectrum of flame front oscillation frequency obtained by Fast Fourier Transform (FFT) (Fig. 5.40). Although the existence of error caused by the discontinuity and limitation of the quartz windows, the general trend should not change qualitatively.

For cases 1, 3 and 5, since there are only stable combustions downstream of the cavity, the flame fronts are maintained at the rear of the cavity. The propagation speeds relative to the combustor are concentrated within the 100 m/s region corresponding with the small spatial movements of the flame front. The corresponding dominant

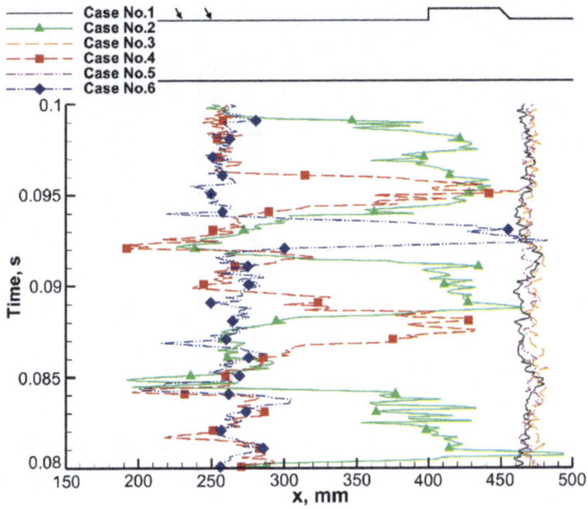

Fig. 5.38 Time history of flame front position in different cases [26]

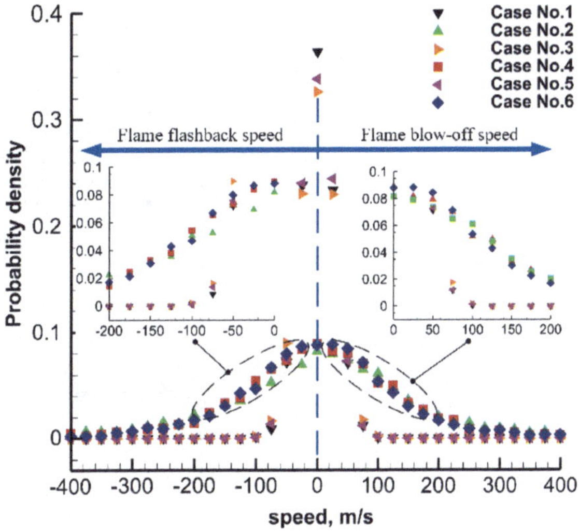

Fig. 5.39 Probability density of flame propagation speed in different cases [26]

frequency cannot be distinguished. For cases 2, 4 and 6, the increasing proportion of flame front surrounding the injector demonstrates that the flames can be held around the injector easily. The combustion oscillation frequency appears as the opposite trend compared with flame front. However, the flame speeds in these cases are similar, both flame flashback and flame blow-off speeds are stable within 300 m/s relative to the combustion chamber.

Figure 5.41 shows the average pressure distribution along the model combustor upper wall. For all cases, the isolator keeps the flow steady, and pressure fluctuations only start at the fifth measurement point. In cases 1, 3 and 5, stable combustions are mainly concentrated in the cavity and the boundary layer downstream of the

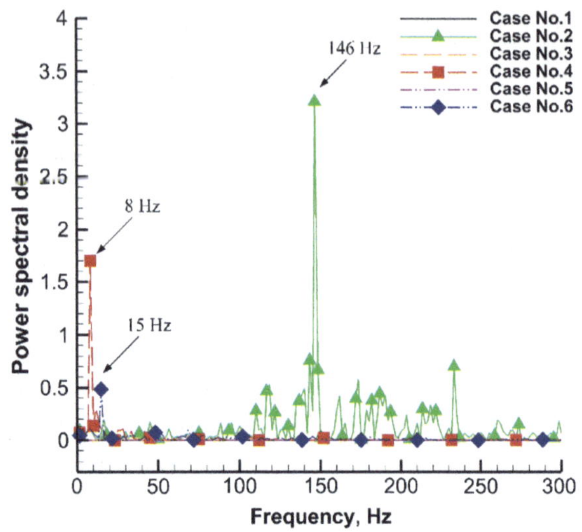

Fig. 5.40 Frequencies spectral density of flame front oscillation in different cases [26]

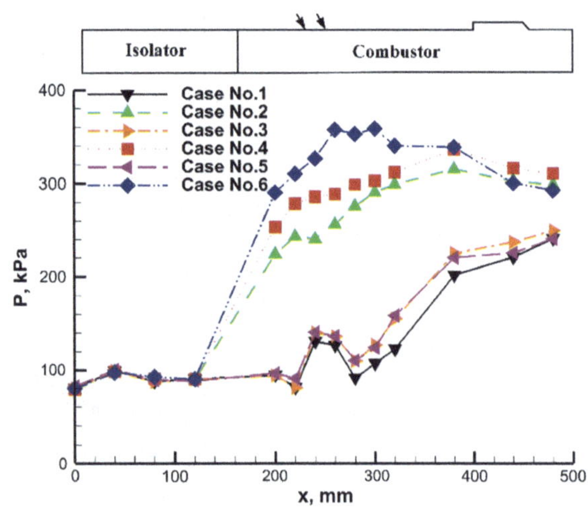

Fig. 5.41 Wall pressure distribution for different cases along the upper wall [26]

cavity, with a peak pressure of about 240 kPa. In contrast, as will be discussed in following section, the larger length-to-depth ratio, stronger cavity aft ramp angle and closer air throttling downstream of the cavity induce flame flashback. The increasing proportion of flame front sustaining near injectors causes the pressure increase near injectors, whose corresponding peak pressure is about 350 kPa, whereas the pressure in cavity and downstream of it is a little bit lower.

The effect of cavity length-to-depth ratio

In this subsection, the effects of the different cavity length-to-depth ratio on combustion oscillation are investigated as the first group. For case 1, although the total injection pressure improves to 2.4 MPa ($\phi = 0.34$), there is no combustion oscillation, only a stable combustion is maintained in cavity shear layer and downstream of the cavity. However, the combustion oscillation accompanying with rapid flame flashback occurs in case 2. As reference analysed [18, 27], We suppose that cavity length will act on the combustion oscillation mainly through the recirculation volume in the cavity and the mass and heat exchange between cavity shear layer and the core flow. With $L = 40$ mm in cavity, the shorter length limits the hear release in the cavity and mass and heat exchange which leads to a stable flame anchored at the cavity. As Luminosity and schlieren images show, the longer cavity length promotes the heat release and the mass and heat exchange between cavity shear layer and the core flow. Compared with case 3, the higher injection pressure leads to the higher global equivalence ratio further the more intense combustion. All above factors will generate intense combustion in cavity and downstream of the cavity, where the flame interacts with boundary layer and gradually occupies the flow channel further forming a thermal throat to trigger flame flashback phenomenon. At flame flashback and flame blown off stages, as schlieren images show, pre-combustion shock trains near the fuel injector are gradually pushed forward, disappear and re-appear upstream. The flame front in case 2 can flashback from cavity to injector, however it cannot be sustained surrounding the injector, and is blown off quickly back to the cavity. Hence, the proportion of flame front surrounding the injector in case 2 is lower than other flame flashback cases. The corresponding dominant frequency is higher than other cases.

The effect of cavity aft ramp angle

In this subsection, the effects of aft ramp angle are studied by the comparative experiments listed in group 2. Compared with case 3, the sharper cavity aft ramp angle in case 4 leads to combustion oscillation. We suppose that cavity aft ramp will act on the combustion oscillation mainly through the impinging shock wave in the cavity aft ramp and the mass and heat exchange between cavity shear layer and the core flow. When the angle turns from 45° to 90°, however, it will maybe result in significant strong impinging shock wave. As Zare-Behtash et al. [28] analyze, the impinging shock occurs over the cavity and on the shear layer, however the interaction results in a similar interaction as with the boundary layer [29], namely the lifting of the shear layer. On the other hand, with sharper cavity aft ramp, a region of reversed flow near shear layer is sometimes created. The roll-up and fall-off derived from the velocity deviation between cavity shear layer and the core flow promote the mass and heat exchange between cavity shear layer and the core flow. From Fig. 5.38, the increasing proportion of flame front surrounding the injector demonstrates that the flames can be held around the injector easily. The same conclusion can be found in the propagation speed spectra (Fig. 5.40).

The effect of air throttling downstream of the cavity

Group 3 investigates the effects of the air throttling locations downstream of the cavity on combustion oscillation. As reference analysed [30], air throttling can enlarge wall boundary and decrease the flow velocity. For case 5, owing to air throttling far away from downstream of the cavity, the separated boundary layer doesn't significantly interact with combustion. There is no thermal throat to introduce flame flashback phenomenon. Hence, only stable flames anchor at the cavity. For case 6, the closer air throttling is loaded near downstream of the cavity. The improved fuel/air mixing in the separated boundary layer leads to the combustion intensity in the cavity and separated boundary layer downstream of the cavity gradually enhanced. The flames occupy the flow channel and form a thermal throat thus triggering flame flashback. It is worth stressing that, by means of air throttling, the boundary layer separation in case 6 is easier and can sustain longer. Hence, case 6 has more proportion of flame front surrounding the injector and lower fluctuation frequency.

5.2.2 Numerical Investigations of Flame Flashback

5.2.2.1 Numerical Condition

In the present study, the fifth-order weighted essentially non-oscillatory scheme developed by Jiang and Shu [31] is used for inviscid fluxes and the second-order centre scheme is used for viscous fluxes. To improve the computational efficiency, time integration is performed using a second-order implicit dual-time-step approach [32], the inner iteration of which is achieved through a lower upper symmetric Gauss-Seidel method. The acoustic CFL number is 0.5.

Geometry and boundary conditions

A sketch of the computational domain is shown in Fig. 5.42. The isolator is set to 200 mm long for better turbulent boundary. For purpose of reducing computational burden, the computational domain has a constant width of 10 mm. For illustrative purpose of effect of wall boundary condition on flame flashback mechanism, a no-slip, no-penetration adiabatic condition or symmetry boundary condition which does not impose boundary layer is set at upper wall, a periodic condition is used on both sides and a no-slip, no-penetration adiabatic condition is imposed at all lower walls. The fuel is injected sonically from two 1-mm-diameter injectors located 110

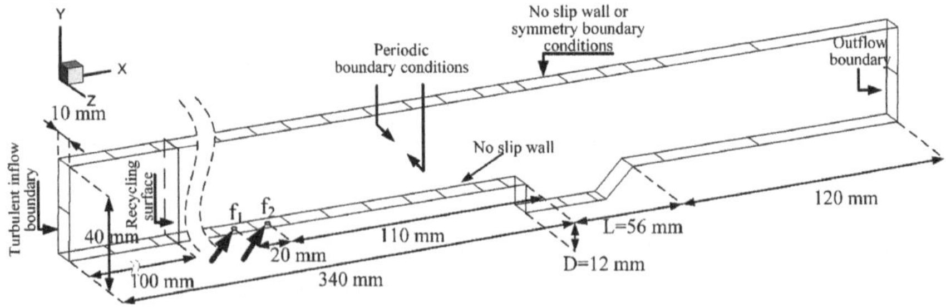

Fig. 5.42 Schematic diagram of the computation domain and boundary conditions [33]

and 130 mm upstream of the cavity leading edge. A turbulent boundary-layer with $\delta_{inf} = 3$ mm is adopted for the present study. This thickness is estimated based on an advanced two-dimensional RANS simulation.

The entire domain is divided into 44 blocks for parallel computing, consisting of 40 blocks in the main flow and 4 blocks in the cavity.

Turbulence models

The governing equations have been well described in the previous literatures [34] and omitted here for brevity. A hybrid LES/RANS method [35] is adopted. This method combines the Spalart-Allmaras (S-A) RANS model [36] which is used for near-wall regions and the Yoshizawa sub-grid scale model [37] which is used for regions away from the wall. While the interaction between the turbulence and chemistry is neglected in the RANS region. The basic assumption in the present hybrid method is that the RANS region is considerably thin (~$0.1\delta_{\text{inf}}$, where δ_{inf} is the inflow thickness). Therefore, only a significantly small portion of the combustion may occur in the RANS region. Neglecting the interaction between the turbulence and chemistry in this region is believed to negligibly influence the results. An assumed sub-grid PDF (Probability Density Function) closure model [38] is used for turbulence-chemistry interaction.

A recycling/rescaling method is used to treat the turbulent inflow condition, which is considered to be a promising way to prescribe time-dependent turbulent inflow conditions for LES or hybrid RANS/LES of spatially developing turbulent flows [39–42]. In the present study, a method similar to that of Xiao et al. [41] is used.

5.2.2.2 Effects of Chemical Kinetic Mechanism

To reduce computation costs and identify the most important chemical paths, the reduced chemical kinetic mechanism of seven species and three steps for ethylene/air combustion [43] (denotes as Original Mechanism) is used and outlined in Table 5.12. Note that accurate prediction of the reaction process and flame temperature is not expected. Nonetheless, at the very least, reasonable macro-effects of heat release on combustion should be obtained.

Figure 5.43 presents the ignition delay time of ethylene/air mixtures obtained using CHEMKIN 4.1 with the reduced chemical kinetic mechanism [44]. This figure

Table 5.12 Reduced chemical kinetic mechanism of seven species and three steps for ethylene/air combustion

Reaction	A (cm^3/mol s)	b	T_a (K)
$C_2H_4 + O_2 \Leftrightarrow 2CO + 2H_2$	2.100E+14	0.0	18015.3
$2CO + O_2 \Leftrightarrow 2CO_2$	3.450E+11	2.0	10134.9
$2H_2 + O_2 \Leftrightarrow 2H_2O$	3.000E+20	−1.0	0.00

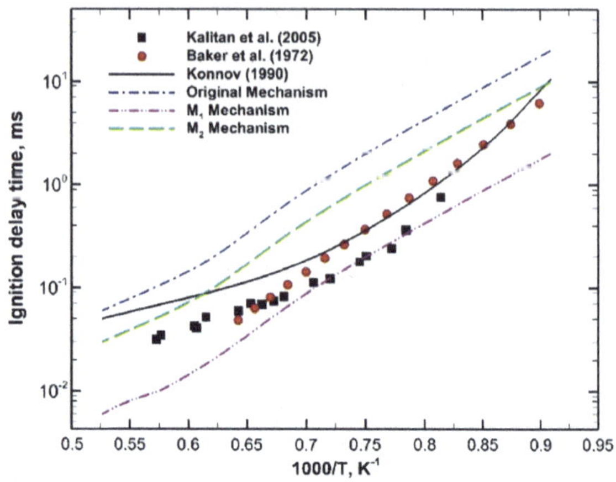

Fig. 5.43 Comparison of experimental ignition delay data with calculated data using different reaction mechanism and calculated ignition delays with the $C_2H_4/O_2/Ar$ mixture at $\varphi = 1$, Ar = 96% , $P \approx 3$ atm

compares the calculated ignition delays of the C2H4/O2/Ar mixture with experimental data reported by Kalitan et al. [45], Baker et al. [46], and Konnov [47], validating the reduced chemical kinetic mechanism. The maximum of pressure gradient is used as the ignition criterion in accordance with the experimental counterpoint. Within a range of 1100–1900 K, the ignition delay time is overestimated or underestimated using the original and the first modified ethylene/air chemical kinetic mechanism (denotes as M1 Mechanism in which the pre-exponential factor is modified as $A_1^* = 10 \times A$, where A stands for pre-exponential factor of Original mechanism), respectively. Hence, a considerable longer or shorter time is required to reach the reaction temperature. In particular, when the temperature decreases, the ignition delay time shows a big difference from the experimental value. The ignition delay time for the second modified ethylene/air chemical kinetic mechanism (denotes as M2 Mechanism in which the pre-exponential factor is modified as $A_2^* = 2 \times A$) remains different from the experimental value; nonetheless, the difference is significantly reduced. Thus, the M2 ethylene/air chemical kinetic mechanism is adopted in the following sections.

5.2.2.3 Effects of Boundary-Layer Conditions

In this section, the boundary-layer effect on flame flashback will be investigated. The boundary, disturbance and flame flashback status are listed in Table 5.13.

Based on the above analysis, it is apparent that the cavity is a key component for promotion of stable combustion. In addition, different boundary conditions affect the flame flashback phenomenon in the combustor with the cavity flameholder. As is apparent from Fig. 5.44, the result of case 1 exhibits a strong flame flashback phenomenon. Owing to the increasing boundary-layer effects, the large separated boundary-layer on the upper wall enhances the main flow compression. The flame mainly develops from the boundary-layer downstream of the cavity. During the flame flashback process, owing to the interaction of the separated boundary-layer and the

Table 5.13 Concise descriptions of two cases

Case	Up wall	Down wall	Disturbance condition	Outcome
No. 1	Turbulent boundary[a]	Turbulent boundary[a]	No disturbance	Flame flashback
No. 2	Symmetric boundary	Turbulent boundary[a]	No disturbance	No flame flashback

[a]Turbulent boundary-layer with 3-mm thickness

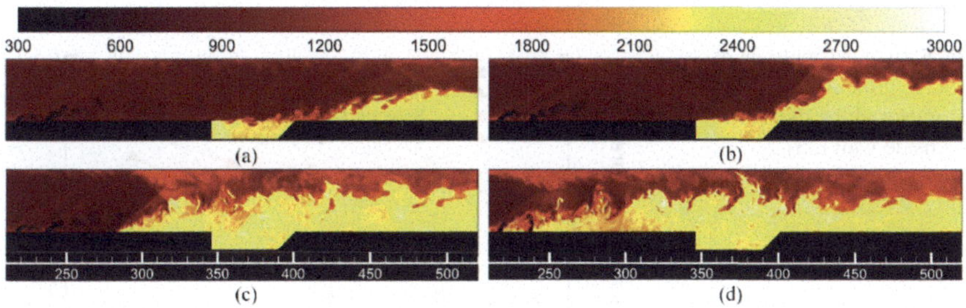

Fig. 5.44 Instantaneous temperature distribution for case 1 (The non-dimensional time interval between two images is $\Delta t = 60 \times D/U_{\infty}$; unit K) [33]

flame, the combustion intensity downstream of the cavity increases. By accumulating the energy from the exothermal reactions, the flame is gradually enhanced, further occupying the flow channel. When the energy exceeds a threshold level, the forming thermal throat triggers flame flashback phenomenon. When the downstream flame spreads to the cavity, the flame in the cavity interacts with the downstream flame and further accelerates itself, until it reaches the injection area. In contrast, there is no boundary-layer in upper wall because of the symmetric boundary condition as presented in Fig. 5.45. Hence, the compression of main flow is too weak to form a thermal choking. A stable combustion, therefore, is maintained downstream of the cavity.

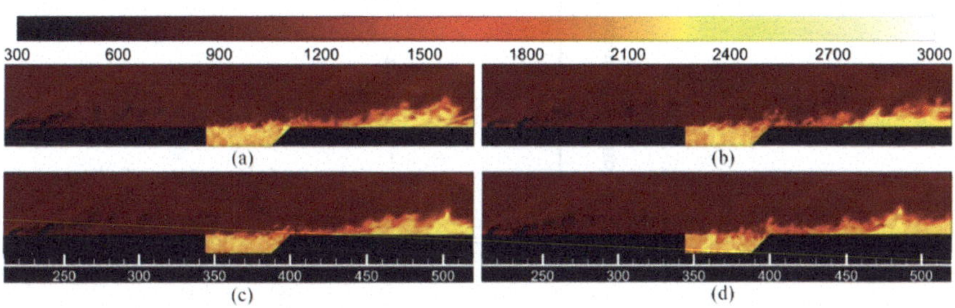

Fig. 5.45 Instantaneous temperature distribution for case 2 ($\Delta t = 60 \times D/U_{\infty}$; unit K) [33]

Although it is known that the boundary condition of case 1 causes the flame flashback phenomenon, the detailed mechanism underlying the flame flashback phenomenon remains unclear, since both temperature and pressure in the flow field differ from those in case 2. In this section, we will clarify what the sensitive parameter is. Where the sensitive region is. Figure 5.46 shows the distributions of average pressure and temperature at different y-z slices along X-axis, it is apparent that the difference of temperature is more obvious than pressure. Figure 5.47 shows the dimensionless temperature distribution along the model combustor lower wall, which also confirms the notable temperature difference downstream of the cavity. Therefore, it can be

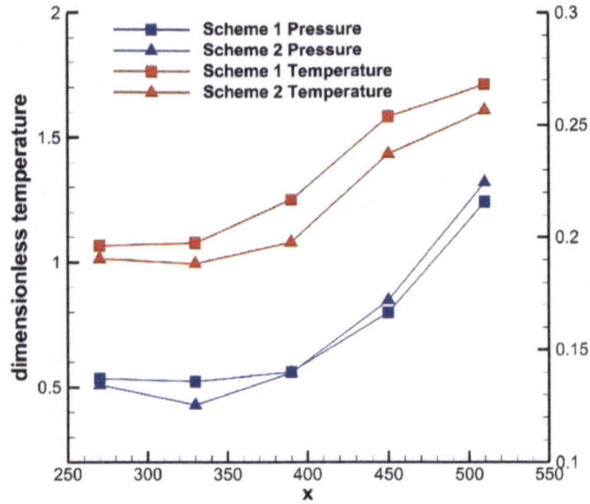

Fig. 5.46 Distributions of average dimensionless parameter along the downside wall [33]

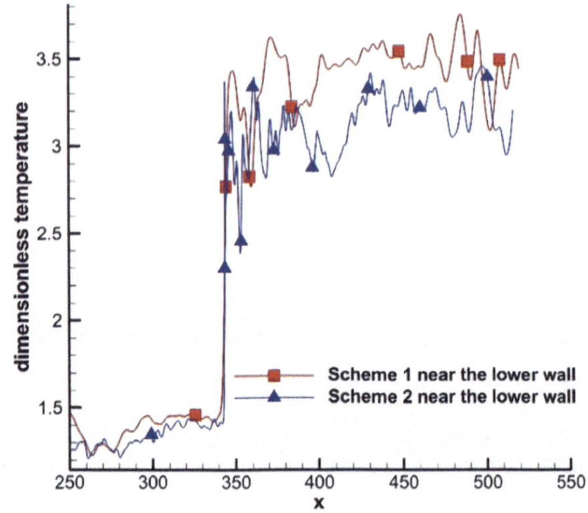

Fig. 5.47 Distributions of dimensionless temperature along downside wall [33]

concluded that the flame flashback phenomenon is sensitive to temperature fluctuation rather than pressure fluctuation, and the boundary-layer downstream of the cavity is the area sensitive to the flame flashback phenomenon.

5.2.2.4 Effects of Thermal Disturbances

Based on the above analysis, it is apparent that temperature is the sensitive parameter and the downstream region of the cavity is the sensitive area for flame flashback phenomenon. In this section, an additional thermal disturbance is loaded at downstream of the cavity to simulate the fluctuation of the local parameters. In addition, the fluctuation locations in the sensitive area should also be investigated. The boundary, disturbance and diffusion conditions, and flame flashback status are listed in Table 5.14.

Figures 5.48 and 5.49 show the instantaneous temperature distribution results for different cases. For case 3, the 30-mm-long thermal disturbance ($T = 2500$ K) is located at 72 mm downstream of the cavity, which is 12 mm upstream of the outflow boundary. For case 4, the 15-mm-long thermal disturbance ($T = 2500$ K) is located at 36 mm downstream of the cavity, which is 72 mm upstream of the outflow boundary. Although the thermal disturbance energy in case 3 is larger than case 4, the residence time of the disturbance is too short to induce the flame flashback phenomenon. This is because the thermal disturbance is close to the outflow boundary and easily blown out of the flow channel. In contrast, once the accumulated energy exceeds a certain threshold as the residence time of the disturbance increases, the small thermal

Table 5.14 Concise descriptions of three cases

Case	Up wall	Down wall	Disturbance and diffusion condition	Outcome
No. 3	Symmetric boundary	Turbulent boundary[a]	Disturbance (30 mm long) located at 72 mm downstream of cavity	No flame flashback
No. 4			Disturbance (15 mm long) located at 36 mm downstream of cavity	Flame flashback
No. 5			Increasing diffusion coefficient downstream of cavity	Flame flashback

[a]Turbulent boundary-layer with 3-mm thickness

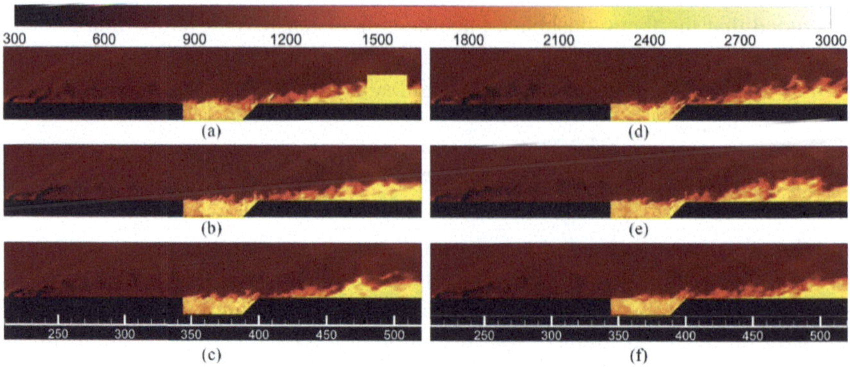

Fig. 5.48 Instantaneous temperature distribution for case 3; Disturbance (30 mm long) located at 72 mm downstream of cavity ($\Delta t = 50 \times D/U_\infty$; unit K) [33]

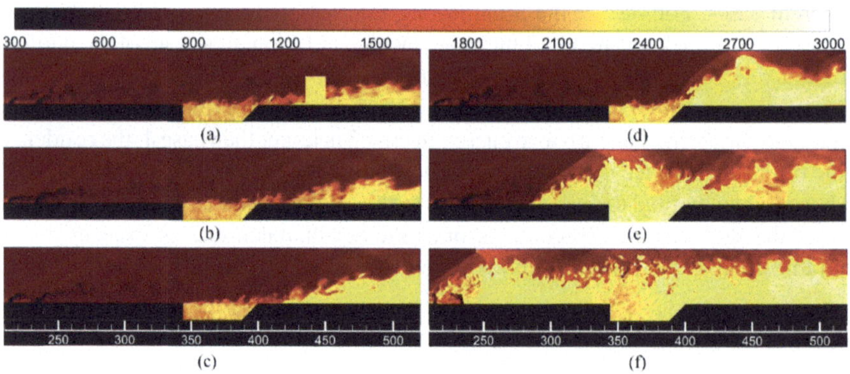

Fig. 5.49 Instantaneous temperature distribution for case 4; Disturbance (15 mm long) located at 36 mm downstream of cavity ($\Delta t = 50 \times D/U_\infty$; unit: K) [33]

disturbance in case 4 is amplified and spreads throughout the entire fluid flow, inducing the flame flashback phenomenon under the same incoming flow and boundary conditions. It can be demonstrated that the thermal disturbance in the sensitive area induces fluctuation of the local parameters, further enhancing the combustion. This also indicates that, with increasing residence time of the thermal disturbance, the combustion fluctuation downstream of the cavity is enhanced. Compared with the temperature distribution in case 1, although a thermal disturbance is loaded, the speed of flame flashback in case 4 is slower than case 1. As a result, when the flame front reaches the injectors location, the flame front in case 4 remains a certain distance from the injectors location.

Adjusting the diffusion coefficient of the species in the sensitive area to improve the local mixing degree is another approach to induce flame flashback. Figure 5.50 shows the temperature distribution results for case 5. The diffusion coefficient is

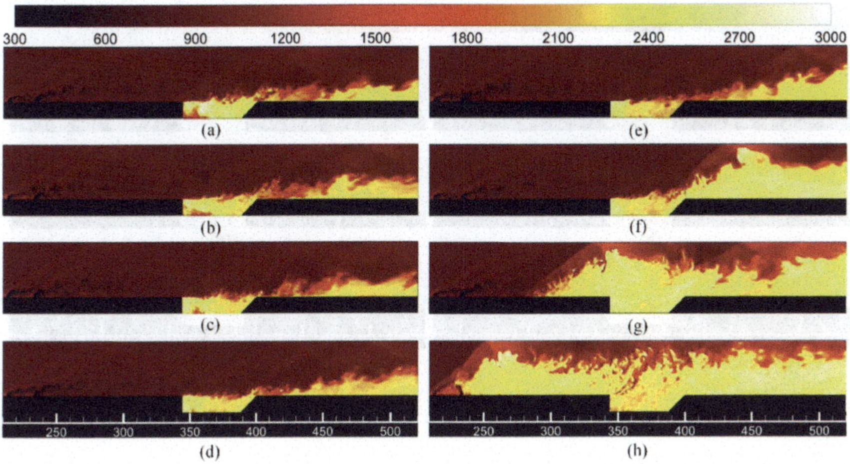

Fig. 5.50 Instantaneous temperature distribution for case 5; Increasing diffusion coefficient downstream of cavity ($\Delta t = 50 \times D/U_\infty$; unit: K) [33]

magnified to seven times bigger than original at downstream of the cavity. At the fuel and air interface, the amplified diffusion coefficient improves the mixing speed. This enhanced mixing efficiency promotes the combustion intensity. In contrast to the intense flame flashback, twice as much time is required for energy accumulation as in case 4, owing to the simply amplifying diffusion coefficient downstream of the cavity.

5.2.2.5 The Detailed Investigation of Representative Case

In the following section, detailed analyses are conducted with different methods to investigate the combustion characteristics of case 4. Figure 5.51 shows the instantaneous heat release distribution with a sonic line. Initially, the heat releases are located in the jet-with-cavity shear layer and boundary-layer downstream of the cavity, concentrating on the sonic line. During the flame flashback process, as the thickness of separated boundary-layer increasing, the heat release rate gradually increases. The heat release mainly distributes on the flame front and propagates forward from downstream of the cavity to injectors. As we can see, part of strong heat release distributes outside of the sonic line, owing to the strong interaction of turbulence and vortex. This strong heat release forms a thermal throat when the flame spreads to the cavity position and further accelerates itself, inducing flame flashback by means of interaction with a large separated boundary-layer. Under this process, the interaction between flame and separated boundary-layer gradually intensifies the combustion intensity.

Firstly, Fig. 5.52 shows the speed distribution and the coloured streamline with pressures. Figure 5.53 gives the instantaneous pressure distribution. Compared with

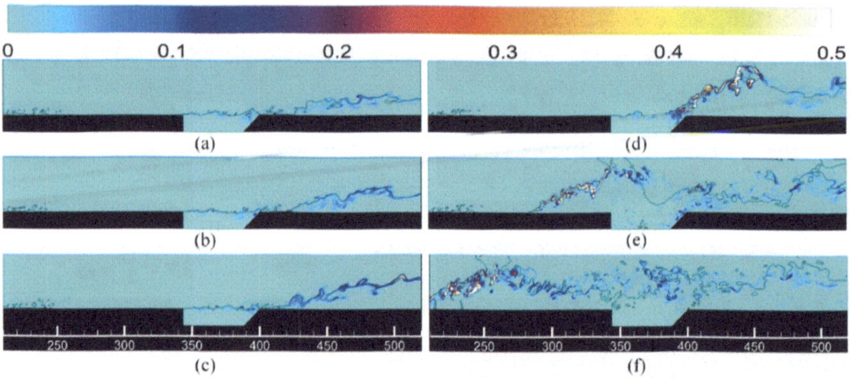

Fig. 5.51 Instantaneous heat release distribution with sonic line for case 4 ($\Delta t = 50 \times D/U_{\infty}$) [33]

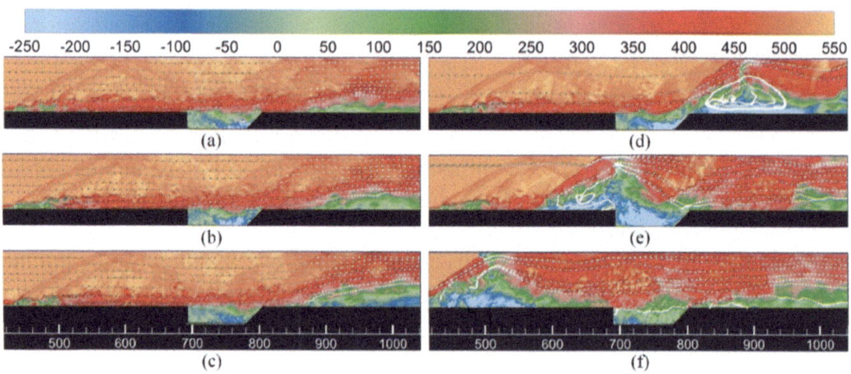

Fig. 5.52 Instantaneous speed distribution and streamline coloured with pressure for case 4 ($\Delta t = 50 \times D/U_{\infty}$; unit: m/s) [33]

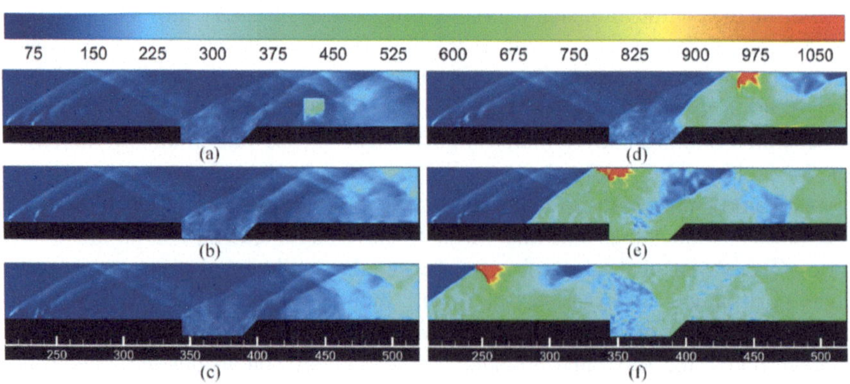

Fig. 5.53 Instantaneous pressure distribution for case 4 ($\Delta t = 50 \times D/U_{\infty}$; unit: KPa) [33]

Fig. 5.49, it can be found that the most of flame is located in the low speed region. During the flame flashback process, the shock wave confines a high-pressure, high-temperature, low-speed region. All these effects lead to higher combustion intensity downstream of the cavity. The back pressure generated from the heat release promotes the boundary-layer to separate. Conversely, the back pressure accompanying the separated boundary-layer gradually occupies the main flow and the forms thermal throat to trigger flame flashback.

As shown in Fig. 5.54, the high pressure appears at downstream of the cavity. Owing to the enhancement of combustion intensity, the pressure peaks which are located at the flame front at the upper and lower wall increase. During the flame flashback process, the magnitude of pressure peaks in upper and lower walls may have the same order of the fuel injection. It is indicated that, since the thermal disturbance enhances the heat release downstream of the cavity, the higher pressure causes the bulk flow to be compressed. It is apparent in Fig. 5.55 that the main distribution of the heat release concentrates on the flame front. The interaction between flame

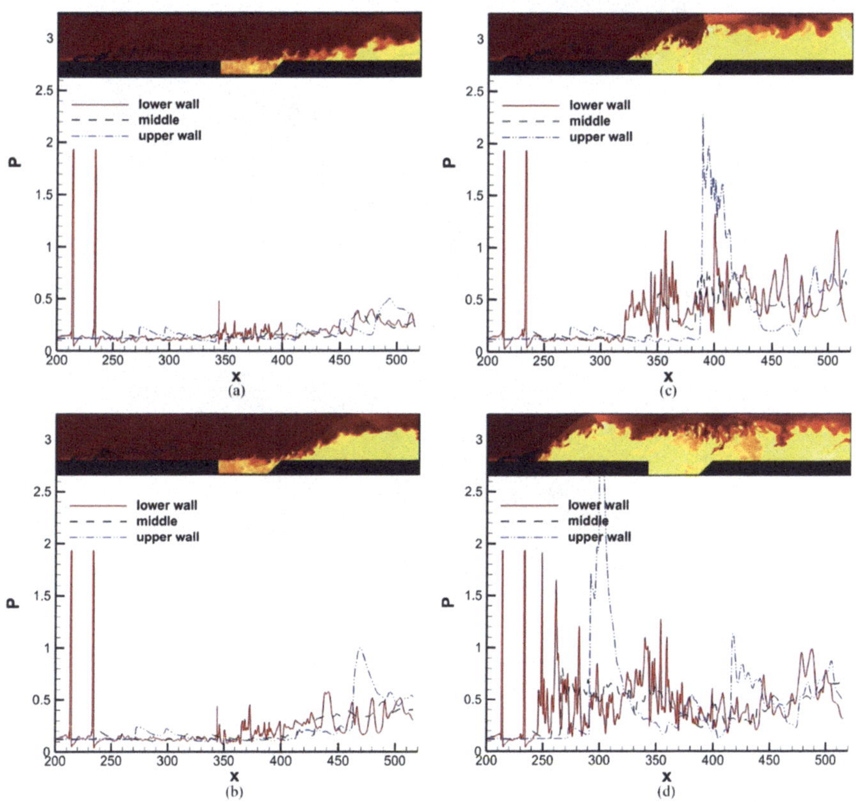

Fig. 5.54 Non-dimensional pressure distributions in three vertical positions for case 4 ($\Delta t = 50 \times D/U_{\infty}$) [33]

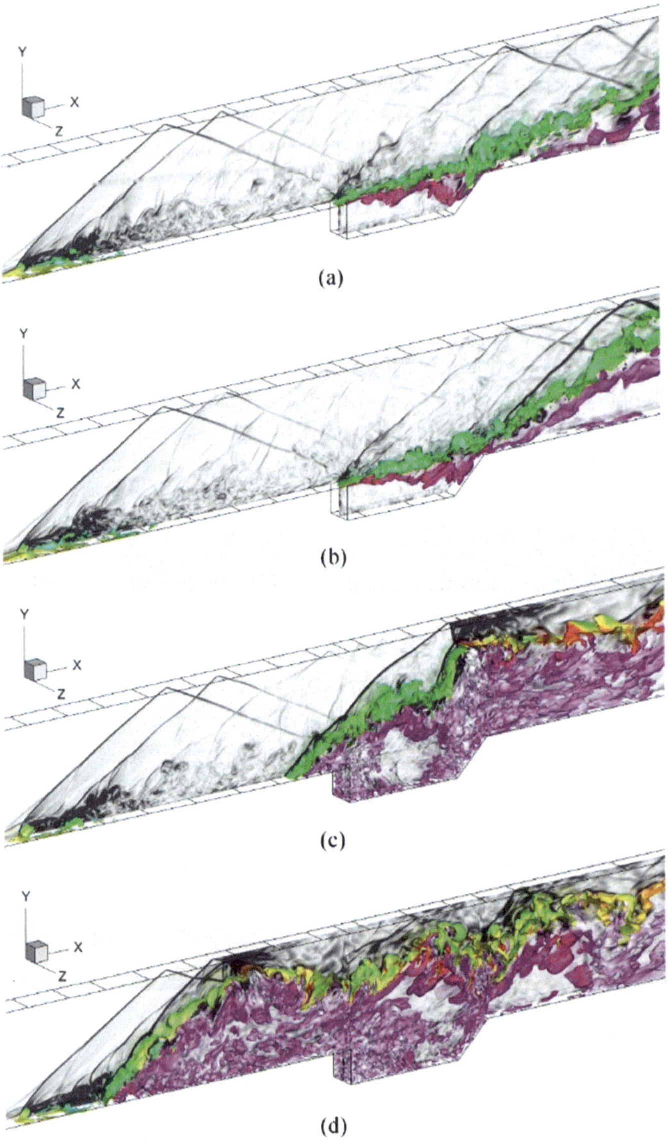

Fig. 5.55 Isolation surfaces of heat release rate, coloured with temperature and density gradients in the centre section for case 4 ($\Delta t = 50 \times D/U_\infty$) [33]

and separated boundary-layer forms a thermal throat when the flame spreads to the cavity position and further to accelerate the flame front, inducing flame flashback phenomenon.

Figures 5.56 and 5.57 illustrate the X-axial slices for the instantaneous temperature contours and the isolation surface of the $\lambda_2 = -0.1$ vortex structure. Initially, the flames are distributed in the jet-with-cavity shear layer and downstream of the cavity, where the subsonic region is located. During the flame flashback process, the flames gradually close even over to the sonic line. It is obvious that a thermal throat accompanied with intense combustion accelerates the flame front to propagate forward. It is apparent from Fig. 5.57 that the hairpin-like vortices near the injectors tense and fracture when they spread to downstream of the cavity, where the subsonic

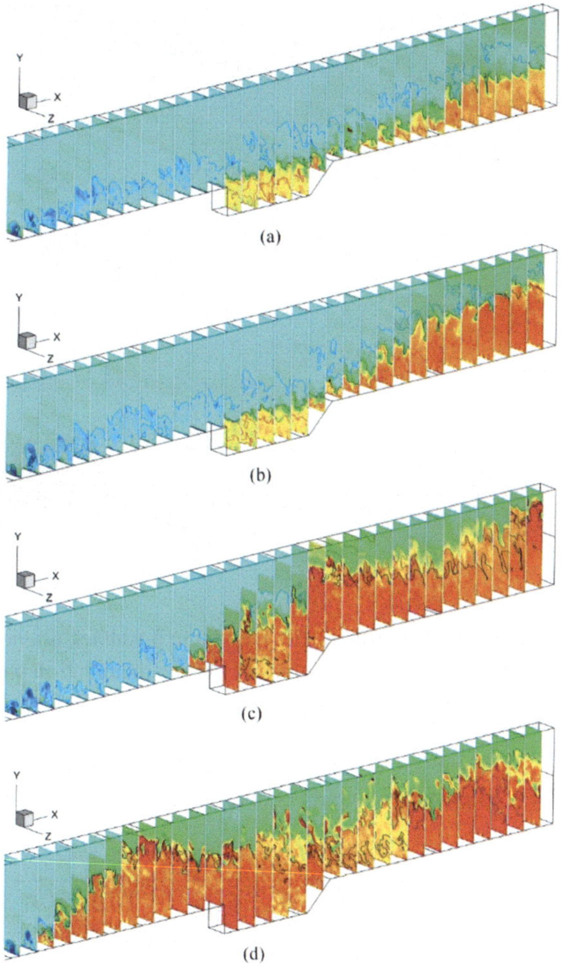

Fig. 5.56 Oblique views of axial slices for instantaneous temperature contours for case 4 (blue line: stochiometric equivalent ratio; black line: sonic line; pink line: stagnation line for flow speed $\Delta t = 50 \times D/U_\infty$) [33]

Fig. 5.57 Isolation surface of $\lambda_2 = -0.1$ vortex structure coloured with temperature for case 4 ($\Delta t = 50 \times D/U_\infty$) [33]

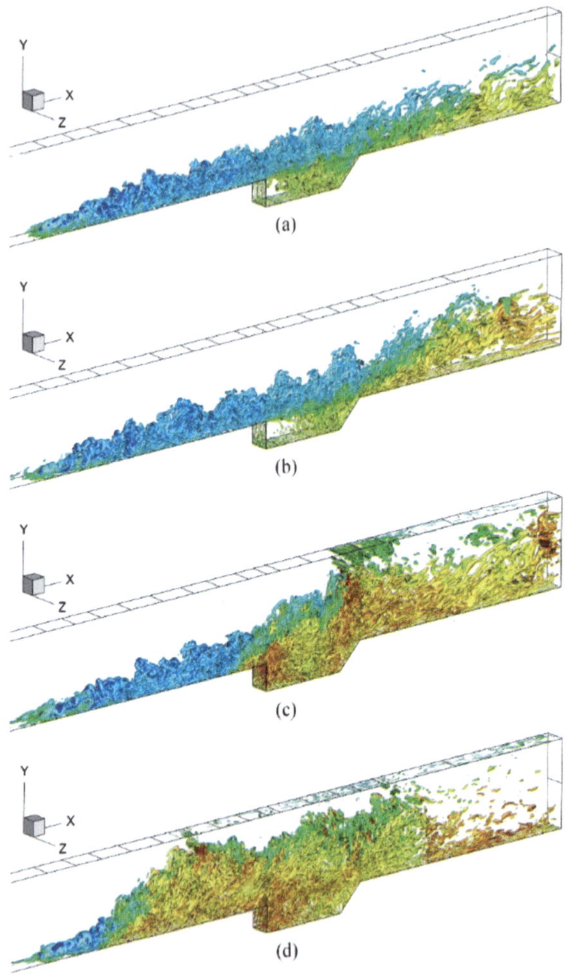

combustion zone is located. The flame flashback phenomenon is a development process of the energy accumulation. During the reaction procedure, the vortices coloured with high temperatures move forward against the flow. The separation zone gradually increases.

As suggested by Bilger [48], we define a mixture fraction to further evaluate the flame index factor, which is derived from C, H, and O elements are as follows

$$Z = \frac{2\left(Y_C - Y_{C,2}\right)/M_C + \left(Y_H - Y_{H,2}\right)/M_H - \left(Y_O - Y_{O,2}\right)/M_O}{2\left(Y_{C,1} - Y_{C,2}\right)/M_C + \left(Y_{H,1} - Y_{H,2}\right)/M_H - \left(Y_{O,1} - Y_{O,2}\right)/M_O}. \quad (5.3)$$

where Y_j and M_j are the elemental mass fractions and atomic masses for the elements carbon, hydrogen and oxygen and the subscripts 1 and 2 refer to values in the fuel and air streams, respectively.

The flame index factor can be defined by combining the mixture fraction and component gradient:

$$I = \frac{Z - Z_{st}}{|Z - Z_{st}|} \frac{1}{2} \left(1 + \frac{\nabla Y_{C_2H_4} \cdot \nabla Y_{O_2}}{\left| \nabla Y_{C_2H_4} \cdot \nabla Y_{O_2} \right|} \right), \tag{5.4}$$

where Z_{st} represents the mixture fraction at the stoichiometric equivalent ratio. From the flame index factor defined by Eq. (5.3), it is apparent that the combustion pattern can be divided into three categories in the combustion process. For fuel-rich premixed combustion, $Z - Z_{st} > 0$, $I_{rich} = 1$. For fuel-lean premixed combustion, $Z - Z_{st} < 0$, $I_{lean} = -1$. For diffusion combustion, $I_{diffusion} = 0$.

The change in the combustion mode in case 4 can be studied using the flame index factor. Figure 5.58 shows the development of the initial flame kernel inside the cavity. The red, yellow and brown areas denote the fuel-rich premixed, the diffusion

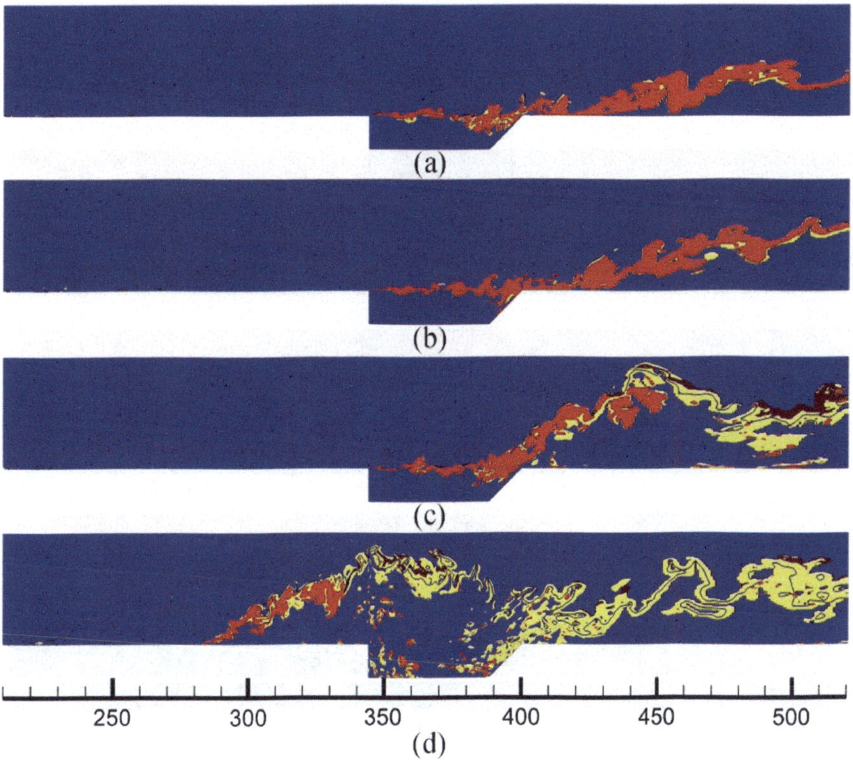

Fig. 5.58 Development of combustion modes for case 4 ($\Delta t = 50 \times D/U_\infty$) [33]

the fuel-lean premixed combustion zones, respectively. The black line denotes the stoichiometric line. Initially, most areas in the cavity and the jet-with-cavity shear layer are located in the fuel-rich premixed combustion zone. The flame downstream of the cavity mainly exhibits the forms of a fuel-rich premixed environment. The flame front spreads near the wall and the flame exhibits a fuel-rich premixed combustion mode. However, the flame located in the shear layer close to the main flow, rapidly changes into the diffusion combustion mode. The flame located in the fuel-rich premixed environment continues to react with the inflowing air, thus the diffusion combustion zone is formed. Additionally, the diffusion combustion zone unceasingly expands, where the vortex structure changes drastically. The entrainment effect can change the fuel or air gradient, which affects the distribution of the lean premixed combustion near the diffusion combustion area.

To quantitatively investigate the experimental and numerical calculation of the flame flashback speeds, a combustion iso-luminosity contour is used to characterise the combustion zone outlines for each flame image. The instantaneous speed of the flame front relative to the wall of the combustion chamber can be calculated using the change of the flame front position and the corresponding time interval. Figure 5.59 shows the combustion distribution at three different times.

Pressure distribution in the X-axial direction is obtained using pressure transducers installed along the centreline of the combustor lower walls. The measured averaged pressure distribution when flame front propagates to the injectors is presented in Fig. 5.60. Compared with the experimental results, the numerical curve of instan-

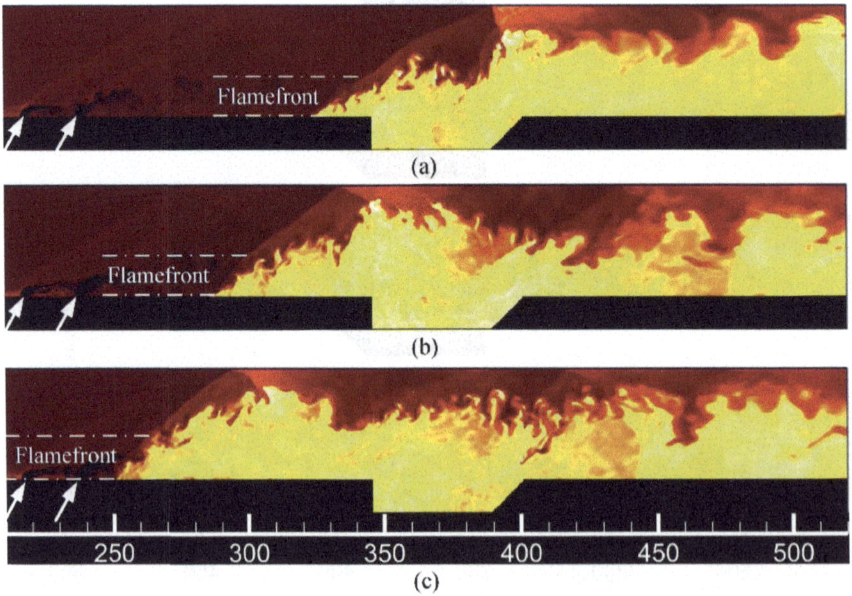

Fig. 5.59 Flame images at three different times with a consistent time step ($\Delta t = 15 \times D/U_{\infty}$) [33]

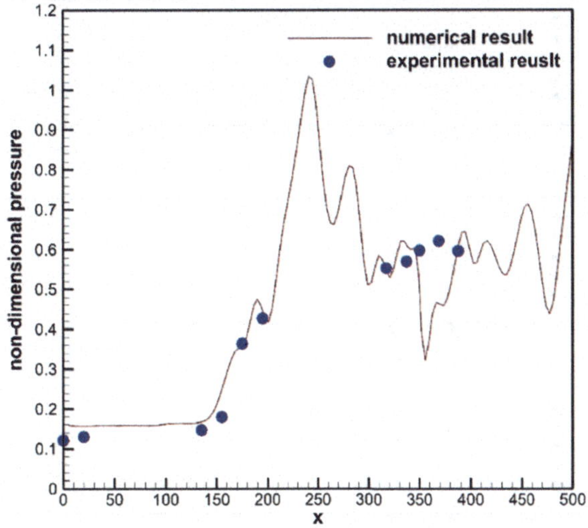

Fig. 5.60 Experimental and numerical dimensionless wall pressure on lower wall [33]

taneous pressure which is smoothed by Tecplot shows relatively good agreements, indicating the reliability of numerical method. The curves succeed to accurately predict the wall pressure upstream of the injectors. The heat-release causes a sharp rise of pressure upstream of the primary injectors. The pressure reaches a plateau in the cavity and rises again due to compression on the ramp wall of the cavity. Because of accumulated energy from the intense heat release, the pressure peak of flame front is slightly higher. These differences further demonstrate that the pressure fluctuations are strongly correlated with the flame flashback.

A series of schematic images are provided in Fig. 5.61 that show the flame flashback process based on the previous analysis in this paper. In typical cavity shear-layer stabilized combustion mode, the flame is anchored in the cavity shear layer, and the main combustion zone is also confined within the shear layer. The strong interactions between flame and boundary-layer greatly enhance the reaction. Then, the combustion fiercely heats the boundary layer downstream of the cavity, which leads to the present of the separated recirculation zone and compresses the bulk flow. In turn, the formed thermal throat induces the flame flashback phenomenon.

5.2.3 Theoretical Analyses

5.2.3.1 Mechanism of Flame Flashback

A simplified combustion opening system model is constructed (Fig. 5.62) based on the following assumptions:

1. The control volume is a tube, open at both ends, and the outer wall of the fluid container is adiabatic. V is the volume and L is the length.

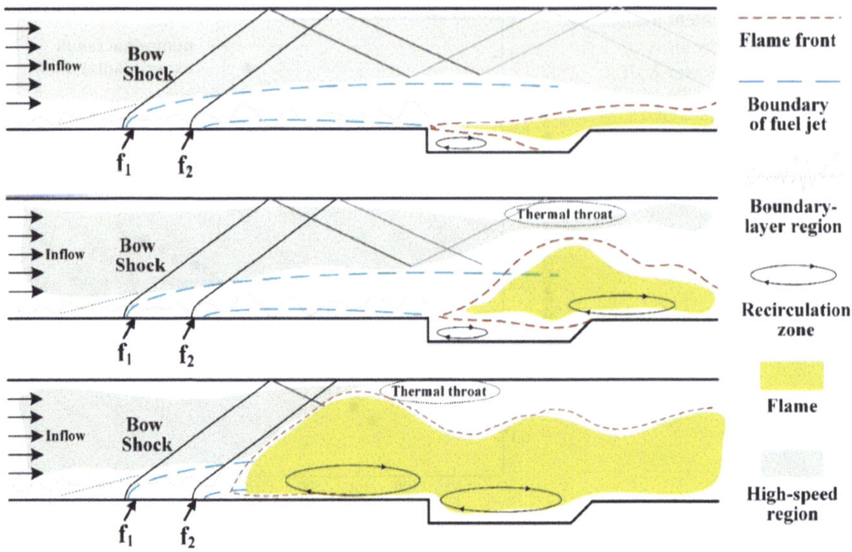

Fig. 5.61 Schematics of flame flashback from the cavity to the injectors for case 4 [33]

Fig. 5.62 Control volume for theoretical analysis

2. The initial temperature is T_0, the initial molarity of the premixed gases is C_0, and the mass flow rate is q (kg/s).
3. The temperature of the control volume rises to T, and the molarity of the premixed gases drops to C after the time period of δ_t. The temperature of the control volumes T and the molarity of the premixed gases C are equal throughout the control volume.
4. There is an n-order reaction model in the control volume.

The energy balance equation is expressed as:

$$\rho V c_p \frac{dT}{dt} = (\Delta H) V C^n k_0 \exp\left(-\frac{E}{RT}\right) - q c_p (T - T_0) \tag{5.5}$$

where ΔH denotes the reaction heat. k_0 and E denote the rate constant and activation energy in Arrhenius equation. The mass balance equation is given as:

$$\frac{dC}{dt} = C^n k_0 \exp\left(-\frac{E}{RT}\right) - \frac{q}{V\rho}(C_0 - C) \tag{5.6}$$

The heat released by the premixed gases in the control volume are used to heat the gases and are released from the system. The product generation is equal to the consumption of the premixed gases. So, the Eqs. (5.4) and (5.5) can be written by:

$$(\Delta H)VC^n k_0 \exp\left(-\frac{E}{RT}\right) = qc_p(T - T_0) \tag{5.7}$$

$$C^n k_0 \exp\left(-\frac{E}{RT}\right) = \frac{q}{V\rho}(C_0 - C) \tag{5.8}$$

Define the dimensionless number: dimensionless heat release rate $\varepsilon_1 = (C_0 - C)/C_0$ refers to the portion of the chemical energy in the premixed combustible gas that has been converted to heat divide the chemical energy that the entire premixed gas has in a unit volume; dimensionless heat dissipation rate $\varepsilon_2 = \left(\rho c_p(T - T_0)\right)/qC_0$ refers to the portion of heat carried out of the product by the product (the portion of the reaction heat used to heat up the product) divide the chemical energy of the entire premixed gas in a unit volume; The dimensionless action time $(\tau_d = \tau_1/\tau_2 = \left(\rho V k_0 C^{n-1}\right)/q)$. Where $\tau_1 = L/v = (\rho V)/q$ is premixed gas residence time in the control volume, and $\tau_2 = 1/\left(k_0 C^{n-1}\right)$ is the time which is required for completely chemical reaction of premixed gas; The dimensionless temperature $\theta = (RT)/E$; and dimensionless total heat release value $\psi = (\Delta HRC_0)/\left(\rho c_p E\right)$. The five dimensionless parameters are inserted into Eqs. (5.6) and (5.7), and the equations are then rearranged to yield

$$\varepsilon_1 = \frac{1}{1 + \frac{\exp\left(\frac{1}{\theta}\right)}{\tau_d}} \tag{5.9}$$

$$\varepsilon_2 = \frac{1}{\psi}(\theta - \theta_0) \tag{5.10}$$

For the purposes of discussion, typical values are obtained using the weighted average method: $C_0 = 0.005$, $\Delta H = 10000$, $E = 80000$, $\rho = 0.7$, $c_p = 2.0$, $T_0 = 1150$, and $\tau_d = 1.2$. τ_2 is calculated using a software package (CHEMKIN [44]).

Figure 5.63 shows that the straight line representing ε_2 cuts the curve representing ε_1 at two points. The lower point is the stable point of combustion, while the upper point is the threshold value for flame flashback. The combined effects of dimensionless heat release and heat dissipation affect the system stability and forms the thermal fluctuation. When the temperature fluctuates over the upper point, the heat release continuously exceeds the heat dissipation and the system temperature cannot self-stabilise. After this stage, the combustion increases widely and the flame in the

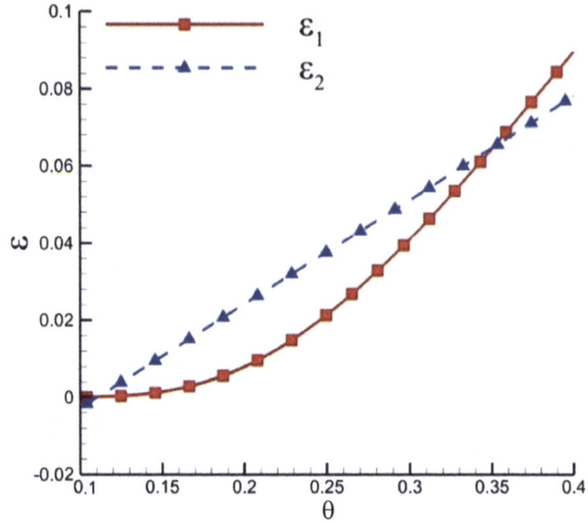

Fig. 5.63 Dimensionless heat release rate ε_1 and heat dissipation rate ε_2 as functions of dimensionless temperature θ [33]

separated boundary-layer spreads into the mainstream, forming a thermal throat and further trigger flame flashback.

Figures 5.64, 5.65, and 5.66 show ε as a function of θ under different conditions. For brief discussion, the x-axis is converted to temperature in following figures. Figure 5.64 shows the effect of the dimensionless initial temperature on dimensionless heat dissipation rate. When the initial dimensionless temperature increases from θ_{01} to θ_{02}, the dimensionless heat dissipation rate ε_2 decreases, leading to the temperature fluctuation threshold decrease. For case 1, due to additional turbulent boundary-layer, the increased compression effect enhances the initial temperature. Hence, the flame flashback phenomenon can be easily triggered under the reduced

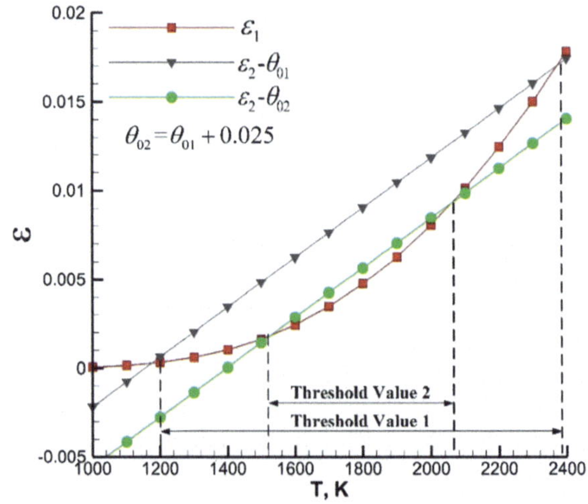

Fig. 5.64 The effect of dimensionless initial temperature on dimensionless heat dissipation rate [33]

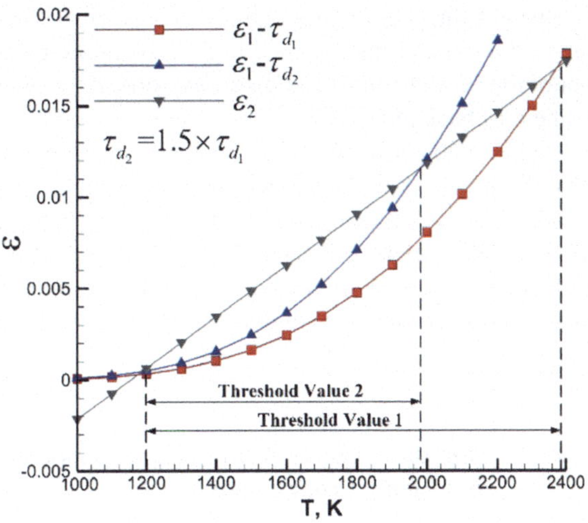

Fig. 5.65 The effect of dimensionless action time on dimensionless heat release rate [33]

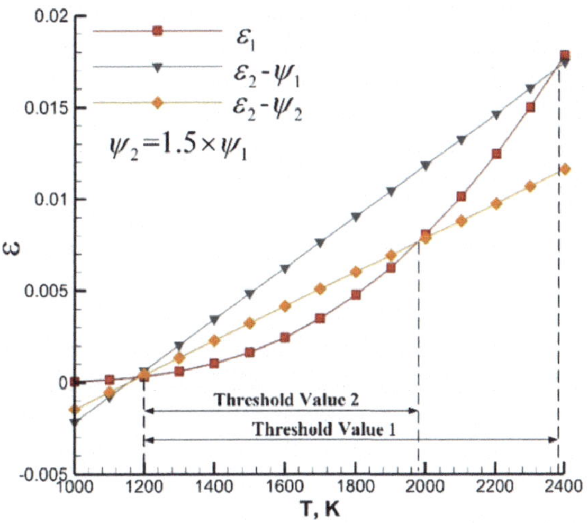

Fig. 5.66 The effect of dimensionless total heat release value on dimensionless heat dissipation rate [33]

temperature fluctuation threshold condition. For case 4, additional thermal disturbance downstream of the cavity increases the temperature fluctuation range. Hence, the fluctuating temperature can easily exceed the upper unstable point, inducing the flame flashback phenomenon.

Figure 5.65 shows the effect of the dimensionless action time on dimensionless heat release rate. The added thermal disturbance in cases 3 and 4 interacting with flame downstream of cavity can prompt boundary layer separation and compressing the bulk flow. This compression effect increases the dimensionless action time of flame which acts on the dimensionless residence time of fuel τ_d. Increasing the dimensionless action time enhances the heat release rate and reduces the temperature fluctuation threshold. Although the energy of the thermal disturbance in case 3 is larger than case 4, the patched high temperature close to outflow is rapidly blown out of the flow channel. The residence time of the disturbance is too short to trigger flame flashback phenomenon.

Figure 5.66 shows the effect of the dimensionless heat value on dimensionless heat dissipation rate. Generally, the improvement of local fuel mixing can change dimensionless total heat release value. For case 5, the amplified diffusion coefficient improves the mixing efficiency at the boundary-layer surface between fuel and air, further decrease temperature threshold value. In addition, the appropriate local equivalence ratio can enhance combustion greatly, increasing the temperature fluctuation range.

5.2.3.2 One Dimensional Analysis in the Competition of Auto-ignition and Flame Propagation

To identify the auto-ignition issue under such high total temperature, the one-dimensional analysis method is introduced in this section. Figure 5.67 shows the streamwise distributions of pressure, temperature and velocity from different longitudinal positions at a certain time instant. The heat release within the flame region can substantially increase the temperature and pressure, compressing the incoming flow and generating shock waves. As the compressing effects spread upstream, temperature and pressure are raised, and velocity is decreased. Near the lower wall, the pressure and temperature rise after the injection (x = 215 mm, x = 235 mm) due to bow shock wave. The abrupt changes are caused by the oblique shock wave. For velocity, the low speed separation and back flow regions appear near lower wall. The velocity in main flow is closer to inflow speed.

As shown in Fig. 5.68, auto-ignition delay times under different temperature and pressure is estimated by CHEMKIN [44] using the seven species and three steps M_2 ethylene/air chemical kinetic mechanism. The auto-ignition delay time rapidly decreases from ~10^{-2} s at 1000 K to ~10^{-5} s at 1800 K, indicating that the auto-ignition behaviour is sensitive to the change of temperature in the range of 1000–1800 K.

To estimate the ignition (flame front) position, as Knop et al. [49] and Colin et al. [50] did, tabulation of auto-ignition delay times based on local pressure and temperature is presented. By defining the rate of auto-ignition progress φ, the auto-ignition delay distance is provided with the local pressure, temperature and velocity data. The auto-ignition delay position x_a is

$$x_a = \int_0^1 \tau_a(x) \cdot \upsilon(x) d\varphi \tag{5.11}$$

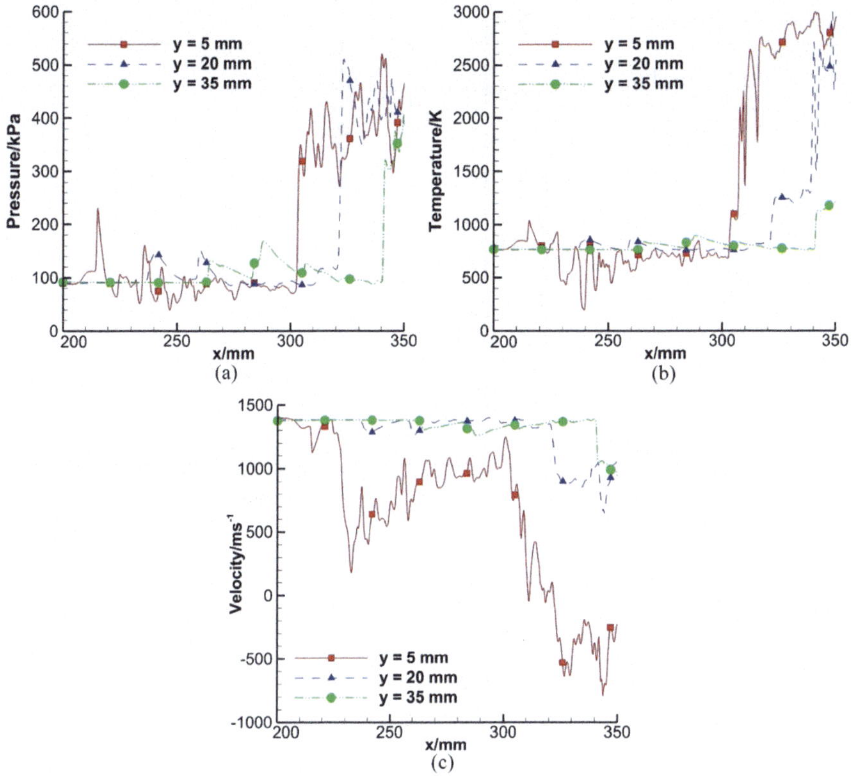

Fig. 5.67 Streamwise distribution of flow parameters at a certain time instant [33]

Fig. 5.68 Auto-ignition delay time [33]

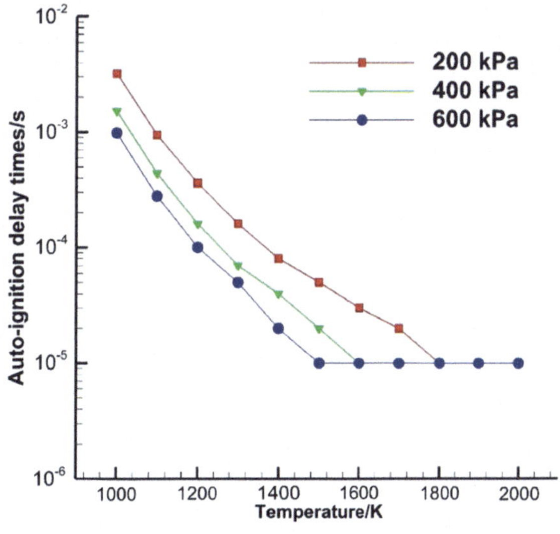

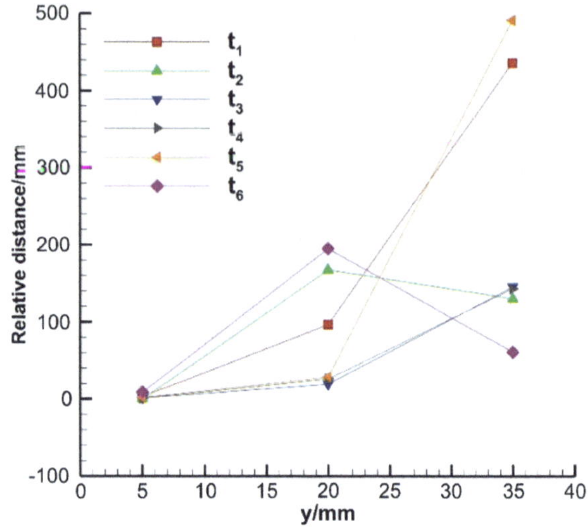

Fig. 5.69 relative distance between auto-ignition position and flame front position at different instantaneous [33]

$\tau_a(x)$ is the local ignition delay time calculated by tabulation of pressure and temperature at point x. $\upsilon(x)$ is the local velocity. The auto-ignition is accomplished and a flame appears when $\varphi = 1$. The integral started at x = 200 mm position, as auto-ignition occurs when the fuel and hot air are mixed. The relative distance (x_r) between auto-ignition position and flame front position x_f is defined

$$x_r = x_a - x_f \tag{5.12}$$

As shown in Fig. 5.69, the estimated relative distances are positive, indicating that the auto-ignition position is located in downstream region of flame front. It can be demonstrated that there are no auto-ignition issues for ethylene under this condition.

5.3 Summary

Flame flashback phenomenon inside an ethylene-fueled scramjet combustor is experimentally investigated in Mach 2.1 facility which simulates Mach 4 flight condition. Experimental and numerical investigations of flame flashback phenomenon have been carried out inside an ethylene-fueled scramjet combustor with a cavity flameholder under the condition of flight Mach 5.5. Some results exhibit quasi-periodic combustion oscillation in the combustion chamber.

1. The experimental results exhibit some factors can separate the boundary layer. The interaction between combustion and separated boundary layer forms a thermal throat which induces thermal chocking, thus induce the flame flashback phenomenon.

2. The flashback develops explosively from the cavity pilot flame at regular intervals. Analysis of the experimental data suggests that the flame flashback is related to flame acceleration similar to deflagration-to-detonation transition.
3. Factors like higher fuel equivalence ratio, sharper injection angle, longer premixing distance, multi-jets, the higher cavity length-to-depth ratio, sharper cavity aft ramp angle, the closer air throttling can induce flame flashback phenomenon. The flame front distributions, flame propagation speeds and quasi-periodic oscillation frequencies obtained by the iso-luminosity contour method shows significant differences in different cases.
4. The centralized injection scheme will result in combustion oscillation in scramjet combustor whose period and intensity will increase as the premixing distance becomes longer. It has been confirmed that distributed injection scheme is an effective method to avoid the combustion oscillation in scramjet combustor, which can induce two parts interacting stable flame.
5. In addition, a simplified combustion opening sys-tem model has been established to analyse combustion oscillation mechanisms, which theoretically demonstrates that above factors can destroy the balance of heat release and dissipation, causing the system cannot self-stabilise once certain temperature fluctuation thresholds in sensitive areas are exceeded. At the same time, the auto-ignition model excludes the possibility of flame flashback generated by auto-ignition effect.

References

1. Wang, H. B., et al. (2015). Mixing-related low frequency oscillation of combustion in an ethylene-fueled supersonic combustor. *Proceedings of the Combustion Institute, 35*(2), 2137–2144.
2. Sun, M. B., et al. (2015). Flame flashback in a supersonic combustor fueled by ethylene with cavity flameholder. *Journal of Propulsion and Power, 31*(3), 976–981.
3. Bychkov, V., et al. (2012). Gas compression moderates flame acceleration in deflagration-to-detonation transition. *Combustion Science and Technology, 184*(7–8), 1066–1079.
4. Valiev, D. M., et al. (2009). Different stages of flame acceleration from slow burning to Chapman-Jouguet deflagration. *Physical Review E, 80*(3), 036317.
5. Dorofeev, S. B. (2011). Flame acceleration and explosion safety applications. *Proceedings of the Combustion Institute, 33*(2), 2161–2175.
6. Ouyang, H., Liu, W., & Sun, M. (2016). Parametric study of combustion oscillation in a single-side expansion scramjet combustor. *Acta Astronautica, 127,* 603–613.
7. Sun, M. B., et al. (2014). Experimental investigation of supersonic model combustor with distributed injection of supercritical kerosene. *Journal of Propulsion and Power, 30*(6), 1537–1542.
8. Fotia, M. L., & Driscoll, J. F. (2013). Ram-scram transition and flame/shock-train interactions in a model scramjet experiment. *Journal of Propulsion and Power, 29*(1), 261–273.
9. Karl, S., et al. (2012). Cfd analysis of unsteady combustion phenomena in the hyshot-ii scramjet configuration. In *18th AIAA/3AF International Space Planes and Hypersonic Systems and Technologies Conference* (p. 5912). AIAA.
10. Larsson, J., et al. (2015). Incipient thermal choking and stable shock-train formation in the heat-release region of a scramjet combustor. Part II: Large eddy simulations. *Combustion and Flame, 162*(4), 907–920.

11. Laurence, S., et al. (2012). Investigation of unsteady/quasi-steady scramjet behavior using high-speed visualization techniques. In *18th AIAA/3AF International Space Planes and Hypersonic Systems and Technologies Conference* (p. 5913). AIAA.
12. Laurence, S., et al. (2011). An experimental investigation of steady and unsteady combustion phenomena in the HyShot II combustor. In *17th AIAA International Space Planes and Hypersonic Systems and Technologies Conference* (p. 2310). AIAA.
13. Laurence, S. J., et al. (2013). Transient fluid-combustion phenomena in a model scramjet. *Journal of Fluid Mechanics, 722,* 85–120.
14. Frost, M. A., et al. (2009). Boundary-layer separation due to combustion-induced pressure rise in a supersonic flow. *AIAA Journal, 47*(4), 1050–1053.
15. Mathur, T., et al. (2001). Supersonic combustion experiments with a cavity-based fuel injector. *Journal of Propulsion and Power, 17*(6), 1305–1312.
16. O'Byrne, S., et al. (2012). *OH PLIF imaging of supersonic combustion using cavity injection.*
17. Ouyang, H., Liu, W., & Sun, M. (2014). Numerical investigation of the influence of injection scheme on the ethylene supersonic combustion. *Advances in Mechanical Engineering, 6,* 124204.
18. Wang, H. B., Wang, Z. G., & Sun, M. B. (2013). Experimental study of oscillations in a scramjet combustor with cavity flameholders. *Experimental Thermal and Fluid Science, 45,* 259–263.
19. Wang, Z. G., et al. (2015). Mixing-related low frequency oscillation of combustion in an ethylene-fueled supersonic combustor. *Proceedings of the Combustion Institute, 35,* 2137–2144.
20. Hu, Z. M., et al. (2008). Numerical study of the oscillations induced by shock/shock interaction in hypersonic double-wedge flows. *Shock Waves, 18*(1), 41.
21. Zhao, G. Y., et al. (2018). Investigations of injection parameters on combustion oscillation in a supersonic crossflow. *Acta Astronautica, 152,* 426–436.
22. Laurence, S. J., et al. (2015). Incipient thermal choking and stable shock-train formation in the heat-release region of a scramjet combustor. Part I: Shock-tunnel experiments. *Combustion and Flame, 162*(4), 921–931.
23. Micka, D. J. (2010). *Combustion stabilization, structure, and spreading in a laboratory dual-mode scramjet combustor.* The University of Michigan.
24. Ali, M., & Islam, A. K. M. S. (2006). Study on main flow and fuel injector configurations for Scramjet applications. *International Journal of Heat and Mass Transfer, 49*(19), 3634–3644.
25. Pudsey, A. S., & Boyce, R. R. (2010). Numerical investigation of transverse jets through multiport injector arrays in a supersonic crossflow. *Journal of Propulsion and Power, 26*(6), 1225–1236.
26. Zhao, G. Y., et al. (2019). Experimental investigations of cavity parameters leading to combustion oscillation in a supersonic crossflow. *Acta Astronautica, 155,* 255–263.
27. Sun, M. B., et al. (2011). Parametric study on self-sustained oscillation characteristics of cavity flameholders in supersonic flows. *Proceedings of the Institution of Mechanical Engineers, Part G: Journal of Aerospace Engineering, 225*(6), 597–618.
28. Zare Behtash, H., et al. (2015). Transverse jet-cavity interactions with the influence of an impinging shock. *International Journal of Heat and Fluid Flow, 53,* 146–155.
29. Babinsky, H., & Harvey, J. K. (2011). *Shock wave-boundary-layer interactions* (Vol. 32). Cambridge University Press.
30. Tian, Y., et al. (2017). Investigation of combustion process of a kerosene fueled combustor with air throttling. *Combustion and Flame, 179,* 74–85.
31. Jiang, G. S., & Shu, C. W. (1996). Efficient implementation of weighted ENO schemes. *Journal of Computational Physics, 126,* 202–228.
32. Wang, H. B., et al. (2014). Numerical study on supersonic mixing and combustion with hydrogen injection upstream of a cavity flameholder. *Heat Mass Transfer, 50,* 211–223.
33. Zhao, G. Y., et al. (2019). Investigation of flame flashback phenomenon in a supersonic crossflow with ethylene injection upstream of cavity flameholder. *Aerospace Science and Technology, 87,* 190–206.

34. Wang, H. B., et al. (2013). Combustion characteristics in a supersonic combustor with hydrogen injection upstream of cavity flameholder. *Proceedings of the Combustion Institute, 34,* 2073–2082.
35. Wang, H. B., et al. (2013). Large-Eddy/Reynolds-averaged Navier-Stokes simulation of combustion oscillations in a cavity-based supersonic combustor. *Internationl Journal of Hydrogen Energy, 38,* 5918–5927.
36. Spalart, P. R., & Allmaras, S. R. (1992). A one-equation turbulence model for aerodynamic flows. In *30th Aerospace Sciences Meeting & Exhibit* (p. 439). Reno, NV: AIAA.
37. Yoshizawa, A., & Horiuti, K. (1985). A statistically-derived subgrid-scale kinetic energy model for the large-Eddy simulation of turbulent flows. *Journal of the Physical Society of Japan, 54*(8), 2834–2839.
38. Wang, H., et al. (2011). A hybrid LES (Large Eddy Simulation)/assumed sub-grid PDF (Probability Density Function) model for supersonic turbulent combustion. *Science China Technological Sciences, 54*(10), 2694.
39. Lund, T. S., Wu, X., & Squires, K. D. (1998). Generation of turbulent inflow data for spatially-developing boundary layer simulations. *Journal of Computational Physics, 140*(2), 233–258.
40. Liu, K., & Pletcher, R. H. (2006). Inflow conditions for the large Eddy simulation of turbulent boundary layers: a dynamic recycling procedure. *Journal of Computational Physics, 219*(1), 1–6.
41. Xiao, X., et al. (2003). Inflow boundary conditions for hybrid large Eddy/Reynolds averaged Navier-Stokes simulations. *AIAA Journal, 41*(8), 1481–1489.
42. Urbin, G., & Knight, D. (2001). Large-Eddy simulation of a supersonic boundary layer using an unstructured grid. *AIAA Journal, 39*(7), 1288–1295.
43. Brindle, A., Boyce, R. R., & Neely, A. J. (2005). CFD analysis of an ethylene-fueled intake-injection shock-induced-combustion scramjet configuration. In *AIAA/CIRA 13th International Space Planes and Hypersonics Systems and Technologies Conference* (p. 3239). Capua, Italy: AIAA.
44. Xu, C., & Konnov, A. A. (2012). Validation and analysis of detailed kinetic models for ethylene combustion. *Energy, 43*(1), 19–29.
45. Kalitan, D. M., Hall, J. M., & Petersen, E. L. (2005). Ignition and oxidation of ethylene-oxygen-diluent mixtures with and without silane. *Journal of Propulsion and Power, 21*(6), 1045–1056.
46. Baker, J. A., & Skinner, G. B. (1972). Shock-tube studies on the ignition of ethylene-oxygen-argon mixtures. *Combustion and Flame, 19,* 347–350.
47. Konnov, A. A. (2009). Implementation of the NCN pathway of prompt-NO formation in the detailed reaction mechanism. *Combustion and Flame, 156,* 2093–2105.
48. Bilger, R. W. (1980). Turbulent flows with nonpremixed reactants. In P. A. Libby & F. A. Williams (Eds.), *Turbulent reacting flows* (pp. 65–113) Berlin, Heidelberg: Springer.
49. Knop, V., Michel, J.-B., & Colin, O. (2011). On the use of a tabulation approach to model auto-ignition during flame propagation in SI engines. *Applied Energy, 88*(12), 4968–4979.
50. Colin, O., Pires da Cruz, A., & Jay, S. (2005). Detailed chemistry-based auto-ignition model including low temperature phenomena applied to 3-D engine calculations. *Proceedings of the Combustion Institute, 30*(2), 2649–2656.

Chapter 6
Flame Behaviors Near Blowoff in Supersonic Flows

Blowoff limits, including rich blowoff limit (RBO) and lean blowoff limit (LBO), are the boundaries between stable combustion and flame blowoff. The blowoff limits usually depend on the conditions of inflow, fuel types, fuel injection patterns, and the configuration of the cavity. In the flame extinction stage, the dynamic behaviors near blowoff are very important to understand the blowoff mechanisms [1]. However, the detailed studies in the supersonic flow are rare in the existing literature.

A comprehensive understanding of flame behaviors near blowoff in scramjet is of great significance for designing the cavity configuration and injection schemes in academic and engineering fields. In this chapter, the blowoff limits of supersonic combustion are analyzed and modeled, the characteristics near flame blowoff as well as the flame blowoff and reignition are analyzed using numerical simulation and experiments.

6.1 Blowoff Limits of Supersonic Combustion

In order to understand the unsteady flame behaviors near blowoff, the physics mechanisms of the blowoff limits were analyzed theoretically in this section. Based on the shear-layer combustion stabilization mode and three-dimensional flowfield structure in the cavity, we analyzed the penetration and mixing model of transverse jet and the entrainment process and the mass exchange between the jet and shear-layer/recirculation zone. Then the Damkohler number and the effective equivalence ratio are redefined for modeling the criteria for blowoff limits.

6.1.1 Physical Interpretation of Blowoff Limits

Hydrogen is injected upstream of the cavity when approaching to blowoff limits. The OH* spontaneous emission photographing and the structure of the reaction zone in

M. Sun et al., *Unsteady Supersonic Combustion*,
https://doi.org/10.1007/978-981-15-3595-6_6

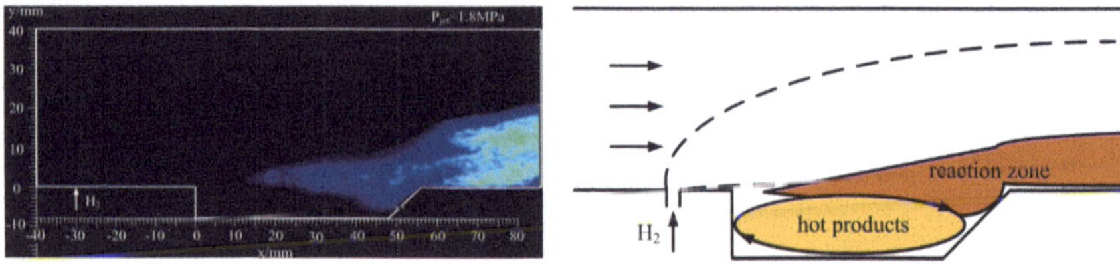

(a) OH* spontaneous emission photographing (b) Schematic of the the structure of the reaction zone

Fig. 6.1 Combustion flow field structure of the cavity shear-layer combustion stabilization mode [2]

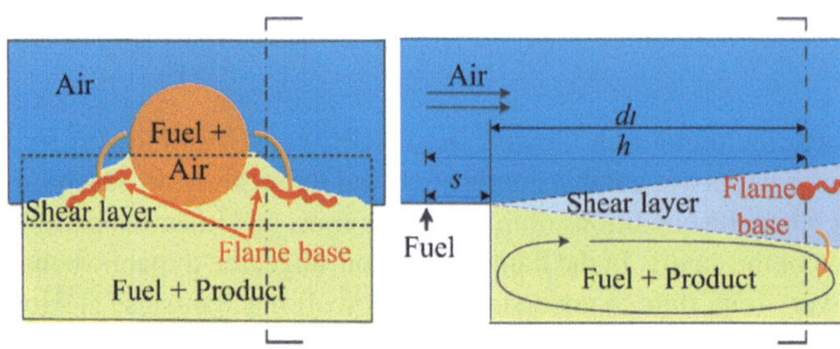

Fig. 6.2 Schematic of the distribution of the components and the combustion process in the cavity at the rich blowoff limit [3]

the cavity shear-layer combustion stabilized mode are shown in Fig. 6.1. There is almost no combustion in the cavity recirculation zone, and the heat release is weak. The recirculation zone is mainly filled with high-temperature products.

According to the observation in Fig. 6.1, it is assumed that the combustion structure can be described as that shown in Fig. 6.2.

This model considers the three-dimensional effect of the flame distribution at the rich blowoff limit, as shown in Fig. 6.2. The cavity consists of the high-temperature products and surplus fuel. The flame base is located in the shear layer below the main stream near the trailing edge of the cavity, with the mixing fraction is approaching to the stoichiometric contour f_s.

Near the rich blowoff limit, the flame lift distance h (the distance from the upstream injector to the flame base) increases as the mass flow rate of the injected fuel increases, so that the flame base moves downstream to the vicinity of the trailing edge of the cavity, thereby achieving rich blowoff limit. There are:

$$h = L_{RZ} + s \tag{6.1}$$

where L_{RZ} is the length of the recirculation zone, which can be considered as the length of the bottom wall of the cavity (L) in the model; s is the distance from the leading edge of the cavity to the injector; d_l is the distance from the flame base to the leading edge of the cavity, there is $d_l = L_{RZ}$ for the rich limit.

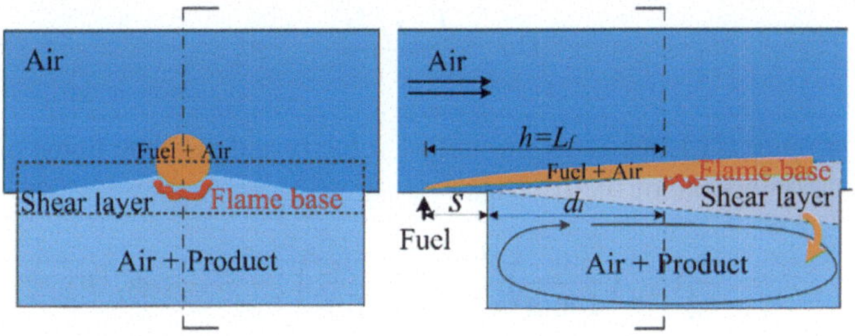

Fig. 6.3 Schematic of the distribution of the components and the combustion process in the cavity at the lean blowoff limit [3]

For the lean blowoff limit, the combustion flow field is shown in Fig. 6.3. There are products and surplus air in the cavity, and the flame base exists in the shear layer below the jet.

Driscoll defined the flammable zone length, the definition is:

$$L_f = \beta_1 \frac{m_F}{\rho_j U_A \delta_L} \tag{6.2}$$

where β_l is the empirical constant, m_F is the fuel mass flow rate, ρ_j is the fuel density at the injector, δ_L is half of the thickness of the shear layer at the position L_f away from the injector.

When the fuel mass flow rate decreases near the lean blowoff limit, the combustible zone retracts upstream. When the length of the combustible zone L_f is closer to the upstream injector than the position where the flame base may be stable, the lean blowoff limit is reached, at this time:

$$L_f = h \tag{6.3}$$

6.1.2 *Modeling of Blowoff Limits*

In order to measure the degree of lean/rich combustion, it is necessary to define an effective equivalence ratio Φ_0, which mainly indicates the characteristic equivalence ratio near the flame base:

$$\phi_0 = \left(m_F^* / m_A^*\right) r_s^{-1} \tag{6.4}$$

In the formula, m_F^* is the effective fuel mass flow rate and m_A^* is the effective air mass flow rate. r_s is the stoichiometric fuel-air mass ratio.

In this section, the Damkohler number is defined by the following formula:

$$Da = \frac{L_{RZ}/U_A}{\tau_f} \tag{6.5}$$

In the formula, U_A is the upstream inflow velocity and τ_f is the flame time scale. The flame time scale of fuel-rich state:

$$\tau_f = \frac{\alpha_0}{S_0^2} C_{T,r}^{-2} \left(\frac{p}{1\ \text{atm}}\right)^{-0.6} \left(1 - \frac{f_s}{f_{RZ}}\right) \frac{L_{RZ}}{L_{RZ}+s} \alpha \phi_0^{-1} \beta_2 \tag{6.6}$$

$$C_{T,r} = \left(Y_{P,RZ} \frac{T_{AD} - T_{OF}}{300\ \text{K}} + \frac{T_{OF} - T_{OA}}{300\ \text{K}}\right) \frac{f_s}{f_{RZ}} + \frac{T_{OA}}{300\ \text{K}} \tag{6.7}$$

The flame time scale of fuel-lean state:

$$\tau_f = \frac{\alpha_0}{S_0^2} C_{T,l}^{-2} \left(\frac{p}{1\ \text{atm}}\right)^{-0.6} \frac{f_s}{f_{up}} \frac{L_{RZ}\eta}{d_l} \alpha \phi_0 \beta_3 \tag{6.8}$$

$$C_{T,l} = \left(Y_{P,RZ} \frac{T_{AD} - T_{OA}}{300\ \text{K}} + \frac{T_{AD} - T_{0,up}}{300\ \text{K}}\right)\left(1 - \frac{f_s}{f_{up}}\right) + \frac{T_{0,up}}{300\ \text{K}} \tag{6.9}$$

In Eqs. (6.6–6.9), S_0 is the laminar combustion rate and α_0 is thermal diffusivity of the fuel under the pressure of 1 atm and the temperature of 300 K. T_{AD} is the adiabatic combustion temperature. T_{0A} is the total temperature of the inflow and T_{OF} is the total temperature of the fuel. $T_{0,up}$ and f_{up} are the total temperature and the mixing fraction of the upper edge of the shear layer at the flame base at the lean blowoff limit, respectively. f_{up} can be calculated through the conversion relation between the mixing fraction and the mole fraction $f_c = \frac{C_c r_m}{c_j + (r_m - 1)C_c}$, C_c and C_j are the molar fraction of the fuel at the center trace and the injector, respectively. For the pure fuel injector, $C_j = 1$. r_m is the fuel-air molar mass ratio. $Y_{P,RZ}$ is the mass fraction of combustion products in the cavity recirculation zone. p is the static pressure of the inflow above the cavity. $\alpha = (1 + \text{Ma}^2(\gamma+1)/2)^{-1}$; $\eta = \gamma + \frac{1-\gamma}{b}(y_c - \delta_l - \delta_{el})$, and it is a velocity attenuation factor, where δ_l is half of the thickness of the shear layer at the flame base and δ_{el} is the offset of the shear layer at the flame base at the lean blowoff limit. y_c is the height of the jet center trace at that position, $\gamma = 1 - c_5(r_j/r_A)0.5(L_f/d)^{-1}$. d_l is the distance between the flame base and the cavity leading edge. s is the distance between the cavity leading edge and the fuel injector; β_0 and β_3 are empirical constants.

The boundaries of rich and lean blowoff limits are given as follows,

Rich blowoff limit:

$$Da = \frac{L_{RZ}/U_A}{\tau_f} = \phi_0 \tag{6.10}$$

Lean blowoff limit:

$$Da = \frac{L_{RZ}/U_A}{\tau_f} = \phi_0^{-1} \tag{6.11}$$

In conclusion, this model can characterize the unsteady flame behaviors near blowoff limits in scramjet.

6.2 Mixing and Combustion Characteristics Near Blowoff

It can be concluded that the flame stabilization behaviors are very sensitive to the parameters near blowoff from Sect. 6.1. In this section, we interpreted the blowoff limits and studied the mixing and combustion characteristics with different injection schemes in low equivalence ratio conditions experimentally and numerically. Mixing efficiencies, combustion efficiencies and the oscillations of the flame base are also analyzed.

The experimental facility is composed of an air heater, a model scramjet engine, a fuel supply system and a measure control system. A Ma = 2.52 nozzle is installed behind the air heater which heats air from room temperature to 1600 K and increases the stagnation pressure of vitiated air up to 1.65 MPa through air/O_2/alcohol combustion to simulate flight Ma = 6 conditions. The total mass flow rate of vitiated air is about 1.0 kg/s. 99.5–99.8% pure C_2H_4 [4] with room temperature is used in all the experiments. The run time is 6–7 s for the air heater and 1.8–2 s for the fuel injection. The non-cooling supersonic combustor walls are directly exposed to the atmosphere. Detailed flow conditions at the nozzle and fuel jet exits are listed in Table 6.1.

The schematic of the test section is shown in Fig. 6.4. The entrance section of

Table 6.1 Inflow and fuel jet conditions of all the test cases [5]

Parameter	T_0 (K)	P_0 (MPa)	Ma	Y_{O_2} (%)	Y_{H_2O} (%)	Y_{CO_2} (%)	Y_{N_2} (%)	$Y_{C_2H_4}$ (%)
Air	1600	1.65	2.52	23.0	7.1	11.6	58.3	0
Fuel jet	300	0.8/1.8	1	0	0	0	0	100

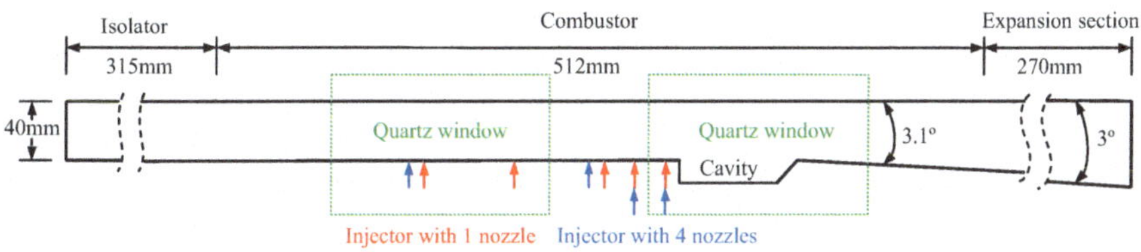

Fig. 6.4 Schematic of the test section and local diagram of the cavity and injectors [6]

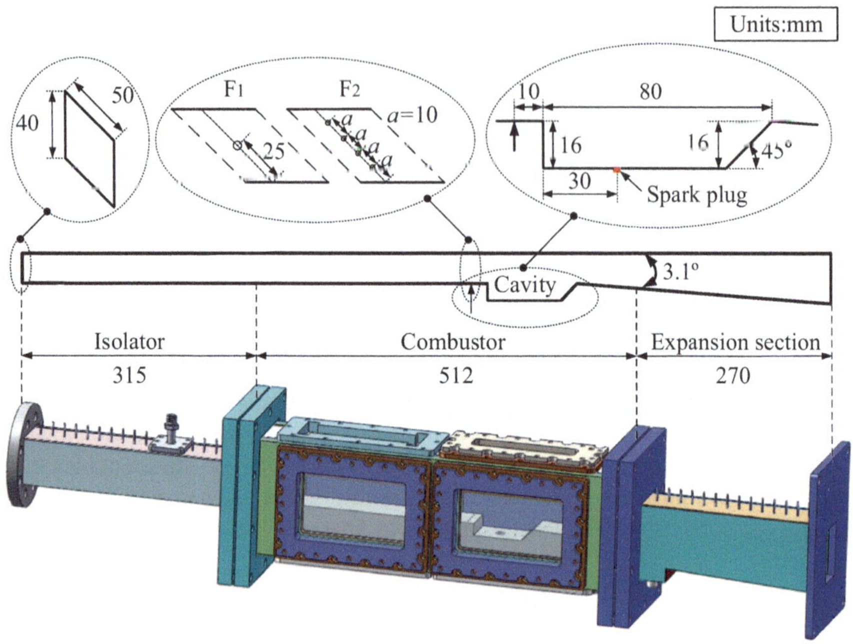

Fig. 6.5 Schematic diagram of the test section [6]

the combustor is a rectangle with a width of 50 mm and a height of 40 mm. Quartz glass windows can be mounted on both sides and top of the combustor to allow optical access. An integrated fuel-injection/flame-holding cavity can be installed on the lower wall of the combustor. The installation scheme is shown in Fig. 6.5. The divergence angle of the combustor lower wall is 3.1°, which starts behind the trailing edge of the cavity. Fuel injection schemes and cavity specific parameters are given in Fig. 6.6. The diameter of pressure tap is 0.5 mm. Two kinds of injection schemes are considered: single-orifice injection and multiple-orifice injection. The diameter for multiple orifices is 1 mm, as shown in Fig. 6.6a, and the diameter for single orifice is 2 mm, as shown in Fig. 6.6b, which can make sure that the global equivalent ratio is the same when the injection pressure is equal. In the experiments, the injection pressure varies from stable combustion states to blowoff states, different injection schemes and distances are considered and C_2H_4 with room temperature is used.

The flame propagation and blowoff process are visualized by high-speed photography. In order to capture the dynamic process of the lean flame blowoff, the camera is set at 20000 fps with a shutter time of 1/20416 s and an aperture of 1.4. Besides, static pressure taps with a sampling frequency of 100 Hz are installed along the lower wall of the combustor, as shown in Fig. 6.6, to obtain the wall-pressure distribution. The pressure transducers have uncertainties of ±0.5% in this section [7].

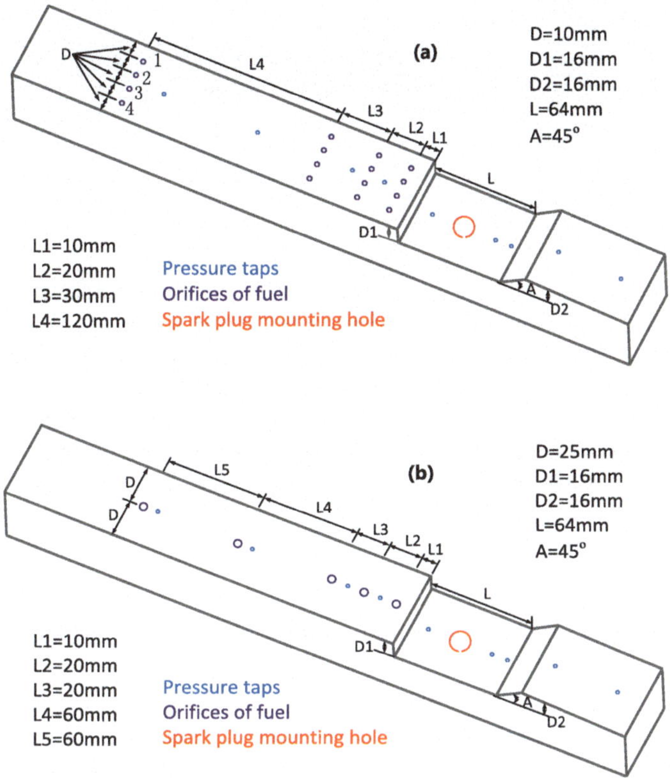

Fig. 6.6 Schematic diagram of cavity and injection configurations [6]

The three-dimensional Reynolds-averaged Navier-Stokes (RANS) equations [8] and the SST k-ω turbulence model [9] are used with density-based (coupled) double precision solver of Fluent for numerical simulation.

A schematic of the computational domain with a cavity studied here is presented in Fig. 6.7. As with the experimental condition, the width of the computational domain

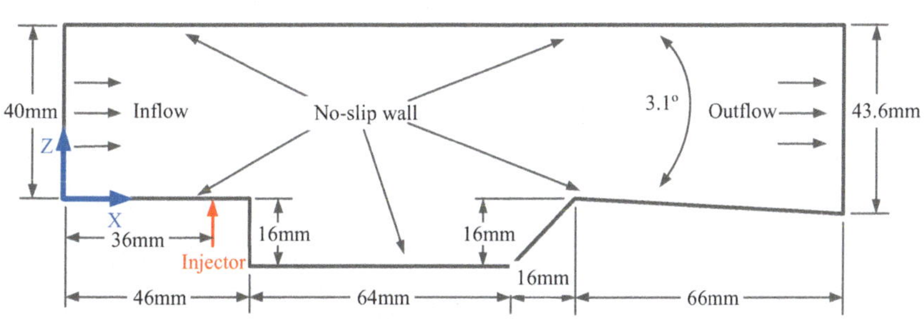

Fig. 6.7 Schematic of the computational combustor with a cavity [6]

Table 6.2 Fuel jet arrangements of all the test cases

	Case 1	Case 2	Case 3	Case 4	Case 5	Case 6
Equivalent ratio	0.08	0.08	0.18	0.18	0.06	0.04
Injection pressure (MPa)	0.8	0.8	1.8	1.8	0.6	0.4
Numbers of injection orifices	1	4	1	4	1	4
Diameter of injection orifices (mm)	2	1	2	1	2	1

is 50 mm and the side walls are set to the solid walls. No-slip and adiabatic boundary conditions are applied at all the solid walls by setting the heat flux to zero. The pressure inlet condition is used for the inflow conditions of the airstream and the fuel. All parameters are set according to the experiments except the species mass fractions of airstream inflow, which are 23% O_2 and 77% N_2. The pressure outlet condition is used for the outlet, but it does not have an effect on the flow field of the combustor for the reason that the flow is supersonic at the outflow. All the test conditions in this section are listed in Table 6.2.

6.2.1 *Interpretation of Blowoff Limits*

In this section, the lean blowoff limits are defined as the minimum injection pressure that can keep continuous combustion. The lean blowoff limits are gained by decreasing the injection pressure at the steps of 0.2 MPa from stable combustion states. The injection pressure is controlled by a pressure reducing valve and an electronic pressure displayer with minimum display unit of 0.1 MPa. At the lean blowoff boundary shown in Fig. 6.8 with the solid lines, the flame can always be stabilized by the cavity as long as the fuel supply is constant. Below the lean blowoff boundary shown in

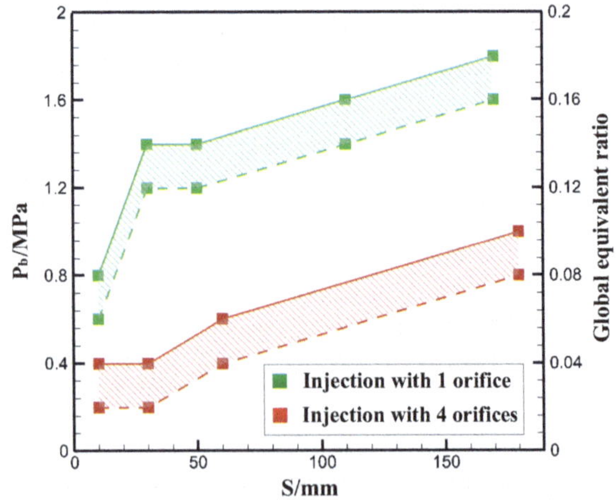

Fig. 6.8 Lean blowoff limits for different injection distances and orifices [6]

Fig. 6.8 with the dashed line, the flame cannot be ignited or will be blown off even if the fuel is served continuously. More accurate lean blowoff boundary should in the area between the dashed line and solid line.

As shown in Fig. 6.8, lean blowoff limits increase with increasing injection distances and which for the single-orifice cases are found to be higher than that for the multiple-orifice cases. The jet penetration deep calculation formula [10] shows that with increasing of injection distance, fuel jet penetration above the cavity becomes larger, and less fuel enters into the shear layer via the interaction between the jet and cavity shear layer. So the lean blowoff limits increase with increasing injection distance. The research of Wang et al. [11] shows that the interaction between the fuel jet and the cavity can promote the transportation of fuel into the cavity recirculation zone. When fuel is injected closer to the cavity leading edge, the interaction between the fuel jet and the cavity is stronger. This may explain why in the single-orifice cases the lean blowoff limits for the injection distance of 10 mm are much lower.

The experimental studies show that the flame behaviors can be different for different injection schemes. The flame is basically steady for the multiple-orifice cases, but the flame shape changes violently and the combustion is quite unsteady for the single-orifice cases. There are no obvious changes in the flame behaviors with increasing injection distance under the present conditions. Only the cases with injection distance of 10 mm are displayed to be analyzed.

6.2.2 Mixing Characteristics

To explain the combustion characteristics in a cavity-based supersonic combustor with different fuel injection schemes, the mixing characteristics should be firstly analyzed. Figures 6.9 and 6.10 show the vorticity magnitude and fuel distributions for different injection schemes under an injection pressure of 1.8 MPa in non reacting flowfield. As depicted in the figures, the origin of coordinates (0, 0, 0) are set at point O, and the spanwise planes are the plane of $y = 25$ mm in the single-orifice case and the plane of $y = 20$ mm in the multiple-orifice case, respectively. The streamlines in the spanwise planes illustrate that the core of the cavity recirculation is deeper inside the cavity in the multiple-orifice case. It indicates that the cavity shear layer develops deeper into the cavity. Besides, it can be seen that more fuel is transported into the cavity in the multiple-orifice case. Moreover, the spanwise distributions of fuel are more uniform because four injection orifices are interval distributed spanwisely. Vorticity gives an indication of the turbulence eddies in the flow field, and it is an important parameter in evaluating the fuel transport and mixing processes [7]. It can be seen that the peak value of the vorticity magnitude concentrates in the shear layer of the cavity and around the fuel jet. The vorticity magnitude decreases along the streamwise. Either in or close to the cavity, the vorticity is much larger in the multiple-orifice case than that in the single-orifice case. From the fuel and vorticity magnitude distributions, it can be expected that multiple-orifice cases have better mixing effects near the fuel injection orifice and above the cavity.

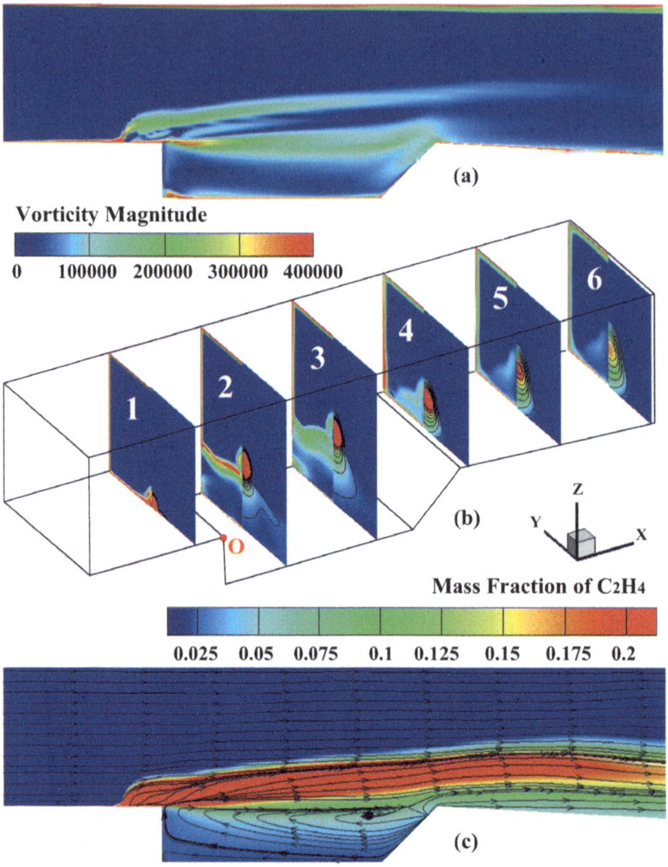

Fig. 6.9 Vorticity magnitude and fuel distributions in the single-orifice case 3. **a** Vorticity magnitude contours in spanwise plane; **b** vorticity magnitude and fuel mass fraction contours in streamwise planes; **c** streamlines and fuel mass fraction contours in spanwise plane [5]

In order to demonstrate the different mixing processes between the multiple-orifice case and the single-orifice case, the iso-surface of stoichiometric mixture fraction colored by vorticity magnitude in cases with an injection pressure of 1.8 MPa for different injection schemes are shown in Fig. 6.11. The iso-surface of the stoichiometric mixture fraction shows that the interface is larger between the fuel jet and the air flow in the multiple-orifice case. Furthermore, the vorticity magnitude in the iso-surface of stoichiometric mixture fraction decreases along the streamwise, and above the cavity it is higher in the multiple-orifice case than that in the single-orifice case. Therefore, the larger interface between the fuel jet and the airflow and the higher vorticity magnitude in the multiple-orifice case may play important roles in the mixing process.

The mixing efficiency is introduced here to evaluate the mixing performance in the combustor. It is defined as the ratio of the mixed mass flow rate of fuel to that of

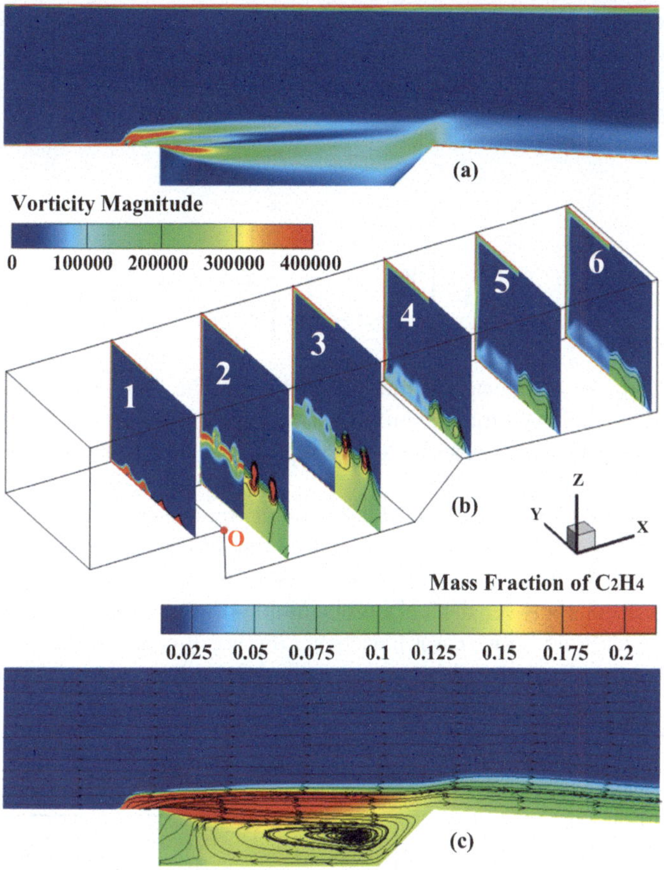

Fig. 6.10 Vorticity magnitude and fuel distributions in the multiple-orifice case 4. **a** vorticity magnitude contours in spanwise plane; **b** vorticity magnitude and fuel mass fraction contours in streamwise planes; **c** streamlines and fuel mass fraction contours in spanwise plane [5]

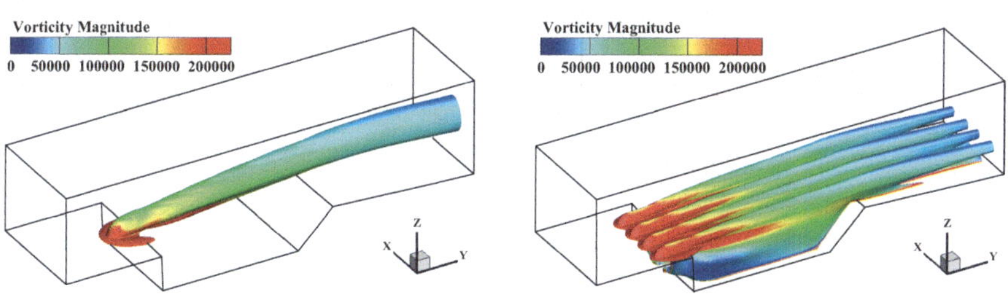

(a) Single-orifice case (Case3)

(b) Multiple-orifice case (Case4)

Fig. 6.11 Iso-surface of stoichiometric mixture fraction colored by vorticity magnitude [5] the total one at the local cross-sectional plane at the given stream-wise location x. The formulas of mixing efficiency can be expressed as follows [9]:

$$\eta_m(x) = \frac{\int_{A(x)} Y_{react}\rho u dA}{\int_{A(x)} Y\rho u dA} \tag{6.12}$$

And

$$Y_{react} = \begin{cases} Y & Y \le Y_{stoic} \\ Y_{stoic}\left(\frac{1-Y}{1-Y_{stoic}}\right) & Y > Y_{stoic} \end{cases} \tag{6.13}$$

where Y, Y_{stoic}, Y_{react} donate the fuel mass fraction, the fuel stoichiometric mass fraction and the fuel mass fraction mixed in a proportion that can react, respectively. ρ and u mean the local density and velocity, respectively. $A(x)$ represents the sectional area at streamwise location x.

The mixing efficiencies calculated based on the time-averaged nonreacting field are shown in Fig. 6.12. It is noticeable that mixing efficiencies in multiple-orifice cases are higher than that in single-orifice cases. It is consistent with the foregoing that the interactions between the fuel jets in multiple-orifice cases promote the mixing process. The multiple-orifice cases have better mixing effects near the fuel injection orifice and above the cavity. Furthermore, the mixing efficiencies of the cases with lower injection pressure are higher. The quicker mixing process may be caused by the lesser total fuel mass with lower injection pressure.

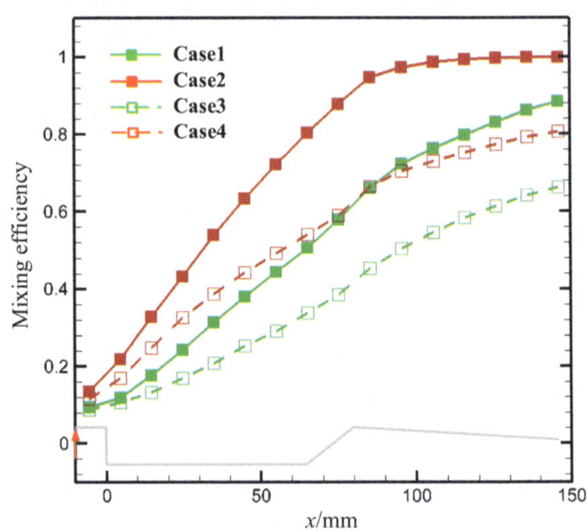

Fig. 6.12 Mixing efficiencies from the nonreacting numerical simulations [5]

6.2.3 Combustion Characteristics

In order to analyze the combustion characteristics, both numerical and experimental results are displayed in this section. Figure 6.13 shows lower wall pressure distributions of case 1, case 2, case 6 and the cold flow without injection. The figure in the top right corner of Fig. 6.13 shows the pressure changes over time in this four cases of position A and indicates that the accuracy of pressure measurements almost have no influence on the interpretation of results in this paper. The rise of the pressure caused by the combustion is slight in all cases as shown in Fig. 6.13. With the more intense combustion, the pressure is higher [12, 13]. With the same injection pressure, the multiple-orifice case appears to have higher pressure rise in the cavity and lower pressure rise downstream of the cavity. It indicates that the flame is more upstream in the multiple-orifice cases. For the cases of lean blowoff boundaries, the pressure rise in the single-orifice cases is higher. This indicates that to maintain the flame exist constantly, the greater combustion intensity is needed for the single-orifice cases. In a word, the lower wall pressure distributions illustrate that the flame is more stable for the multiple-orifice cases.

Figure 6.14 shows the calculated temperature and fuel distributions for multiple-orifice cases and single-orifice cases, respectively. For the multiple-orifice cases, the temperature of the whole cavity volume is relatively high, the spanwise distribution is more uniform and the fuel jet is closer to the cavity. So when the combustible mixture passes the cavity, it can be ignited more readily. For the single-orifice cases, the high-temperature region only exists in the latter part of the cavity volume, the fuel concentrates in the jet wake and it is more far from the cavity, so it is harder to ignite the combustible mixture. The results of the calculation well explain the experimental

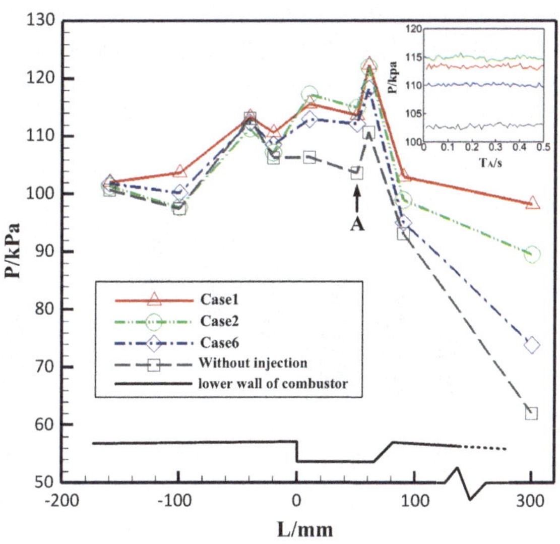

Fig. 6.13 Lower wall pressure distributions of the combustor in experiments [6]

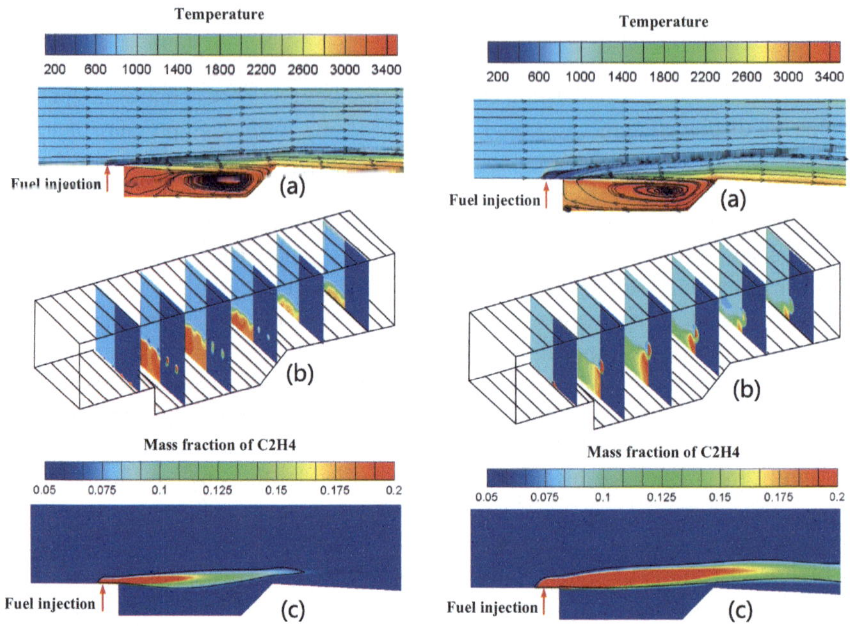

(a) Single-orifice case (Case1) (b) Multiple-orifice case (Case2)

Fig. 6.14 Temperature and fuel distributions contours from the reacting numerical simulations. **a** Streamlines and temperature contours in spanwise plane; **b** temperature contours in streamwise planes; **c** stoichiometric line and fuel mass fraction contours in spanwise plane [6]

phenomenon that the lean cavity flame for the multiple-orifice cases is more stable than that for the single-orifice cases.

Figure 6.15 shows the temperature and mass fraction of C_2H_4 contours from the reacting numerical simulations with a fuel injection pressure of 1.8 MPa. In the single-orifice case, the fuel concentrates near the center line of jet stream and there

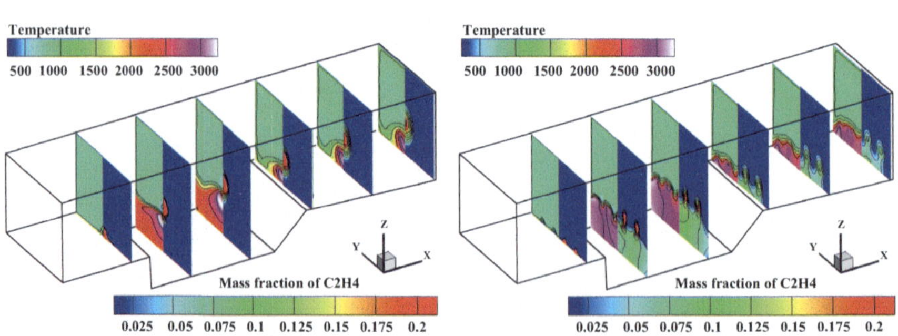

(a) Single-orifice case (Case3) (b) Multiple-orifice case (Case4)

Fig. 6.15 Temperature and fuel distributions contours from the reacting numerical simulations [5] is little fuel in the cavity shown in Fig. 6.15a. Moreover, the temperature is higher than 3000 K in only a little region that below the fuel jet while it is lower than 2500 K in most regions of the cavity. By contrast, it can be seen from Fig. 6.15b, the temperature in almost all the regions of the cavity is higher than 2500 K and the most area is higher than 3000 K shown in the multiple-orifice case. Therefore, the temperature in the cavity in the multiple-orifice case is higher than that in the single-orifice case in reaction flow conditions. In addition, there is still much fuel in the cavity in the multiple-orifice case except the fuel that concentrates near the center line of jet stream like the single-orifice case. Therefore, the cavities are filled up with high-temperature mixture of fuel and combustion products in the multiple-orifice case, however, the cavities are filled with high-temperature mixture of inflow gases and combustion products in the single-orifice case.

To analyze the flame stabilization process in the combustor quantitatively, the combustion efficiencies are calculated along the axial direction. It is defined as the ratio of the burned mass flow rate of fuel at the local cross-sectional plane at the given streamwise location to the total mass rate of the ethylene jet that injected from the injection orifices. This parameter reveals the fuel consumption by the chemical reaction along the streamwise direction and a larger combustion efficiency could also represent a more stable flame stabilization process in the combustor [8]. The combustion efficiency is calculated as follows [14]:

$$\eta_c(x) = \frac{\dot{m}_{jet} - \int_{A(x)} \rho Y u dA}{\dot{m}_{jet}} \tag{6.14}$$

Here $\dot{m}_{jet}$ represents the total mass rate of the ethylene jet that injected from the injector.

The combustion efficiencies calculated based on the time-averaged reacting field are shown in Fig. 6.16. It is obvious that multiple-orifice cases have higher combustion efficiencies in the same streamwise position. Moreover, the slopes of the combustion efficiencies curves above the cavity are much higher in multiple-orifice cases. It reveals that more fuel is burned above the cavity in multiple-orifice cases and it is consistent with the foregoing that temperature in the cavity in multiple-orifice cases is higher than that in single-orifice cases in reaction flow conditions. It is owing to the nonreacting flow characteristics that more fuel is transported into the cavity and the mixing efficiencies are higher in multiple-orifice cases. Also, the cases with lower injection pressure have higher combustion efficiencies than that with higher injection pressure and they may due to that the cases with lower injection pressure have higher mixing efficiencies.

Figure 6.17 shows flame luminosity images for multiple-orifice with fuel injection pressures of 0.4 and 0.8 MPa. It can be seen from the images that the flame is mainly stabilized in the cavity shear layer. Decreasing the injection pressure, the flame becomes weaker and smaller, though it still remains stable. When the lean blowoff limit is approached, the flame only exists below the cavity mouth. Figure 6.18 shows flame luminosity images for single-orifice with fuel injection pressures of 0.8

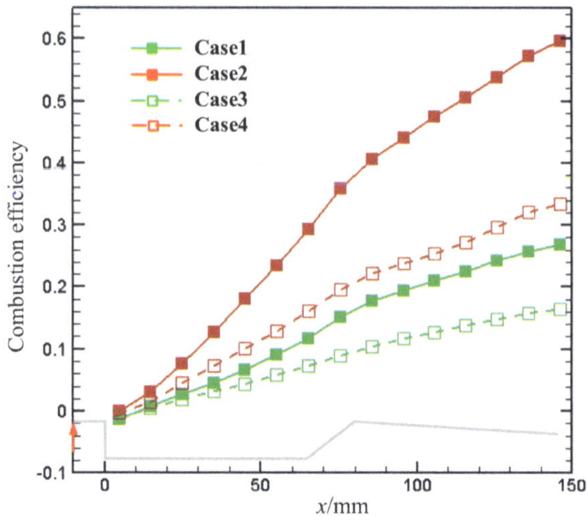

Fig. 6.16 Combustion efficiencies from the reacting numerical simulations [5]

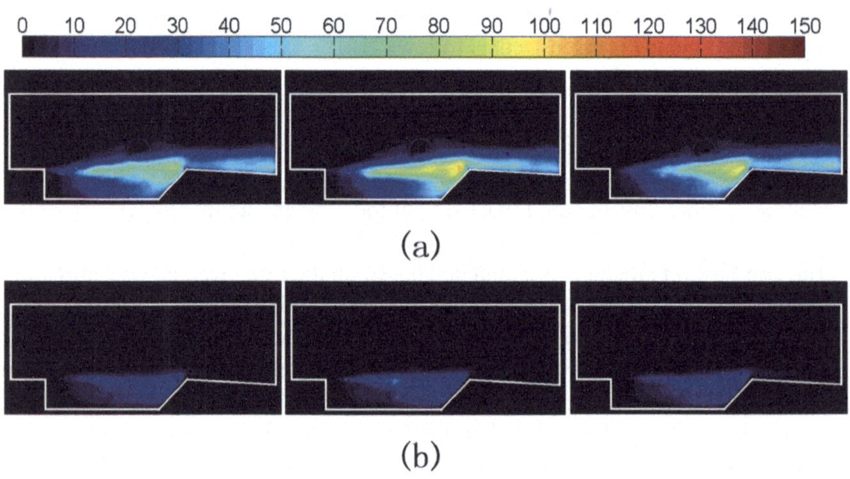

Fig. 6.17 Flame luminosity images for multiple-orifice cases with different injection pressures. The time intervals of adjacent images are 0.05 ms. **a** Case 2, **b** case 6 [6]

and 1.8 MPa. It can be seen that the flame is also stabilized in the cavity shear layer. Decreasing the injection pressure, the flame becomes more unstable, and local extinction takes place intermittently in the cavity shear layer. Compared to that for multiple-orifice cases, the robustness of the lean flame for single-orifice cases seems relatively poor.

Figure 6.19 depicts the time-averaged flame luminosity images in the combustor from the experiments, which are calculated by ten thousand images that represent

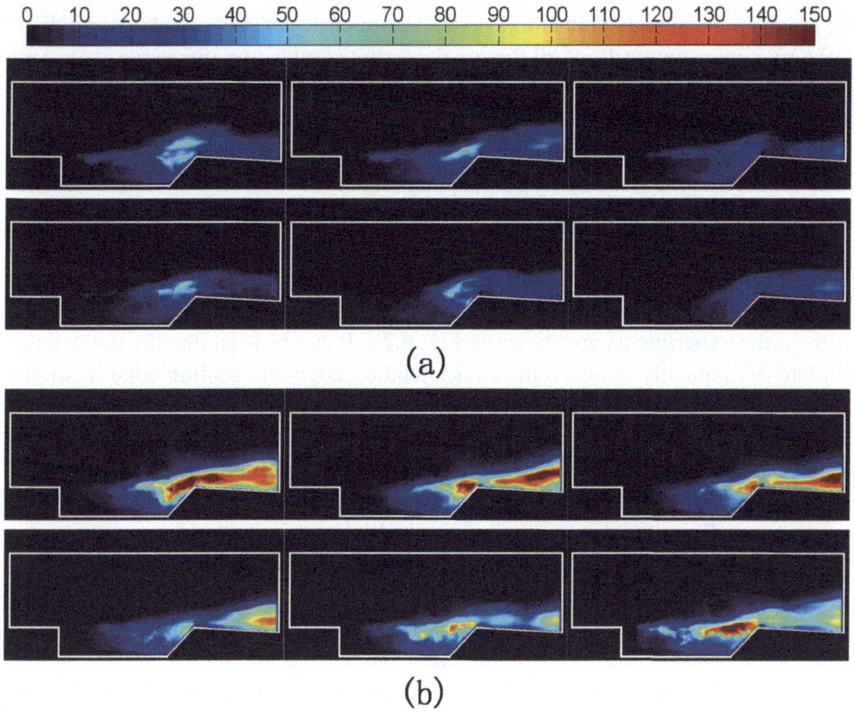

Fig. 6.18 Flame luminosity images for single-orifice cases with different injection pressures. The time intervals of adjacent images are 0.05 ms. **a** Case 1, **b** case 3 [6]

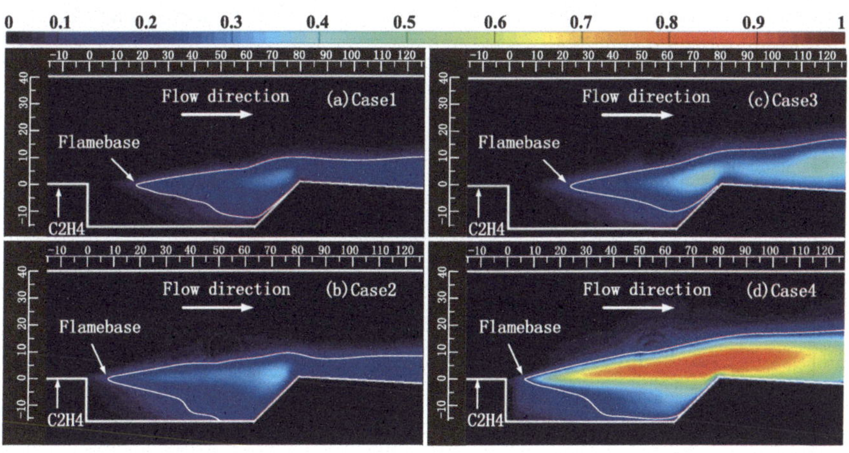

Fig. 6.19 Time-averaged flame luminosity images from the experiments [5]

500 ms from the stable combustion stage. The intensity of each pixel is normalized by the maximum intensity of all the pixels during combustion processes [15]. And, these images are displayed in false-color to provide good contrast. The white line is the iso-luminosity line, the averaged gray of each time-averaged flame luminosity image, and it is used to characterize the combustion zone outline. Furthermore, the flame base is defined as the leading axial position of the combustion zone. It is obvious that the combustion zones are deeper in the cavity in multiple-orifice cases and the flame bases are closer to the cavity leading edge. Moreover, the flame intensities with higher injection pressure are stronger due to the higher equivalence ratio.

To analyze the unsteady combustion process, histories of flame base oscillations [9] from the experiments are shown in Fig. 6.20. It can be seen that the flame bases oscillate dramatically between the cavity leading edge and trailing edge in single-orifice cases (Case 1 and Case 3). However, the flame bases oscillate slightly in the cavity shear layer just downstream of the leading edge in multiple-orifice cases (Case 2 and Case 4). Moreover, the oscillation amplitudes under the injection pressure of 0.8 MPa are larger than those under the injection pressure of 1.8 MPa in multiple-orifice cases. Even so, the oscillations in multiple-orifice cases are much weaker than those in single-orifice cases. As a result, with the advantages of closer fuel jets to the cavities, higher mixing efficiencies and temperature in the cavities, the flame bases in multiple-orifice cases will be more stable. However, these conditions are not available in single-orifice cases, so the flame bases oscillate dramatically.

Figure 6.21 shows frequency spectra of flame base oscillations obtained by Fast Fourier Transform (FFT) [9, 16] from the experiments. The flame base shows a widely distributed spectra of oscillations, indicating that the flame does not couple with the acoustics within this frequency range. Also, the power spectral density of flame base oscillations are large in the frequency 0–1 kHz. However, there is not a fixed dominant frequency in all cases.

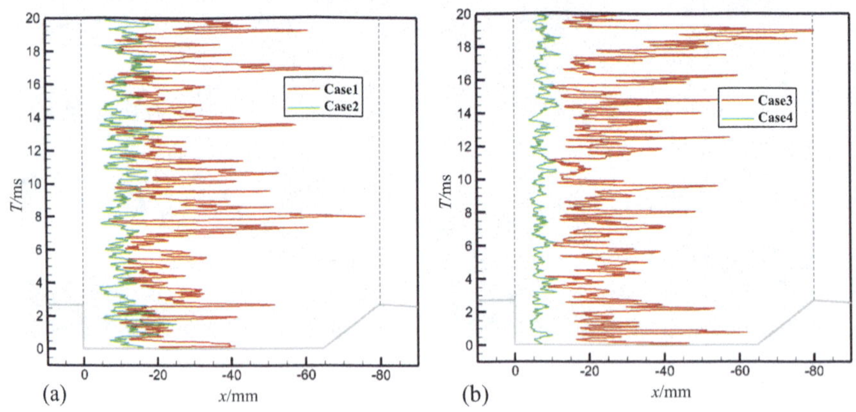

Fig. 6.20 Histories of flame base oscillations from the experiments [5]

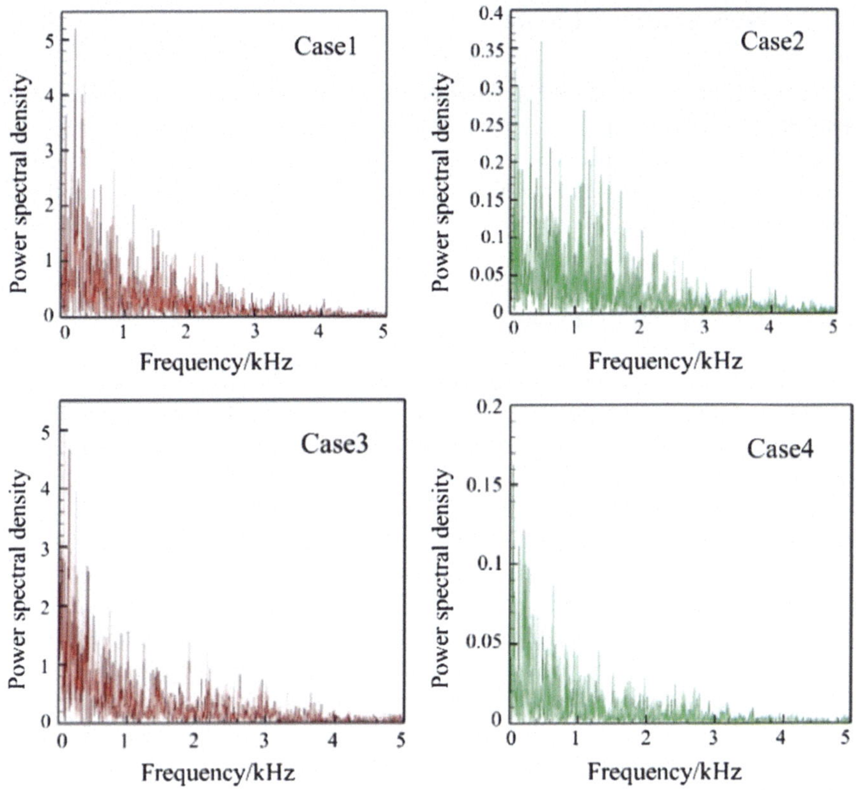

Fig. 6.21 Frequency spectra of flame base oscillations from the experiments [5]

6.3 Near Blowoff Flame Dynamics

In subsonic flows, the dynamic behaviors near flame blowoff are different from the stable combustion. In scramjet combustors, due to the complex structures induced by the interaction between jet and cavity, the mechanisms of dynamic behaviors near flame blowoff are not clear. Very lean or rich flames may exist at certain stages or in some local regions even if the global equivalence ratio is close to unity. Thus, the near-blowoff characteristics may be crucial to the stabilization of the entire flame. The characteristics of the flame blowoff process need to be clarified, and it is necessary to understand the mechanism. Therefore, this section studied the blowoff process of the non-premixed flame and the premixed flame.

6.3.1 Non-premixed Flame

Lean blowoff characteristics of an ethylene-fueled model scramjet combustor with cavity flameholder are investigated experimentally and numerically. Near-blowoff dynamics and blowoff process for the single-orifice cases are captured and analyzed. The experimental facility and the numerical analysis method are the same as those described in Sect. 6.2.

6.3.1.1 Blowoff and Reignition

At the lean blowoff boundary shown in Fig. 6.8, the robustness of flames for the single-orifice cases are poor and the cavity shear-layer flames disappear intermittently. Since all the dynamic processes are similar, case 1 is taken for detailed analysis.

The time of the first image in Fig. 6.22 is defined as 0.00 ms. The flame in the shear layer becomes weaker from 0.00 to 0.15 ms. The possible reason is that the shear layer moves down due to the disturbance of the inflow, and which is not fully discussed here and ought to be investigated further. The unsteady characteristics bring about a large stretch of flames [4]. When the stretch rate near the trailing edge is larger than that the flame can withstand, the flame is extinguished. So the flame is partially extinguished there and the flame is broken down into two parts at 0.20 ms. Then the flame downstream of the cavity is blown off and the other part of the flame shrinks into the cavity at 0.25 ms. Later, the flame in the cavity spreads downstream of the cavity from 0.30 to 0.45 ms, but the flame at 0.45 ms is not as strong as that at 0.00 ms. At 0.50 ms, the flame is locally extinguished and broken down again, similar to that at 0.25 ms. However, the flame is much weaker, so it cannot reignite the mixture in the cavity shear layer immediately. Due to the weakening of the flame around the cavity, the cavity shear layer lifting caused by combustion is weakened and the interaction between the shear layer and the trailing edge is enhanced. As a result, more fuel is entrained into the cavity recirculation zone, and the combustion

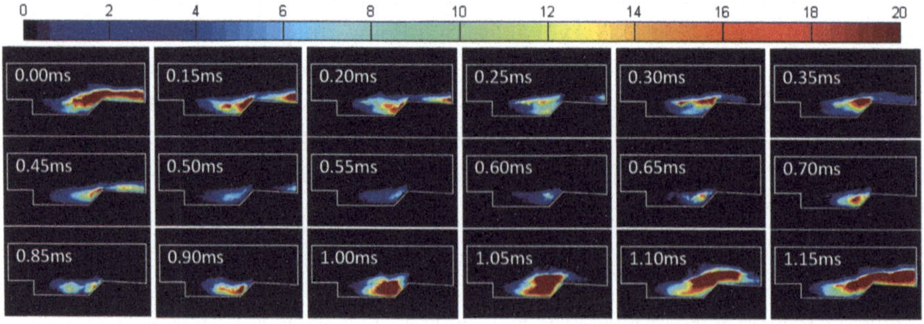

Fig. 6.22 Local extinction and reignition of the cavity flame with the injection pressure 0.8 MPa for single-orifice case 1 [6]

within the cavity is intensified. From 0.65 to 1.00 ms, the flame becomes stronger and larger. Later, the flame spreads into the main stream and becomes similar to that at 0.00 ms.

6.3.1.2 Blowoff Process

Flames for the multiple-orifice cases are always stable and no blowoff process is found in our experiments when the fuel supply is constant. When the injection pressure is lower than the lean blowoff boundary, spark ignition cannot ignite the mixture. Although the flame of hydrogen is used to ignite the ethylene, the flame of ethylene is extinguished while the hydrogen is cut off. The blowoff processes without cutting off the fuel are observed only in the single-orifice cases.

The blowoff process of the cavity flame is shown in Fig. 6.23 and the time marked in the diagram is the moment before the flame is extinguished. Like the process from 0.00 to 0.25 ms in Fig. 6.22, the flame goes through a similar process from 0.60 to 0.40 ms in Fig. 6.23. Since the injection pressure in Fig. 6.23 is lower than that in Fig. 6.22, there are less fuel in the shear layer and there are fewer energy sources in the cavity recirculation zone. After the flame shrinks into the cavity, the mixture in the shear layer cannot be reignited. From 0.35 ms to the time when the flame is extinguished completely, the flame moves to the front half part of the cavity after a short stay in the latter part of the cavity and then disappears.

Comparing the two processes in Figs. 6.22 and 6.23, it is found that the cavity flames are partially extinguished near the trailing edge firstly, and then the flames shrink into the cavities in these two cases. If the injection pressure is higher, there is always a flame in the latter part of the cavity and the mixture in the shear layer can be reignited soon. If the injection pressure is low enough, after the flame shrinks into the cavity, the flame cannot reside in the latter part of the cavity steadily but moves to the front half part of the cavity. The flame in the front half part of the cavity probably cannot reignite the mixture in the shear layer and after a while it is extinguished completely. From the above analysis, the blowoff process of cavity flames can be divided into five steps—the shear layer flame becomes weaker, the flame is partially extinguished near the trailing edge, the flame shrinks into the latter part of the cavity,

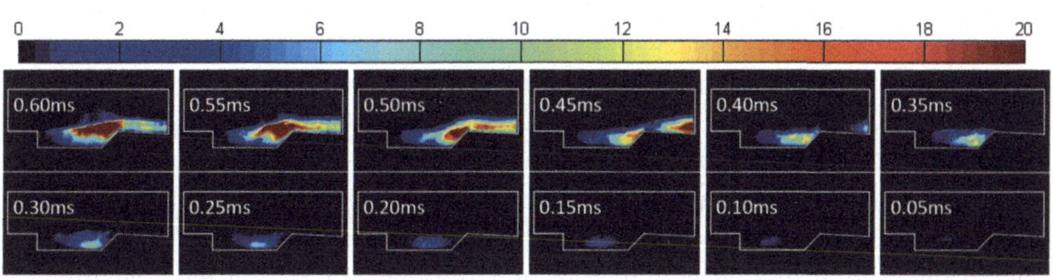

Fig. 6.23 Blowoff process of the cavity flame with the injection pressure 0.6 MPa for single-orifice case 5 [6]

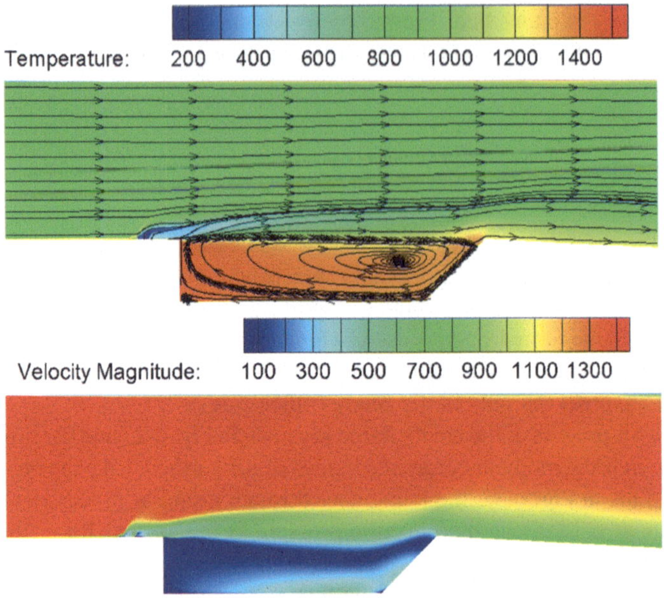

Fig. 6.24 Temperature and velocity magnitude distributions for single-orifice case in nonreacting flows [6]

the flame moves to the front half part of the cavity and the flame is extinguished completely.

As shown in Fig. 6.24, comparing the three regions that flames can exist, there are the highest velocity and the lowest static temperature in shear layer, the highest static temperature and the lowest velocity in the front half part of the cavity, and the temperature and velocity are moderate in the latter part of the cavity. In the blowoff process of the cavity flame, it transfers from the shear layer to the region that is more combustible.

6.3.2 *Premixed Flame*

Near-blowoff characteristics of cavity-stabilized premixed flames in supersonic flows are investigated. It is crucial to understand the flameholding process and important for the design of flameholding schemes in scramjet combustors.

In this section, numerical simulations are carried out to investigate the near-blowoff characteristics of cavity-stabilized combustion in supersonic flows. In order to avoid the complicated effects of fuel injection and mixing on the near-blowoff physics, premixed flames are considered in the present section. The equivalence ratio of the freestream is changed to obtained stable, near-blowoff or blowoff conditions, respectively. The flow and reaction structures under different conditions are discussed, and the near-blowoff behaviors and blowoff process of cavity flames in supersonic flows are characterized.

6.3.2.1 Geometry and Nonreacting Cavity Flow

A cavity with depth of D = 8.9 mm and length-to-depth ratio of L/D = 3 that was experimentally studied by Gruber et al. [17] is considered. The supersonic freestream is an unheated flow of Mach 2.0, and the total pressure and total temperature are 690 kPa and 300 K, respectively. These conditions are similar to those of Gruber et al. [17] except that the Mach number is changed. The schematic of the computational cavity is shown in Fig. 6.25. A no-slip, no-penetration adiabatic condition is imposed at all walls and a zero-gradient condition is enforced at the outflow boundary. For supersonic inflow boundary, just physical boundary conditions are employed. For the fifth-order WENO scheme, three extra points (called ghost points) are needed for the requirements of the spatial derivatives near boundaries.

The grid points are respectively 331 × 91 and 151 × 91 for the regions outside and inside the cavity, which is chosen based on grid sensitivity analysis. The time-averaged temperature contours overlapped with streamlines for nonreacting cavity flow is shown in Fig. 6.26. It is observed that the temperature within the cavity is close to but slightly lower than the stagnation temperature of the freestream. Basically, there exist two eddies within the cavity recirculation zone (RZ): a primary eddy (PE) in the aft region of the cavity volume; and a secondary eddy (SE) between the PE and the cavity front wall.

Figure 6.27a shows the pressure history at the cavity aft corner together with the histories of mass-averaged temperature within the cavity, and Fig. 6.27b shows the frequency spectra of the data in Fig. 6.27a. Three kinds of temperature are considered here, i.e. temperature within the cavity, $\langle T_{cav} \rangle$, temperature within the PE, $\langle T_{PE} \rangle$, and temperature within the SE, $\langle T_{SE} \rangle$. The mass-averaged temperature within the cavity is defined as:

$$\langle T_{cav} \rangle = \frac{\iint_{cavity} \rho(x, y) T(x, y) dxdy}{\iint_{cavity} \rho(x, y) dxdy} \tag{6.15}$$

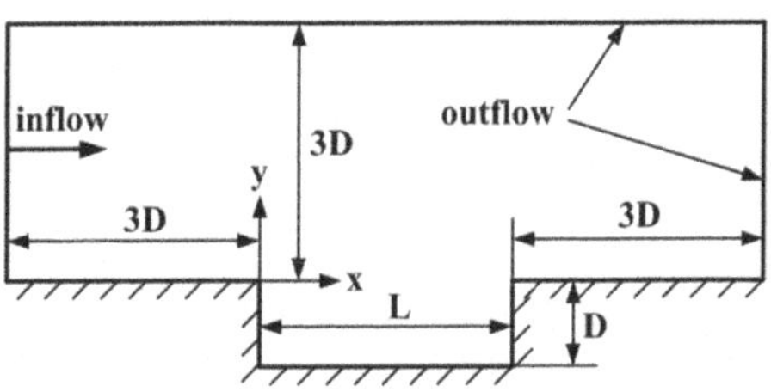

Fig. 6.25 Schematic of the computational cavity [18]

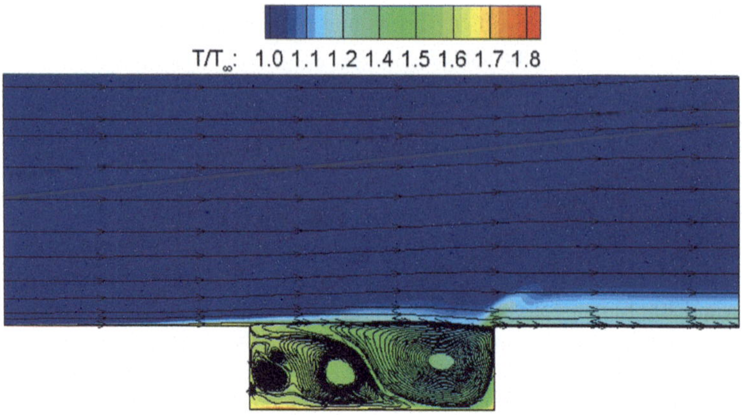

Fig. 6.26 Average temperature contours overlapped with streamlines for nonreacting cavity flow [18]

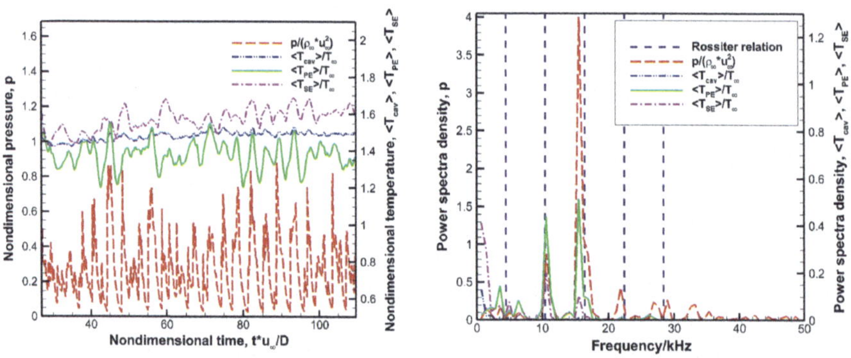

(a) Pressure and temperature histories

(b) FFT results

Fig. 6.27 Pressure and temperature oscillations for nonreacting cavity flow [18]

The definitions of $\langle T_{PE} \rangle$ and $\langle T_{SE} \rangle$ are similar except that their domain of integration respectively include the first third and last third of the cavity volume.

It is observed that the pressure oscillations are quite strong but the temperature oscillations are relatively weak under nonreacting conditions. The calculated dominant frequencies, i.e. 10.5 and 15.5 kHz, compare well with the second and third modes predicted by the modified Rossiter relation [19]:

$$f_n = \frac{u_\infty}{L} \frac{n - \alpha}{M_\infty \Big/ \sqrt{1 + \left[(\gamma_\infty - 1)/2\right] M_\infty^2} + 1/K_v} \tag{6.16}$$

where f_n is the resonant frequency corresponding to the nth mode; u_∞ is the free-stream velocity; L is the cavity length; $\alpha = 0.25$ and $K_v = 0.57$. That is, the pressure and temperature fluctuations under nonreacting conditions mainly result from the self-excited oscillations of the cavity.

6.3.2.2 Impact of Equivalence Ratio on the Flowfield

Firstly, a case with equivalence ratio $\Phi = 0.5$ is run and stable combustion is obtained. Then the equivalence ratio is either increased or decreased to obtain different combustion regimes. Here, the results of four cases will be presented and analyzed, i.e. $\Phi = 0.55, 0.5, 0.45$ and 0.4. $\Phi = 0.55$ and 0.5 correspond to stable conditions while $\Phi = 0.45$ and 0.4 respectively represent near-blowoff and blowoff conditions. At the beginning, a simulation is run for $\Phi = 0.5$. When the flow reaches a periodic state, the other three simulations for $\Phi = 0.55$, 0.45 and 0.4 are initiated using the flowfield of $\Phi = 0.5$ and then the four simulations are continued. The histories of mass-averaged temperature within the cavity are shown in Fig. 6.28.

It is observed from Fig. 6.28 that, under stable conditions, i.e. $\Phi = 0.55$ and 0.5, there exist moderate temperature fluctuations in the cavity, which may be caused by the essentially unsteady characteristics of the cavity shear layer and recirculation zones within the cavity. When the lean blowoff limit is approached by decreasing the equivalence ratio to $\Phi = 0.45$, the temperature fluctuations within the cavity are greatly intensified, accompanied by local extinction and reignition, as will be shown later. When the equivalence ratio is further decreased to $\Phi = 0.4$, a final blowoff occurs with nearly monotonic decrease of the cavity temperature. Detailed analysis of these processes will be presented later.

Figure 6.29 and Fig. 6.30 respectively show the time-averaged heat release rate (HRR) and hydrogen mass fraction contours overlapped with streamlines for $\Phi =$

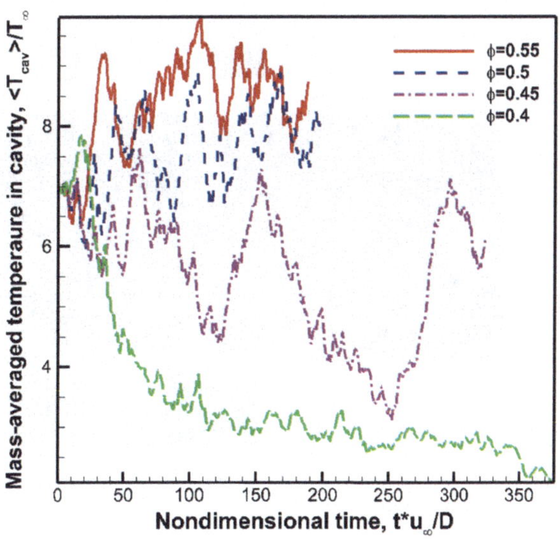

Fig. 6.28 Time histories of mass-averaged temperature within the cavity for different equivalence ratios [18]

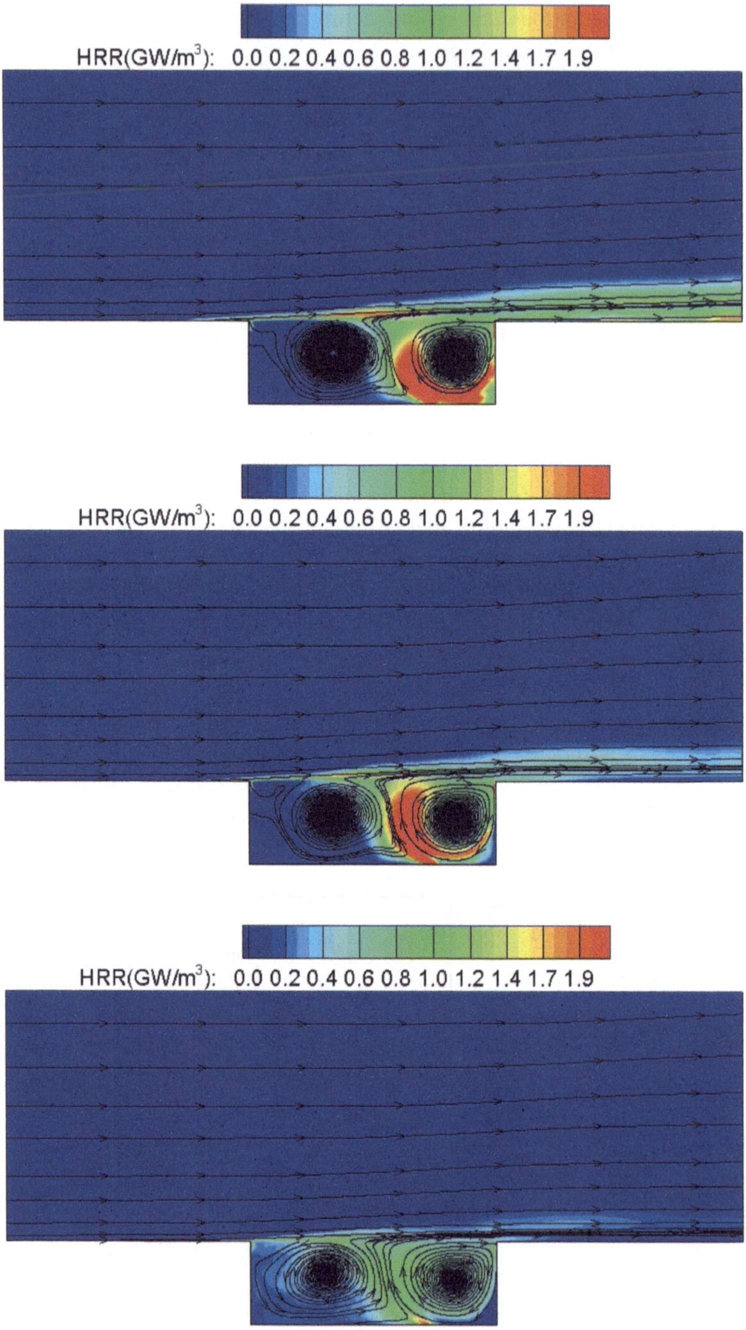

Fig. 6.29 Average heat release rate contours overlapped with streamlines, from top to bottom $\Phi = 0.55$, 0.5 and 0.45 [18]

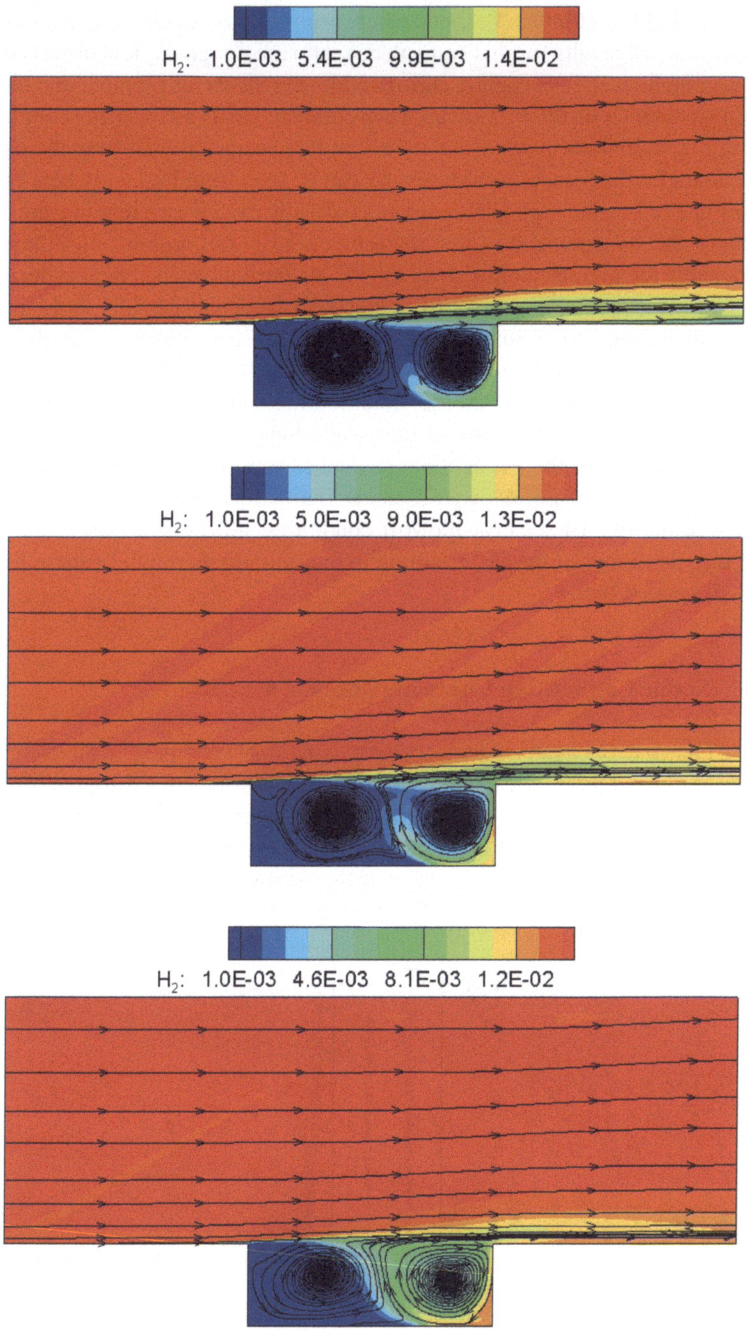

Fig. 6.30 Average fuel mass fraction contours overlapped with streamlines, from top to bottom $\Phi = 0.55$, 0.5 and 0.45 [18]

0.55, 0.5 and 0.45. As for $\Phi = 0.4$, since there is no periodic state during the blowoff, the time-averaged results for this equivalence ratio are not given. It is observed that, in the reacting flow, the PE slightly shrinks while the SE seems plumper and directly comes into contact with the freestream when compared to its nonreacting counterpart shown in Fig. 6.26. However, the cavity RZ pattern seems insensitive to the equivalence ratio under the present conditions. As can be seen from Fig. 6.29, heat-release reactions marked by the high HRR contours basically exist in two regions: the cavity shear layer and its extension in the downstream near-wall regions; and the cavity RZ, especially the regions within/along the PE. As the equivalence ratio is decreased from 0.55 to 0.45, the combustion is obviously weakened, especially in the regions downstream of the cavity. Within the RZ, the heat-release reactions also tend to move from the PE to the regions between the PE and SE with decreasing equivalence ratio. It can be seen from Fig. 6.30 that the unburned fuel enters the RZ from the cavity trailing edge and then is transported upstream along the PE. More unburned fuel is accumulated around the PE as the equivalence ratio is decreased. Under stable combustion conditions, i.e. $\Phi = 0.55$ and 0.5, unburned fuel only survives in the boundary of the PE. Under near-blowoff conditions, however, unburned fuel may intrude into the center of the PE, resulting in intermittent thorough extinction in the PE as will be seen below.

6.3.2.3 Stable Combustion Characteristics for $\Phi = 0.5$

Here, we take $\Phi = 0.5$ as an example to analyze the stable combustion characteristics of cavity-stabilized premixed flames. Figure 6.31a shows the pressure history at the cavity aft corner together with the histories of mass-averaged temperature within the cavity, and Fig. 6.31b shows the frequency spectra of the data in Fig. 6.31a. When compared with Fig. 6.27a, it is found that the pressure oscillations are weakened

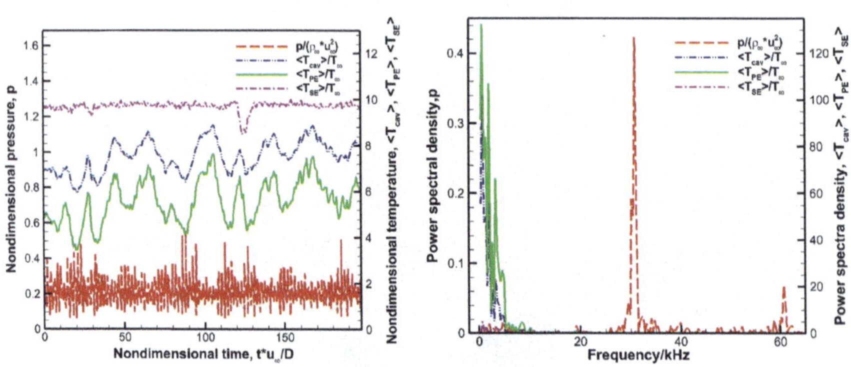

(a) Pressure and temperature histories (b) FFT results

Fig. 6.31 Pressure and temperature oscillations for $\Phi = 0.5$ [18]

while the temperature oscillations within the cavity are intensified. Further observation shows that the amplified temperature oscillations mainly result from those in the PE while the temperature oscillations in the SE are comparable to those under nonreacting conditions. Furthermore, it is observed that the pressure oscillations shift to higher frequencies while the temperature oscillations shift to lower frequencies. The weakened pressure oscillations can be explained by the changed behaviors of the cavity shear layer. It is well known that the energy of the cavity oscillations in the Rossiter mode [19] basically comes from the impingement of the high-speed shear layer on the cavity rear wall. When combustion is stabilized in the cavity, heat release can push the shear layer towards the freestream as can be seen from the streamlines in Figs. 6.26 and 6.29. As a result, the impingement of the shear layer and the cavity rear wall is attenuated and thus the energy supplied to the oscillation system is decreased. However, the reason for the shift of the pressure oscillation energy to higher frequencies is not clear.

Figure 6.32 and Fig. 6.33 respectively show the instantaneous heat release rate (HRR) and hydrogen mass fraction contours overlapped with streamlines at four typical instants. It is found that the reaction layer in the cavity shear layer is thin and intermittently wrinkled by the large-scale vortices. However, the reaction within the RZ is relatively distributed, and the position and amount of heat release to a large extent determined by fuel distribution, as can be seen from Fig. 6.33. Therefore, the temperature oscillations within the cavity are attributed to the fluctuating heat release there and the accompanied intermittent leakage of low-temperature, combustible mixture from the cavity shear layer into the RZ. Under the stable conditions, the low-temperature, combustible mixture entering into the RZ burns soon as it is transported along the PE due to the hot and activated environment. In general, the fuel is completely consumed before it is entrained again by the cavity shear layer and the PE is never filled up with cold mixture. This is a key condition for stable combustion since the hot and activated RZ is crucial to the stabilization of the shear-layer flame.

6.3.2.4 Near-Blowoff Behaviors for $\Phi = 0.45$

Decreasing the equivalence ratio to $\Phi = 0.45$ leads to intermittent/local extinction of the cavity flame. Figure 6.34a shows the pressure history at the cavity aft corner together with the histories of mass-averaged temperature within the cavity, and Fig. 6.34b shows the frequency spectra of the data in Fig. 6.34a. When compared with Fig. 6.34a, it is found that the temperature in the PE, $\langle T_{PE} \rangle$, decreases greatly and sometimes even approaches the nonreacting value shown in Fig. 6.27a, indicating total extinction in the PE. Meanwhile, the temperature within the SE, $\langle T_{SE} \rangle$, may also dramatically decrease at some stages, which is caused by the intrusion of cold mixture into the SE, as will be seen later. The pressure oscillations are slightly augmented especially in extinction period of the PE. As has been pointed out above, the amplitude of the pressure oscillation is mainly determined by the impingement between the shear layer and the cavity rear wall. Once the heat release in the RZ is

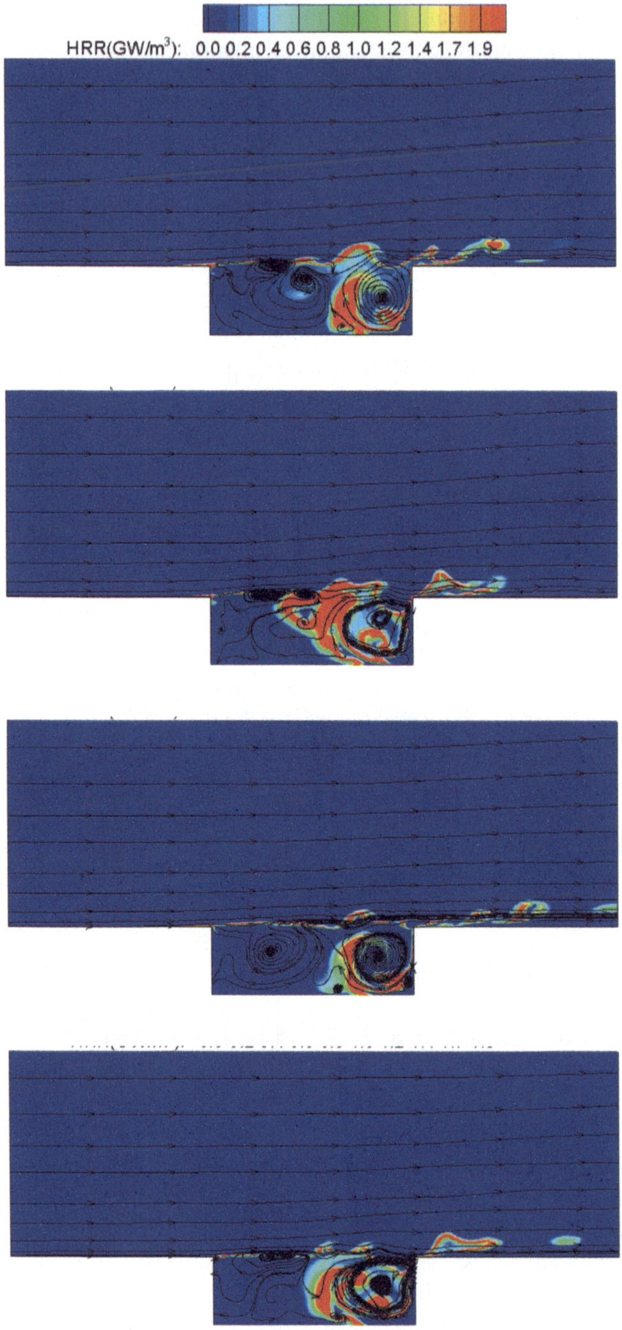

Fig. 6.32 Instantaneous heat release rate (HRR) contours overlapped with streamlines for $\Phi = 0.5$ [18]

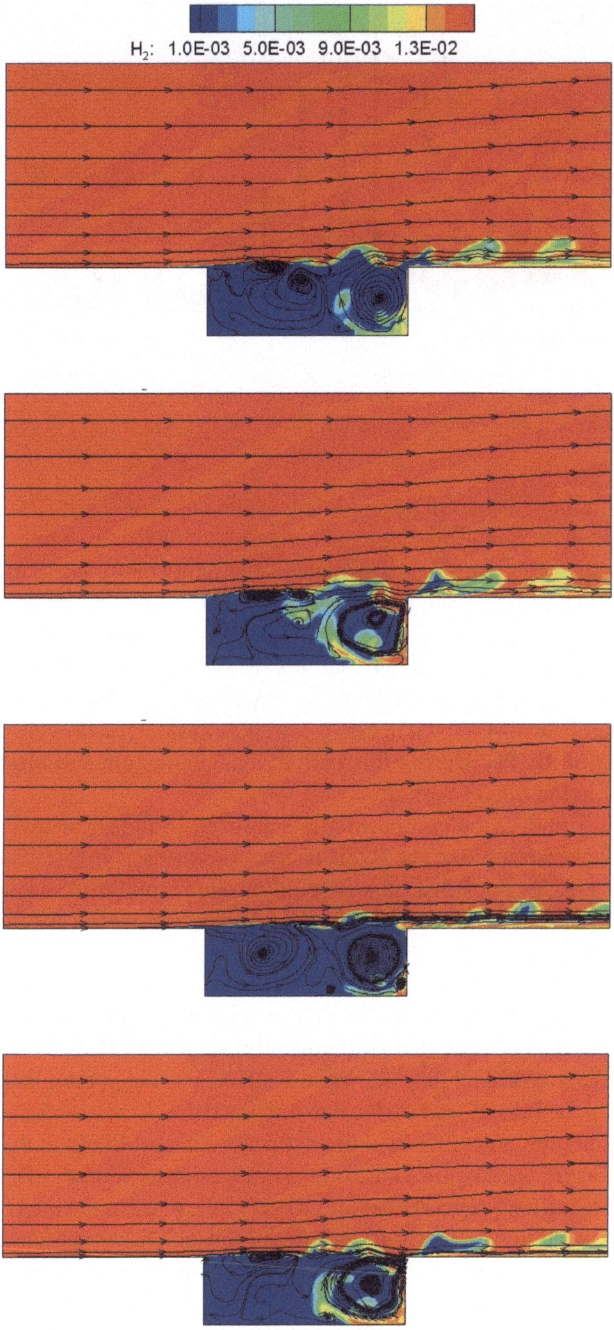

Fig. 6.33 Instantaneous fuel mass fraction contours overlapped with streamlines for $\Phi = 0.5$ [18]

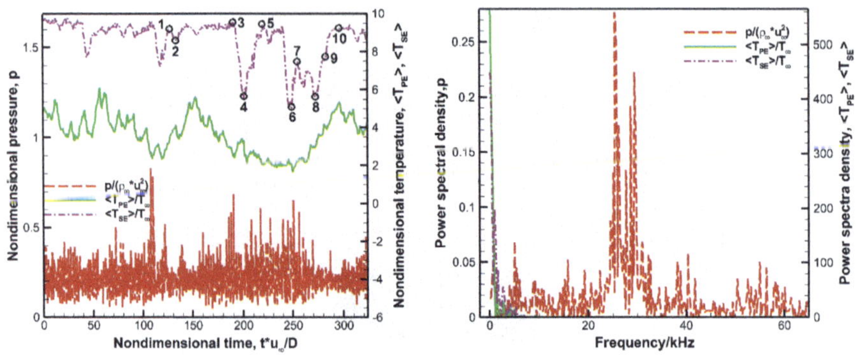

(a) Pressure and temperature histories (b) FFT results

Fig. 6.34 Pressure and temperature oscillations for $\Phi = 0.45$ [18]

reduced, this impingement will be amplified, leading to stronger self-excited oscillations. Since there exist intermittent/local extinction of the cavity flame, the pressure oscillation frequencies are more distributed. The temperature oscillations are still located around O(1 kHz), similar to those for $\Phi = 0.5$, but with greater amplitude.

Figure 6.35 and Fig. 6.36 respectively show the instantaneous heat release rate (HRR) and hydrogen mass fraction contours overlapped with streamlines for $\Phi = 0.45$. These snapshots are taken at the time corresponding to the numbered markers in the temperature history shown in Fig. 6.34a and shows a typical cycle of extinction and reignition of the PE. During this period, local extinction and reignition occur intermittently in the SE.

At instants 1 and 2, the combustion is stabilized in the shear layer and in the PE, and the fuel entering into the RZ is nearly consumed as it is transported along the PE and the PE is not filled up with cold mixture, similar to that observed for $\Phi = 0.5$. At instants 3 and 4, much cold mixture transported by the PE is left unburned when it passes the regions between the PE and SE. Then a portion of this cold mixture is re-entrained by the shear layer, leading to local extinction in the downstream part of the cavity shear layer. As a result of this local extinction in the shear layer, more cold mixture subsequently enters the PE via the impingement between the shear layer and the cavity rear wall. Accordingly, more cold mixture is accumulated in the PE and leads to local extinction there. At instant 5, the PE is filled up with cold mixture, resulting in entire extinction in the PE and in the shear layer adjacent to the PE. Meanwhile, the combustion regions within the RZ are pushed towards the SE and heat-release reactions mainly take place at the interface between the cold, combustible mixture in the aft region of the cavity volume and the hot, activated products in the front region of the cavity volume. At instants 6–8, the cold, combustible mixture enters the SE and burns there, acting as an ignition source. The shear layer adjacent to the PE is intermittently ignited by this ignition source. However, the PE is not successfully ignited yet at these instants. At instant 9, the PE is finally reignited and the combustible mixture accumulated there begins to be

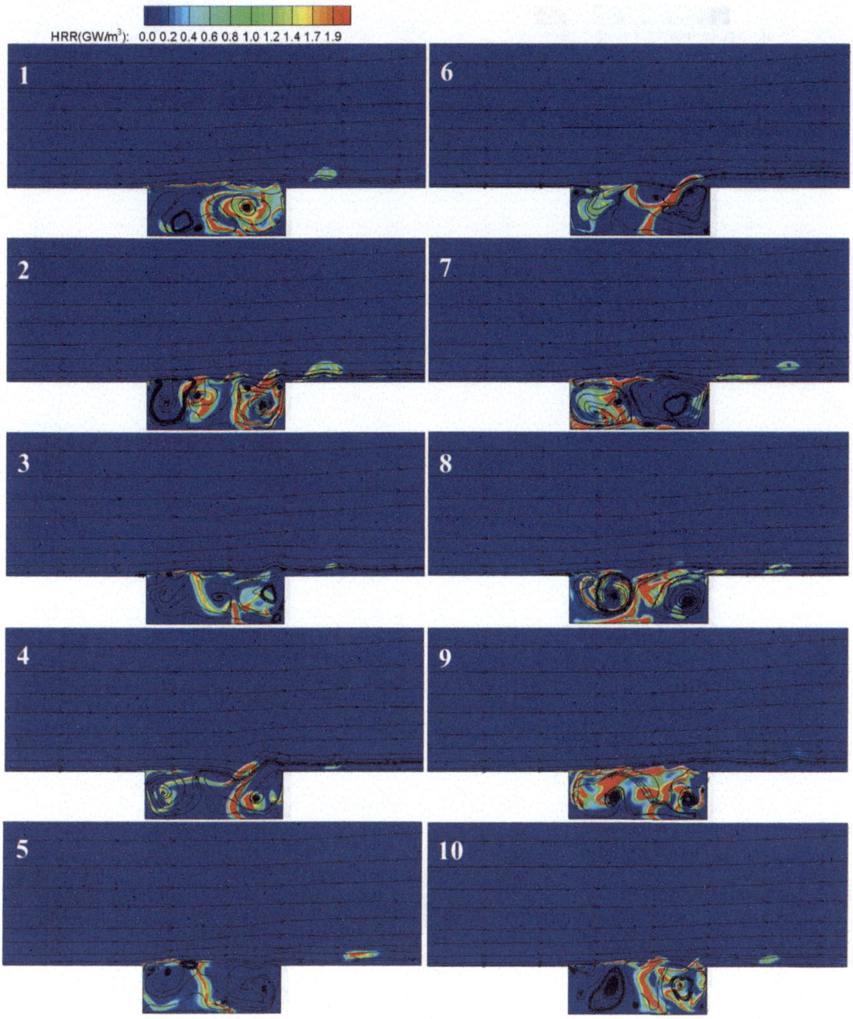

Fig. 6.35 Instantaneous heat release rate (HRR) contours overlapped with streamlines for $\Phi = 0.45$ [18]

consumed. At instant 10, the fuel in the SE runs out and the heat-release reactions there cease. The combustion is again stabilized in the cavity shear layer and in the PE.

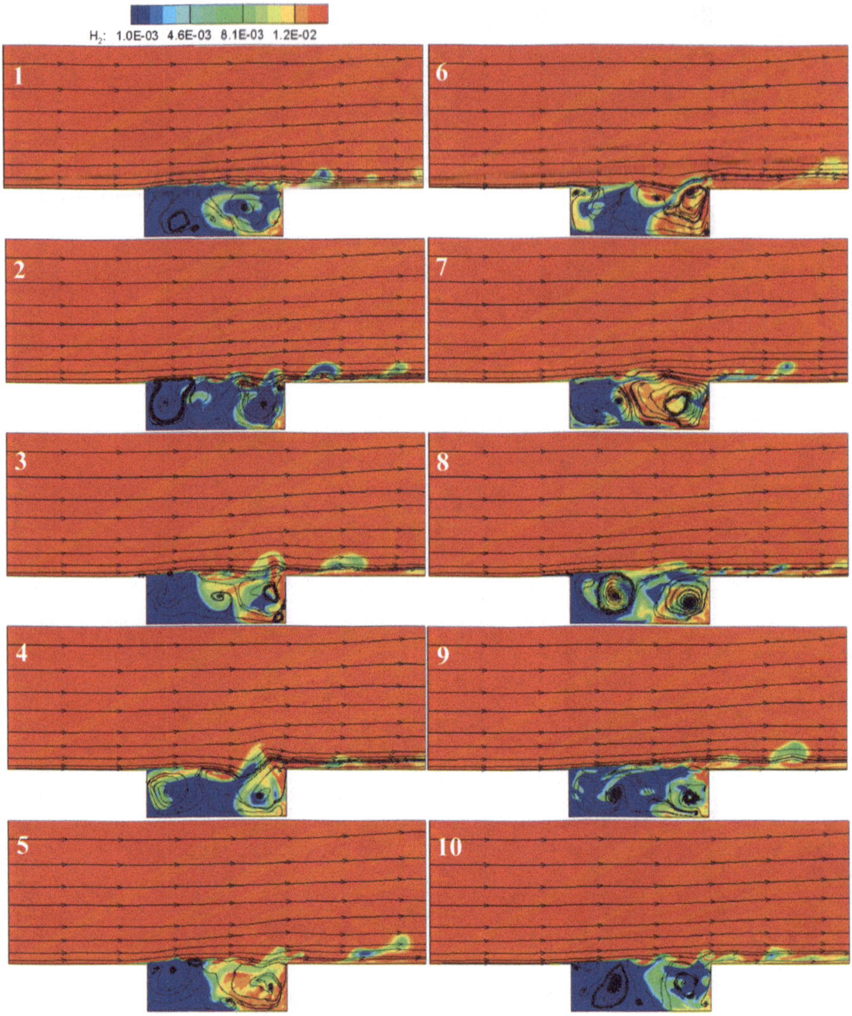

Fig. 6.36 Instantaneous fuel mass fraction contours overlapped with streamlines for $\Phi = 0.45$ [18]

The local extinction and re-stabilization process described above indicates that the SE seems to have better flameholding performance than the PE does. This fact may be explained as follows. Firstly, the PE is mainly driven by the high-speed shear layer while the SE is driven by the PE, so the velocity in the SE is lower, which is beneficial to flameholding. Secondly, the SE usually has higher temperatures, as can be seen from Figs. 6.27a, 6.31a and 6.34a, which is useful to accelerate chemical reactions. Thirdly, the flowfield in the SE is more stable than that in the PE, which also favors combustion stabilization. In fact, the re-stabilization of combustion in the cavity shear and in the PE is similar to the forced ignition process in scramjet

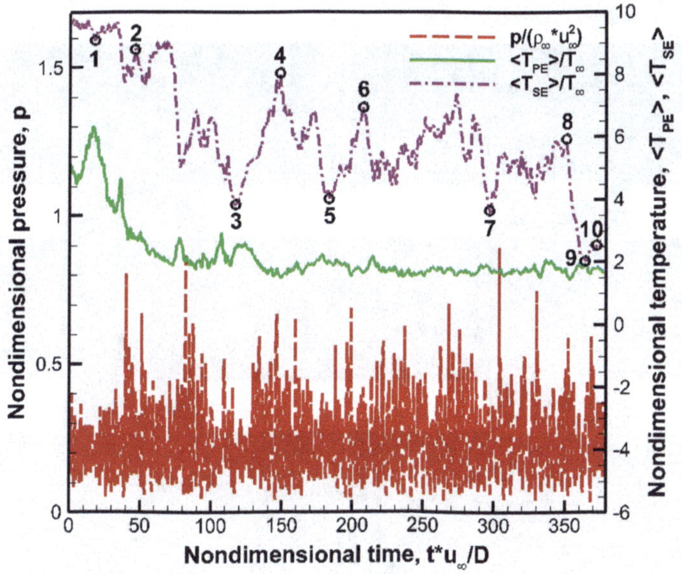

Fig. 6.37 Pressure and temperature histories for $\Phi = 0.4$ [18]

combustors. Thus, it is better to locate the ignition source in the SE, provided the same ignition energy and free-stream conditions.

6.3.2.5 Blowoff Process for $\Phi = 0.4$

Further decreasing the equivalence ratio to $\Phi = 0.4$ leads to the entire blowoff of the cavity flame. Figure 6.37 shows the pressure history at the cavity aft corner together with the histories of mass-averaged temperature within the cavity. It is found that the temperature in the PE, $\langle T_{PE} \rangle$, decreases nearly monotonously in the blowoff process while the temperature within the SE, $\langle T_{SE} \rangle$, falls with apparent fluctuations. The amplitude of the pressure oscillations rapidly approaches to the nonreacting one shown in Fig. 6.27a once the PE is extinguished.

Figure 6.38 and Fig. 6.39 respectively show the instantaneous heat release rate (HRR) and hydrogen mass fraction contours overlapped with streamlines for $\Phi =$ 0.4. These snapshots are taken at the time corresponding to the numbered markers in the temperature history shown in Fig. 6.37 and shows the combustion and flow structures during the blowoff process. Similar to those observed for $\Phi = 0.45$, the PE and the shear layer adjacent to it are firstly extinguished, and then the cold mixture enters into the SE and burns there. However, the heat-release reactions taking place in the SE cannot ignite the PE unlike that for $\Phi = 0.45$. The residual flame in the SE may survive for quite a while but is finally extinguished, leading to the entire blowoff of the cavity flame.

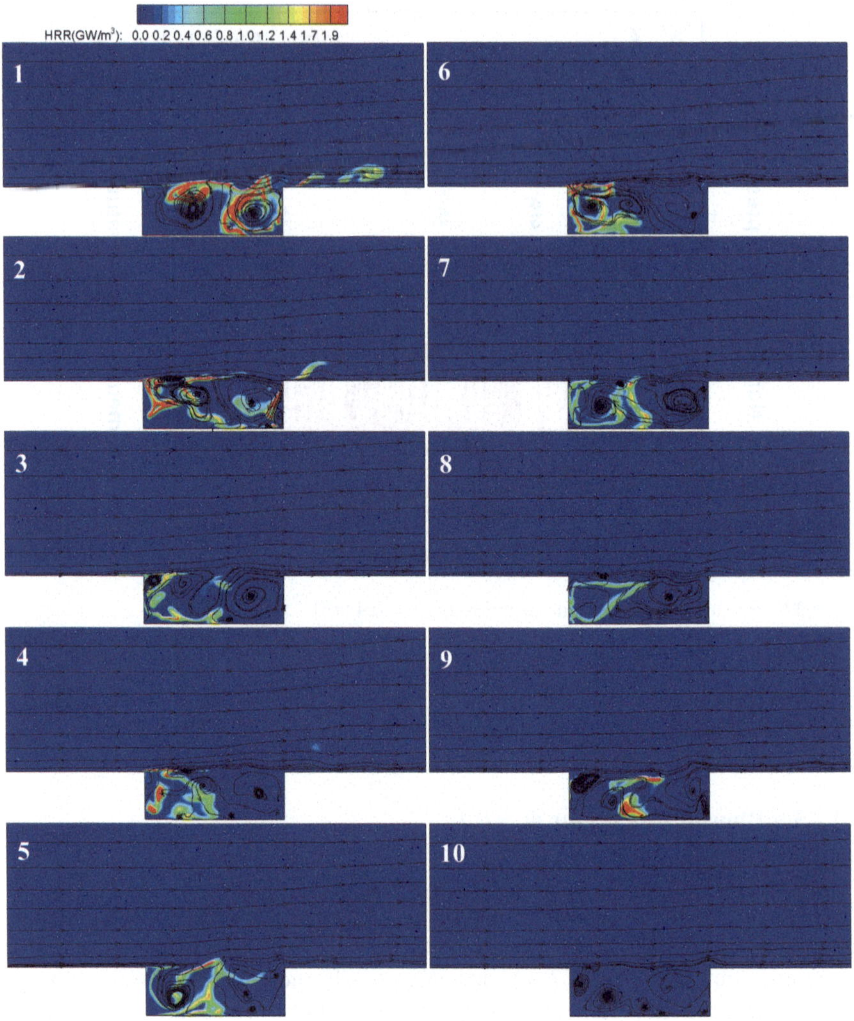

Fig. 6.38 Instantaneous heat release rate (HRR) contours overlapped with streamlines for $\Phi = 0.4$ [18]

In sum, blowoff of the cavity flame is found to occur in multiple steps—local extinction in the cavity shear layer and in the PE, entire extinction of the PE and its adjacent shear layer, local extinction in the SE, and a final blowoff. Under certain conditions (e.g. $\Phi = 0.45$ in the present section), the cavity flame can persist with certain levels of local/intermittent extinction, whereafter the residual flame in the SE may reignite the entire flame. Further decreasing the equivalence ratio (e.g. $\Phi = 0.4$ in the present section) leads to extinction of the residual flame in the SE and entire blowoff of the cavity flame.

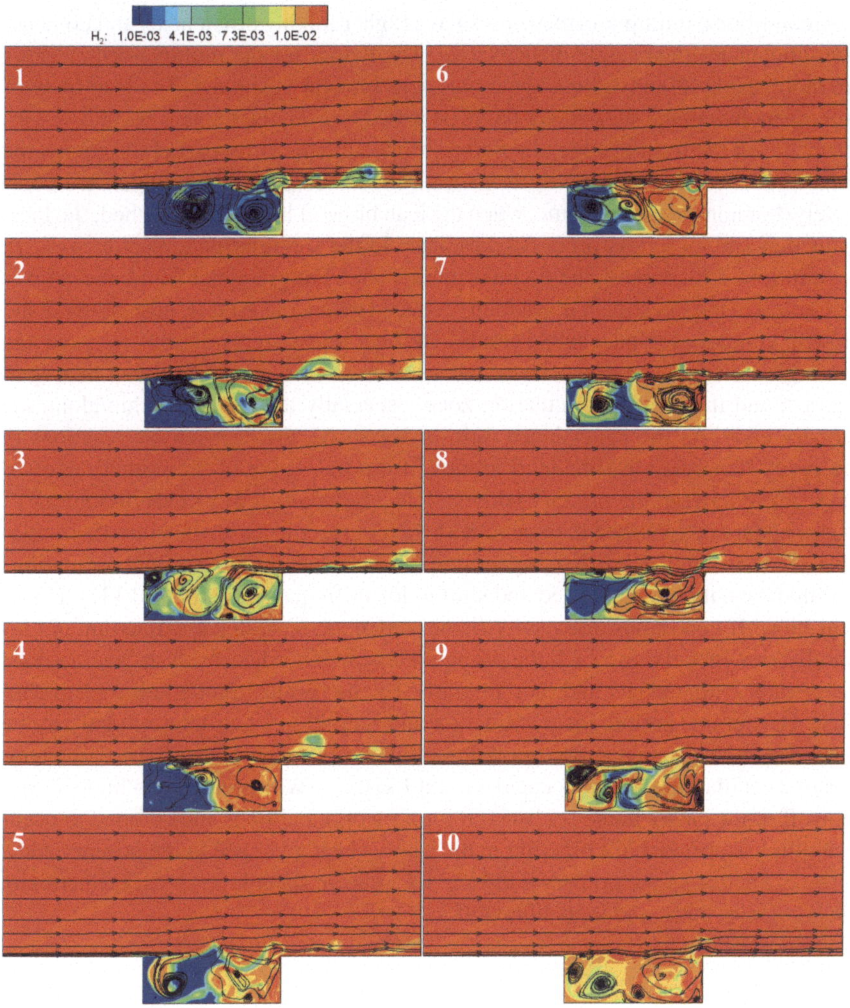

Fig. 6.39 Instantaneous fuel mass fraction contours overlapped with streamlines for $\Phi = 0.4$ [18]

6.4 Summary

This chapter systematically studies the blowoff mechanisms of the flame in the scramjet combustor with transverse injection and a cavity as a flameholder. The blowoff limits model describing the injection of fuel upstream of the cavity in the supersonic flow is theoretically established. Using the assumption of the flame base in the shear layer, the blowoff limits model is correlated with flow condition, injection parameters and cavity configuration. Then the characteristics of mixing and combustion near blowoff limits are studied. In both nonreacting flow conditions and reacting flow conditions, multiple-orifice cases have higher mixing efficiencies and combustion efficiencies than single-orifice cases. The flames for the multiple-orifice cases are more stable. Finally, flame dynamics near blowoff is studied. The lean blowoff process including the shear-layer flame becomes weaker, the flame is partially extinguished near the trailing edge, the flame shrinks into the latter part of the cavity, the flame moves towards the cavity front wall and is subsequently extinguished completely. For non-premixed flame, when the lean blowoff limit is approached, the local extinction and reignition process takes place in the single-orifice injector cases. In the multiple-orifice cases, the flame is more stable. For premixed flame, the equivalence ratio of the freestream is gradually reduced to obtained stable, near-blowoff and blowoff conditions. Under stable conditions, heat-release reactions basically exist in two regions: the cavity shear layer and its extension in the downstream near-wall regions; and the cavity recirculation zone, especially the regions within/along the primary eddy. As the equivalence ratio is decreased, the heat-release reactions in the cavity recirculation zone tend to move from the region of the primary eddy to the regions between the primary eddy and the secondary eddy or even to regions of the secondary eddy. When compared with the nonreacting case, the pressure oscillations are weakened and shift to higher frequencies while the temperature oscillations within the cavity are intensified and shift to lower frequencies around 1 kHz. Under near-blowoff conditions, the cavity flame may experience local extinction and re-stabilization process, in which the primary eddy and its adjacent shear layer may be completely extinguished and then reignited by the residual flame in the secondary eddy. During this period, local extinction and reignition occur intermittently in the secondary eddy. The pressure oscillation frequencies are more distributed. The temperature oscillations are still located around 1 kHz, but with greater amplitude. Entire blowoff of the cavity flame is observed for blowoff conditions and is found to occur in multiple steps—local extinction in the cavity shear layer and in the primary eddy, entire extinction of the primary eddy and its adjacent shear layer, local extinction in the secondary eddy, and a final blowoff.

References

1. Wang, Z. G., Wang, H. B., & Sun, M. B. (2014). Review of cavity-stabilized combustion for scramjet applications. *Proceedings of the Institution of Mechanical Engineers, Part G: Journal of Aerospace Engineering, 228*(14), 2718–2735.
2. Wang, H. B. (2012). *Combustion modes and oscillation mechanisms of cavity-stabilized jet combustion in supersonic flows*. National University of Defense Technology.
3. Wang, Z. G., Yang, Y. X., Liang, J. H., et al. (2014). Analysis and modeling of blowout limits of cavity flame in supersonic flows. *SCIENTIA SINICA Technologica, 44*(9), 961–972.
4. Chaudhuri, S., Kostka, S., Tuttle, S. G., et al. (2011). Blowoff mechanism of two dimensional bluff-body stabilized turbulent premixed flames in a prototypical combustor. *Combustion and Flame, 158*(7), 1358–1371.
5. Song, X. L., Wang, H. B., Sun, M. B., et al. (2019). Mixing and combustion characteristics in a cavity-based supersonic combustor with different injection schemes. *Acta Astronautica, 159,* 584–592.

6. Song, X. L., Wang, H. B., Sun, M. B., et al. (2018). Experimental study of near-blowoff characteristics in a cavity-based supersonic combustor. *Acta Astronautica, 151,* 37–43.
7. Cai, Z., Liu, X., Gong, C., et al. (2016). Large Eddy simulation of the fuel transport and mixing process in a scramjet combustor with rearwall-expansion cavity. *Acta Astronautica, 126,* 375–381.
8. Cai, Z., Wang, Z. G., Sun, M. B., et al. (2016). Effect of combustor geometry and fuel injection scheme on the combustion process in a supersonic flow. *Acta Astronautica, 129,* 44–51.
9. Wang, H. B., Wang, Z. G., Sun, M. B., et al. (2014). Experimental and numerical investigation of cavity-based supersonic flow and combustion. *Proceedings of the Institution of Mechanical Engineers, Part G: Journal of Aerospace Engineering, 228*(5), 781–798.
10. Torrez, S. M. (2012). *Design refinement and modeling methods for highly-integrated hypersonic vehicles*. University of Michigan.
11. Wang, H. B., Wang, Z. G., Sun, M. B., et al. (2013). Combustion modes of hydrogen jet combustion in a cavity-based supersonic combustor. *International Journal of Hydrogen Energy, 38*(27), 12078–12089.
12. Cai, Z., Yang, Y. X., Sun, M. B., et al. (2016). Experimental investigation on ignition schemes of a supersonic combustor with the rearwall-expansion cavity. *Acta Astronautica, 123,* 181–187.
13. Zhang, C. L., Chang, J. T., Ma, J. X., et al. (2018). Effect of Mach number and equivalence ratio on the pressure rising variation during combustion mode transition in a dual-mode combustor. *Aerospace Science and Technology, 72,* 516–524.
14. Wang, H. B., Wang, Z. G., Sun, M. B., et al. (2013). Nonlinear analysis of combustion oscillations in a cavity-based supersonic combustor. *Science China Technological Sciences, 56*(5), 1093–1101.
15. An, B., Wang, Z. G., Yang, L. C., et al. (2017). Experimental investigation on the impacts of ignition energy and position on ignition processes in supersonic flows by laser induced plasma. *Acta Astronautica, 137,* 444–449.
16. Zhao, G. Y., Sun, M. B., Song, X. L., et al. (2019). Experimental investigations of cavity parameters leading to combustion oscillation in a supersonic crossflow. *Acta Astronautica, 155,* 255–263.
17. Gruber, M., Baurle, R., Mathur, T., et al. (2001). Fundamental studies of cavity-based flameholder concepts for supersonic combustors. *Journal of Propulsion and Power, 17*(1), 146–153.
18. Wang, H. B., Li, P. B., Zhang, Y. X., et al. (2017). Numerical study on near-blowoff characteristics of cavity-stabilized premixed flames in supersonic flows. In *21st AIAA International Space Planes and Hypersonics Technologies Conference* (p. 2160).
19. Heller, H., & Bliss, D. (1975). The physical mechanism of flow-induced pressure fluctuations in cavities and concepts for their suppression. In *2nd Aeroacoustics Conference* (pp. 75–491). AIAA.

The manufacturer's authorised representative in the EU is Springer Nature Customer Service Centre GmbH, Europaplatz 3, 69115 Heidelberg, Germany. If you have any concerns regarding our products, please contact ProductSafety@springernature.com

Printed and bound by CPI Group (UK) Ltd, Croydon, CR0 4YY

30/09/2026

02225556-0009